Using Activity Domain Theory for Managing Complex Systems

Lars Taxén
Linköping University, Sweden

Information Science REFERENCE

INFORMATION SCIENCE REFERENCE

Hershey · New York

Director of Editorial Content:	Kristin Klinger
Senior Managing Editor:	Jamie Snavely
Assistant Managing Editor:	Michael Brehm
Publishing Assistant:	Sean Woznicki
Typesetter:	Michael Brehm, Michael Killian
Cover Design:	Lisa Tosheff
Printed at:	Yurchak Printing Inc.

Published in the United States of America by
Information Science Reference (an imprint of IGI Global)
701 E. Chocolate Avenue
Hershey PA 17033
Tel: 717-533-8845
Fax: 717-533-8661
E-mail: cust@igi-global.com
Web site: http://www.igi-global.com/reference

Library of Congress Cataloging-in-Publication Data

Taxen, Lars, 1944-
Using activity domain theory for managing complex systems / by Lars Taxen.
p. cm.
Includes bibliographical references and index.
Summary: "This book offers a new approach toward managing the coordination of complex system development tasks, combining a deep understanding of concrete, every-day conditions of the telecom industry with innovative theoretical development of the Activity Domain Theory"--Provided by publisher.
ISBN 978-1-60566-192-6 (hbk.) -- ISBN 978-1-60566-193-3 (ebook) 1. System theory. 2. Network analysis (Planning) 3. Management information systems. 4. Workflow--Management. 5. Industrial management. I. Title.

T57.6.T39 2010
384.01'15--dc22

2009038606

British Cataloguing in Publication Data
A Cataloguing in Publication record for this book is available from the British Library.

Table of Contents

Section 2
The Theoretical Trail

Section 3
The Activity Domain Theory

Foreword

"Why make it simple when you can do it so beautifully complicated?" This answer to a student is attributed to one of the classical German philosophers of the 19th century. Some readers may think that the author of this book would reply in a similar way. The theoretical discourse of this book, which is partly inspired by some of these philosophers, is indeed a tough piece of cerebral aerobics. However, it results in the wonderful simplicity of understanding complex issues, which makes the reader feel amply rewarded.

Mastering the mindboggling complexity of creations such as modern telecom systems can not rely on shortcuts and silver bullets. On the basis of extensive professional experience and reflection the author convincingly demonstrates that coordinating the development of such systems needs to build on well-grounded theories and thoughtful application. The successful development and diffusion of the "anatomic-centric" approach to project coordination within the Ericsson telecommunications company, where the author was so deeply involved, testifies to the importance of this contribution.

In the area of complex systems development, thoughtful project management is a key factor for innovation, for bringing together system capabilities to actually working systems and for taking them to the customer. The critical question then is: How can managers and practitioners conceptualize and understand the central ingredients of successful project management in this and similar fields? In the extant literature there is a plethora of tools for advanced planning and scheduling, for system decomposition and modularization, for reducing interdependencies and avoiding errors. But there is also a growing criticism of these approaches. A number of these studies have criticized mainstream models of project management for an over-emphasis on the role of planning and scheduling and highlighted the need for developing models that take into account the need for flexibility and adaptability. These studies have singled out the importance of fitting project management to the situation and working out contingency formulae as critical for firm-level competitive advantage. This critique, however, tends to be overly general in character and lack grounded suggestions for effective managerial practices and coordination mechanisms which are needed to make complex system development at all possible.

This work pursues a different route, different both from the traditional planning road, and the alternative "flexibility" route, where everything is open for negotiation. By applying rigorous theoretical analysis it brings a new depth to the art and science of complex system development in general and to project management practices in particular. As the author forcefully argues, nothing is as practical as a good theory. The theories discussed and analyzed here do indeed lead to very practical results such as new forms of representing and expressing interdependencies, new means of creating shared understanding, new ways of communicating system characteristics and integrating complicated project activities into systems which function as predicted and can be delivered as promised.

The reader is invited to a rewarding ride through research and experimentation, ending up in highly useful new insights.

May 2009
Christian Berggren
Professor in Industrial Management & Program Director for KITE – Knowledge Integration & Innovation in Transnational Enterprises
Linköping University, Sweden

Christian Berggren *is Professor in Industrial management at the University of Linköping, Sweden and director for the KITE research program, Knowledge Integration and Innovation in Transnational Enterprise (http://www.liu.se/kite). Christian Berggren has published extensively on innovation and project management within the automotive, electro-technical and telecommunications industries. His recent papers include "The integrator's new advantage - reassessing outsourcing and production competence in the global telecom industry", European Management Journal, 2008, 26, 5, 314-324; "Technological capabilities and late shakeouts: Industrial dynamics in the advanced gas turbine industry", Industrial and Corporate Change, 2008,* 17, 335-392*; "Lagomizing, Organic Integration, and Systems Emergency Wards: Innovative Practices in Managing Complex Systems Development Projects, Project Management Journal, 2008, 39, S111-S122 and "Rethinking Project Management Education: Socials twists and Knowledge Co-production", International Journal of Project Management, 2008, 26, 3, 286-296. Berggren's current research focuses on the role of knowledge integration capabilities for realizing innovative solutions by combining complex technologies.*

Preface

The sociologist Kurt Lewin once said that "there is nothing so practical as a good theory". As a general motto, this view is no doubt acknowledged among practitioners and academics. Industrial accomplishments should be informed by insightful theories that in turn may be further elaborated from practical experience.

However, when taking a closer look at how this motto is practiced in industry and academia, things become more intricate. My experience, as someone who has spent many years in both camps, is that theories are still by and large the concern of academia, and practice the concern of industry. When academia approaches practice, it is usually after something interesting has taken place in industry. For example, the emergence of Enterprise Resource Planning (ERP) systems took place several years before academia began researching this major upheaval in organizations. Moreover, the main knowledge interest is to *explain* the ERP phenomenon using various informing theories. Accounts for the use of theories to *influence* the trajectory of ERP systems are rare in academia.

From the other vantage point, industry shows limited interest in theories unless they give immediate pay-back in ongoing development tasks. As early as 1995, David Parnas expressed concerns about software engineering conferences:

The sad fact is that most Engineers actually writing code do not come to these conferences. They also do not read IEEE Transactions on Software Engineering. However, if we want to influence the way that software is written, we have to admit that we are doing something wrong. (Parnas, 1995, p. 30)

Unfortunately, we still seem to be doing something wrong. Industry feels that academic research is irrelevant since – so it is believed – academia has no real understanding of the magnitude and complexity of the practical challenges industry is facing. Practical applications emanating from theories used in academia are mostly seen as vehicles for advancing theory development and thus of little relevance to practitioners.

In contrast, the message of this book is that theories can indeed make a difference in demanding practical settings. The book communicates a story where theory and practice mutually have influenced each other. The theory is what I have coined the Activity Domain Theory (ADT), and the practice is the development of telecommunication systems at Ericsson, a major supplier of telecommunication equipments all over the world.

The subject area of the book is, as the title indicates, the interplay between coordination, complex projects and ADT. The motivation for focusing on this arena is as follows. Our technological capabilities to develop products have increased dramatically. This enables us to produce ever more complex systems like telecom networks, airframes, cars, weapons systems, and so on. The complexity is augmented by the many different types of technologies used; in particular the increasing utilization of embedded control software.

However, our ability to cope with the complexities that these technological developments raise has not increased at the same pace. There is a gap between the possibilities that technology offers us and our capability to take advantage of this development. This is perhaps most evident in, for example, the poor track record of developing large software systems. The Standish Group classifies such projects as successful, challenged and failed. Although there has been some increase in the rate of successful projects, still in 2001 about half of the projects were considered challenged, i.e. completed with overrun budget, over the time estimated, and with fewer functions than specified (Standish Group, 2001). 23% of the projects were outright failures and only 28% were considered successful. Thus, there is an urgent need to improve the track record of complex projects.

Coordination is at the heart of complex system development since such endeavors are possible only by the purposeful coordination of socially organized work. There is a pressing need to understand the nature of coordination in order to improve the way coordination is carried out. However, theoretical efforts to define coordination (e.g. Kraut & Streeter, 1995; Malone & Crowston, 1994) seldom reach into the unruly domains of industrial practice. When things become concrete and detailed, esoteric and abstract theories will come off badly. Thus, the focus on coordination is motivated by the need to operationalize this concept, i.e., to make it useful in practice.

Why do we need to theorize about coordination? After all, well-working and efficient products appear on the market at an ever increasing pace. It seems that industry is doing pretty well without dwelling into intricate theories. However, the story of failed and target overrun projects shows that things are not all that well. Either we take the position that such failures are inevitable, considering the magnitude and complexity of these projects, or we seek to improve the way we run our projects guided by, hopefully, well-informed theories.

As many other scholars, I am convinced that the root cause of failed projects are to be found, not primarily in the complexity of the technology, but in our social and cognitive capabilities to manage the complexity of such projects. Social and cognitive aspects are inherent features of ADT, which implies that this theory should be well armed to guide the coordination of complex development tasks. The reason why I dare to state this rather bold claim is that I worked for more than 30 years in different roles at Ericsson. This provided me with a down-to-earth appreciation of the concrete problems the development of a telecom system poses. Rather late in my professional life I spent ten years at the university, first as a Ph.D. student and later as an associate professor. This enabled me to reflect and theorize about my experiences from Ericsson. In this sense, the book is a synthesis of a personal meandering between practical and theoretical camps, something that should provide a good basis for practical theory development. It is up to the reader to judge whether I have succeeded.

THE REALITY OF THE TELECOM INDUSTRY

The telecommunication industry in general and Ericsson in particular can be seen as a paradigmatic example of the complexity many product development organizations face today. The ensemble of telecommunication systems has been called the world's largest machine; an immense network of interacting nodes, each performing some kind of utility like setting up calls, keeping track of the position of a mobile, providing charging functions, etc. The network is constantly evolving. New types of services such as live TV directly in the mobile are implemented as the capacity of the network increases. In addition, there is a legacy of existing equipments and networks that must be considered when making changes in the network. Another source of complexity is the many different technologies used such as radio, software, hardware, mechanical, and optical ones.

Figure 1. An integration plan of one node in the 3rd generation of mobile systems (Taxén & Svensson, 2005. Used with permission)

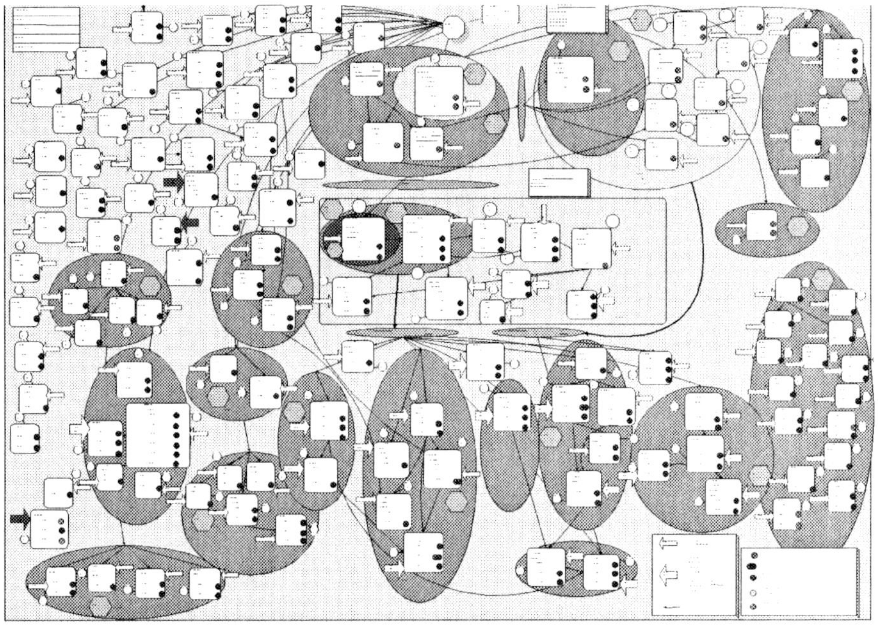

In the telecommunications market there is a fierce competition among systems providers like Ericsson, Nokia™ and others. This competition has been heightened by the deregulation, leading to volatile requirement specifications of systems and pressure to shorten lead-times and costs. In addition, frequent re-organizations are carried out such as outsourcing, acquisitions, and the establishment of partnerships.

As an example of the complexity underlying a mobile phone call we may take a look at the image in Figure 1. The image, which should be read from top to bottom, is called an "integration plan". It is used as an instrument for coordinating development tasks (square white boxes) in one of the nodes in the network. Each task, which is called a "work package", provides a specific functionality to the node. The thin lines mark dependencies between the packages, indicating which packages must be ready in order for other packages to function properly. Thick arrows show the datum for integration and verification of the packages. Small dots indicate the status of a package such as "in design", "in test", "delayed", "ready", etc. The ovals signify basic services like registering the location of the mobile, calling the mobile, answering a mobile call, etc. In most cases the functionality is provided by software, where the total number of source code lines may be in the order of millions. In other cases, complex integrated hardware chips are used. The project may take more than a year to execute and can involve several thousand persons at development sites all over the world.

It is beyond doubt that the coordination of projects like the one in Figure 1 is a challenging task that needs to be supported by information systems (ISs). In order to implement such systems, there must be a sufficient consensus of how coordination should be conceived. At a general level, it is fairly easy to agree about what coordination items like, for example, a requirement is. However, as the level of detail increases, disagreements begin to surface. How should a requirement be identified? What kind of status values does a requirement have? What attributes characterize a requirement? How should requirements be related to other things such as customers, products, test cases, etc? In fact, the experience from Ericsson testifies that attaining common understanding about coordination is an overwhelming endeavor that takes precedence over other tasks. How to accomplish this endeavor is the recurrent theme of this book.

THE DEVELOPMENT OF THE ACTIVITY DOMAIN THEORY

The emergence of ADT can be seen as a result of two "trails" that met around 1998: a practical and a theoretical one. The practical trail provided me with a thorough understanding of the immense problems of coordinating huge, globally distributed development projects. The theoretical trail rendered me with theoretical instruments by which the practical experiences could be understood and subsequently be reflected back into guidelines for improving practice.

The Practical Trail

The molding of the practical trail occurred roughly between the late 1990s and 2003 when Ericsson was involved in the transition from the 2nd to the 3rd generation of mobile systems. This was a major challenge for Ericsson. In order to meet these challenges, Ericsson worked out a development method, which I will refer to as the *anatomy-centric approach* throughout this book. The origin of this approach dates back to the early 1990s, when Ericsson was involved in delivering the 2nd generation mobile systems to the Japanese market. The method wandered back and forth over many years in the company, sometimes heavily used, and sometimes more or less forgotten until it became compulsory for it to be used in all subprojects in the 3G development.

The anatomy-centric approach put quite new demands on the management and coordination of the 3G projects. One of the most critical tasks was to make sure that work packages implementing various functionalities were designed, tested and delivered to system integration according to the project plan. Typically, between 5,000 to 10,000 items needed to be managed with respect to revision, status, dependencies, etc.

In 1996 I was taking part in a work aiming at supporting the coordination of anatomy-centric projects. Early on, it was clear that specific IS support had to be developed for such projects. I became responsible for the pre-studies and subsequent implementation of such a system at a site in Stockholm. In early 1996 I happened to meet a lonely looking salesman at a construction fair. It turned out that the salesman was the retailer for Matrix, a Product Data Management system that was then fairly new on the market (MatrixOne, 2008). He showed in a couple of minutes the ease by which a particularly demanding data model construct could be implemented in Matrix. I became impressed with this demonstration, and managed to persuade Ericsson to take a closer look at this system. The system turned out to have the qualities we wanted, above all ease of changing the implementation.

During 1997, Matrix was prototyped in some pilot projects and 1 May 1999, the first usage of Matrix in the 3G projects was a fact. This was the start of an astonishing development. Between 1999 and 2003 other projects followed, both at Stockholm and other Ericsson sites. At its peak, around 140 projects and subprojects were using similar, albeit distinct, applications built on the same Matrix IS platform.

A striking feature of the coordination support developed was the establishment of separate applications built on Matrix at different sites. Although each application turned out to provide highly efficient coordination support, they were completely differently constructed, both in terms of information managed and functionality offered. There was no apparent reason why they should differ, since all sites were engaged in the same 3G development enterprise. However, any attempts to consolidate the applications more or less failed. This and other equally perplexing experiences make up the substance of the practical trail.

The Theoretical Trail

The origin of the theoretical trail can be traced back to a long personal interest in philosophy in general and in dialectical thinking, the Marxian concept of praxis, semiotics, and meaning in particular. With such a backpack it was near at hand to reflect on the daily practice where I spend most of my working hours. The philosophical "spectacles" enabled me to see taken-for-granted values and conceptions from other perspectives. For example, an outspoken value at Ericsson was to "develop what the customer requires", or – expressed differently – "The customer is always right". The problem, however, was that the customer seldom knew precisely what he wanted and often changed his mind in the midst of the development of a particular system. It appeared to me that the real process was much more interactive and dynamic than expressed by prevailing tenets. Customer and supplier together worked out a common view of the system, sometimes in conflict but more often in cooperation.

So, the theoretical trail had to some extent been cleared alongside with my practical, daily work. However, this trail would have remained dormant if it hadn't been for a company controversy over the choice of the Matrix system. The Corporate IT organization at Ericsson stipulated the use of another system, which however did not have the necessary capabilities – above all flexibility to change the implementation in an easy and straightforward way.

The controversy over the platform intensified during 1998, and at a certain point in time, Matrix was on the brink of being thrown out in favor of the corporate system. Along with that my ideas and the promising results we had achieved so far would disappear. In order to save this work I turned, more or less in despair, to Linköping University with a research proposal. After some initial hesitation, the proposal was accepted, and so my academic career started, rather late in life – I was 54 years old then. At that point I was quite convinced that a promising initiative to improve coordination support was for ever gone. However, in May 1998, the project manager for one 3G project got interested in our prototype and decided to use this in his project. From that moment on, the tide changed and Matrix became a key component in the 3G development endeavor.

Thus, I could elaborate the theoretical trail alongside my work at Ericsson until 2003 when I was required to leave the company together with many others as a result of the telecom crisis. So, from that moment the practical trail faded away while I could continue to refine the theoretical trail in my research. In retrospect, the constellation of idiosyncratic circumstances that gravitated around 1998 seems almost inconceivable. The transition from 2G to 3G at Ericsson, my position as responsible for the IS support in one project, the encounter with the Matrix platform, the conviction and support of a single project manager, my interest in philosophy – all these elements coincided in time and space to form the backbone of the ADT-approach and consequently the material for this book.

THE MAIN FEATURES OF THE ACTIVITY DOMAIN THEORY

There is no lack of theories that address social and technological issues. This is perhaps most evident in researching ISs. For example, the Association for Information Systems provides a list of more than 50 theories used in IS research (IS theories, 2008). So, why bother about yet another theory? The main reason, I believe, is that ADT differs in some crucial aspects from most other theories proposed.

- ADT originated and evolved within the telecom industry. Thus, it has been sculptured from and scrutinized by concrete industrial needs. Most other theories originate from outside the industry, and are appropriated or adjusted to fit the particular circumstances of that industry. For example, the Actor Network Theory (e.g. Latour, 1992) originated in social sciences and has been used by a number of researchers to analyze the impact of ERP-systems.

- ADT is operational in industrial settings. By operationalization, I mean that the theory can be expressed in elements that can be manipulated, measured or observed in a particular situation in order to influence this situation. This means that the theory is indeed "practical" in the sense that Kurt Lewin meant.
- A common thread in ADT is the concept of *meaning*. Meaning is considered intrinsically related human action, and only meaningful sensory impressions can be informative and acted upon. Consequently, a key issue in coordinating human actions is the construction of common understanding about how coordination should be conceived.
- The focus on meaning implies that human cognition plays an important role in ADT. A key tenet in ADT is that the properties of human cognition constitute the way we apprehend our socially constructed reality.
- ADT states that common understanding is constructed in the social *practices* (e.g. Schatzki, Knorr Cetina, & von Savigny, 2001). A practice has been defined as a coherent set of human actions characterized by a commonly understood object towards which coordinated actions are directed (Wartofsky, 1987).
- The inclusion of both human cognition and social practices means that ADT integrates individual cognition, social organization and technological-material aspects of human action into a coherent whole.

These features are reflected in the structure of ADT. Its key construct is the *activity domain*, which can be seen as a particular form of social practice where coordinating aspects are at the fore. The motivation for the existence of the activity domain is that it supplies products or services that meet some social need. In order to do so, the actors in the domain work on some work object, the outcome of which fulfills the need.

The work object molds the inner structure of the activity domain, including the meaning actors associate with the object and other related items needed in order to produce the outcome. For example, the activity domain of flying an airplane will differ quite radically from that of building a house in terms of what the work object is, what things are important, what methods work or don't work, what rules and norms are valid, what tools or instruments are used, etc.

In spite of the different realities constructed in activity domains, ADT claims that certain deep structures coined *activity modalities*, can be found in every domain. These modalities denote cardinal dimensions along which meaning is constructed, and they can ultimately be traced back to the way our cognitive system works. Stated differently, ADT claims that the social reality constructed in activity domains reflects the way we as humans apprehend the world. For example, since our minds discern a temporal dimension, we create "temporal" artifacts like calendars and clocks that help us to coordinate our actions.

ORGANIZATION OF THE BOOK

From the background sketched in the previous sections, it is relatively straightforward to organize the book in four parts, each consisting of a number of chapters (see Figure 2). The practical and theoretical trails leading up to the ADT are described in Section 1 and Section 2 respectively. These parts should be

Figure 2. Organization of the book

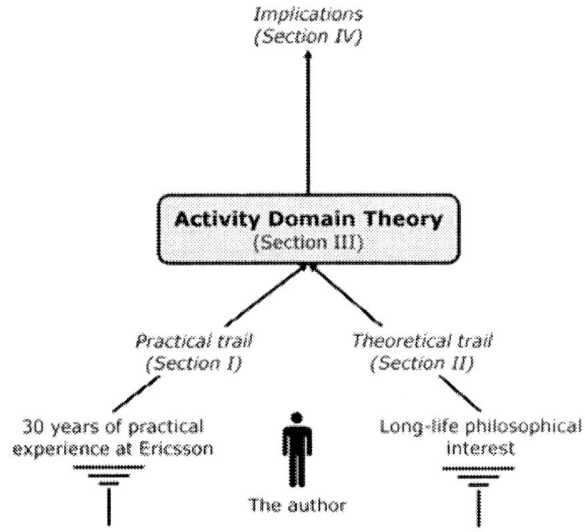

read as paving the way for the ADT, which is presented in Section 3. Section 4 discusses implications of the theory in a number of areas, with a focus on the coordination of complex systems. With that a full circle is closed and we may approach practice and theory with, hopefully, some new insights.

Stakeholders

Since the book aims at a balanced approach between theory and practice, the following main stakeholder groups can be identified:

- **The P-type:** *"Just do it."* This group consists of action oriented practitioners, people that consider theories as irrelevant for achieving results. Research carried out at universities is regarded with great suspicion by the P-type person: nothing practical ever comes out of the ivory towers in academia.
- **The Pt-type:** *"What's happening here?"* This group includes practitioners that reflect on what they are doing and try to find ways to improve practice. Innovative new products and methods often emerge from this group. Still, these innovations occur outside academia; there is no or little interaction with research.
- **The Tp-type:** *"Can my theory make an impact in practice?"* Here we find theoreticians with a practical inclination; persons in academia active in applied sciences such as IS development, computer science, software engineering, etc. These persons want theories to be relevant for practice. However, practical applications developed are mostly used as means for theory development, and these applications are in general less complex compared to ones found in product-developing industries.
- **The T-type:** *"I think, therefore I am."* In this group we find theoreticians without any practical application in mind. Practice is regarded as dirty business that should not ever be let inside the academic fortress. This group corresponds to the P-type group in industry; no interaction whatsoever between academia and industry is the basic attitude.
- **The PT-type:** *"Practice and theory are two sides of the same coin."* Persons in this group are, I dare say, quite rare today. They combine the interests of the Pt and Tp person types.

Figure 3. The stakeholder focus Section 1

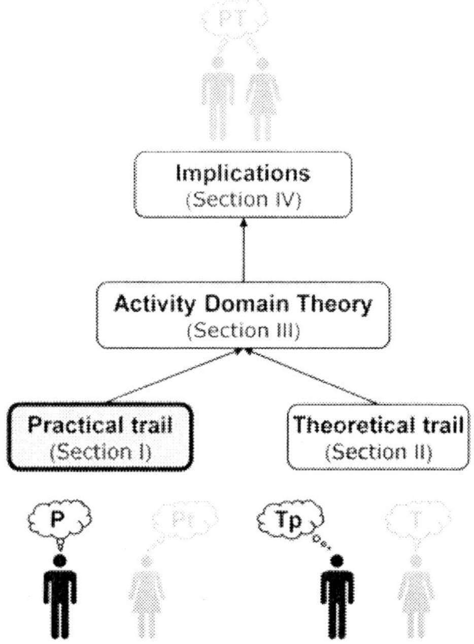

I foresee that people from these, admittedly caricatured, groups will be interested in different parts of the book as follows.

Section 1: The Practical Trail

Relates the empirical background of the ADT. The stakeholder focus for Section 1 is on the P and Tp type of stakeholders (see Figure 3). Practitioners may be interested in how the coordination of the 3rd generation of telecom systems occurred at Ericsson; what kinds of obstacles had to be overcome and what results were obtained. This part should also be of interest to theoreticians who are interested in how a theoretical understanding might grow out of practical experiences rather than from appropriation of other theories for practical purposes[1].

The chapters in Section 1 are as follows.

The Dawn of the Activity Domain Theory
An historical account of events, observations and experiences from the Ericsson 3G development projects between 1997 and 2003 that gradually shaped the ADT.

Reflections
Attempts to make sense of the events reported in the previous chapter. Reports on other observations collected over many years at Ericsson that contributed to the formation of ADT. These observations concern the dilution of the business process concept; the difficulties of establishing common vocabulary in information modeling projects; the chaotic IS/IT architecture in the organization; and the fragmented responsibilities of processes, information architectures, IS/IT and business rules, resulting in glitches between these areas. Summarizes the insights and needs from the practical trail:

Figure 4. The stakeholder focus Section 2

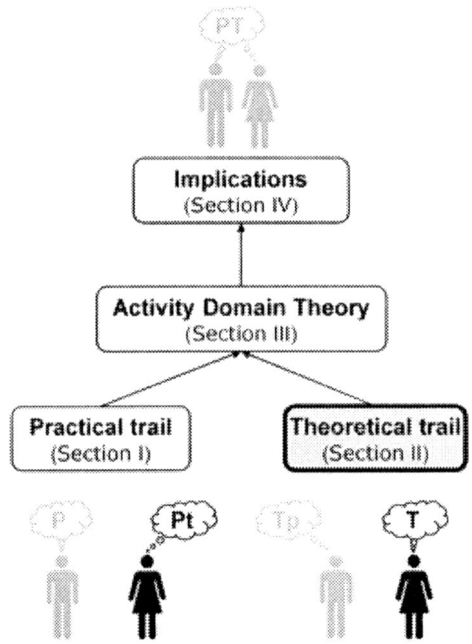

- **Common understanding:** The 3G development experience made it painstakingly clear that achieving common understanding is a difficult task that is easily overlooked. Thus, a methodical way of addressing this issue should be up front on the agenda.
- **Integrating construct:** Many observations pointed to the need of some organizational construct that could integrate elements like IS/IT, business processes, information structures, and corporate business rules. Taking one of these elements, for example, business processes, as the basis for organizational analysis and change initiatives appeared to be insufficient since all elements obviously were interdependent and, consequently, could not be sensibly treated one at a time. Gradually, an insight grew that some kind of *practice* construct (e.g. Schatzki, Knorr Cetina, & von Savigny, 2001) was a viable candidate for this purpose. The practice idea was later elaborated into the *activity domain,* which subsequently became the integrating construct in ADT.
- **Contextualization:** Other observations indicated that situational and contextual aspects must be given more attention in the organizational discourse. This was most conspicuous during the 3G development, where different sites developed site-specific coordination support applications; each of which provided previously unmatched IS support. Thus, by delimiting the scope of commonality, it seemed that the efficiency of IS application development tasks increased dramatically.
- **Recurrent patterns:** A closer look at the most frequent organizational artifacts indicated that these were manifestations of more basic, underlying patterns that appeared over and over again. These patterns came across as being of separate kinds but nevertheless interdependent and mutually constituting each other. Subsequently, these patterns were elaborated into the *activity modalities* in ADT.
- **Enactment:** Organizational artifacts must be what Orlikowski (2000) calls *enacted* in order to become useful in a particular organization. The enactment view "starts with human action and

examines how it enacts emergent structures through recurrent interaction with the technology at hand" (Orlikowski, 2000, p. 407). Some experiences made me aware of the fact that the historical development of organizational artifacts like information models, process models, etc., must be taken into account. Not until these artifacts are regarded in a historical perspective, is it possible to appreciate the enactment efforts behind their particular appearance at a certain moment in time.

Section 2: The Theoretical Trail

Describes the theoretical roots of the ADT. The focus of Section 2 is on the Pt and T type of stakeholders with a focus on the T-type (see Figure 4). Obviously, theoreticians will have an interest in the theoretical background of ADT. However, I also want to address practitioners with a reflective mind, primarily through examples from everyday life and industrial settings.

The chapters in Section 2 are as follows.

The Philosophical Roots

Provides an account of the Marxian philosophical cornerstones that have influenced the ADT: the concept of *praxis*, the *dialectical relation*, the *epistemology of praxis*, and the *dialectical method*. In the concept of praxis, the Marxian tradition tries to capture the fundamental nature of human activity. The dialectical relation conceives related phenomena as mutually constituting each other. The epistemology of praxis concerns the nature of trustworthy and actionable knowledge. The dialectical method is the Marxian approach towards investigating the underlying essence of phenomena that we experience as immediately given in everyday life. Together, these cornerstones provide an alternative conceptualizing of organizational phenomena.

Activity Theory

Reviews three additional cornerstones for ADT from the Russian theory of Activity (AT) (e.g. Kaptelinin & Nardi, 2006): the concept of *activity* (frames the context in which individual actions are meaningful), the concept of *mediation* (humans always put something between themselves and their object of action), and the AT view on *meaning* (how we make sense of reality).

Section 3: The Activity Domain Theory

Provides an account of the ADT as a synthesis of the empirical and theoretical trails, and conceptualizes ADT as an elaboration of AT where coordination is in focus. Describes the *activity domain* as a work setting (team, group, business unit, etc.) whose existence is *motivated* by its capability to produce some *outcome*. The outcome is achieved through the actions of socially organized *actors*, who transform a *work object* into the outcome. The outcome of one domain may be the *prerequisite* for another domain, which means that the activity domain construct is *recursive* and *scalable*.

In order to transform the work object, actors *enact resources* consisting of means and skills. Enactment in the context of ADT refers to the fact that the potential capabilities of means become resources only when actors have drawn these into the social fabric of the activity domain, and collectively learnt how to use them. Thus, capabilities of means and actors must always be related to the work object and motive in order to become useful resources in the domain.

A key tenet in ADT is that society as constructed in human activity reflects or is congruent with the structure of the human cognitive system. A corollary of this tenet is that humans have innate capabilities to coordinate their actions. These capabilities are characterized by the *activity modalities contextual-*

Figure 5. The stakeholder focus Section 3

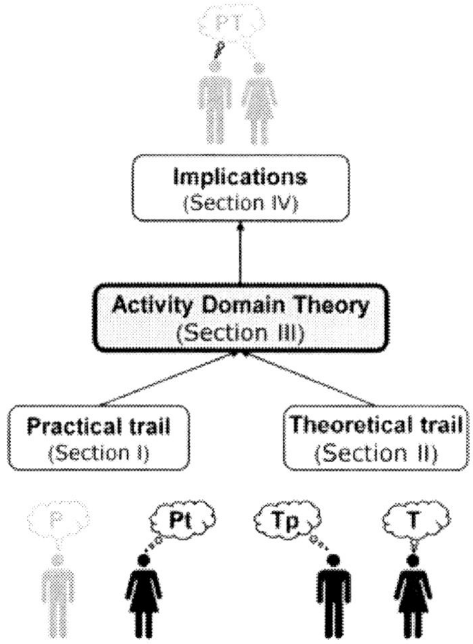

ization, spatialization, temporalization, stabilization, and *transition*. Modalities are manifested both as societal imprints in the form of artifacts, institutions, norms, etc., and as individual, cognitive imprints in the human mind and body. For example, a coordinative means like a calendar is a manifestation of the temporalization modality. In order to become a coordinative resource, however, actors must learn to use it and interpret it in a common way.

An important consequence of the ADT perspective is that sense making is inherently related to the activity domain. In other words, different activity domains enact by necessity different communal meanings[2] or taken-for-granted ideologies about what is meaningful.

The stakeholder focus of Section 3 is similar as for Section 2; however with more emphasis on the Pt type since an important part of ADT is the operationalization of the theory (Figure 5).

The chapters in Section 3 are as follows.

Cognitive Grounding
Reviews indications from cognitive neuropsychology and linguistic sciences that support the claim of activity modalities as basic features of the human cognitive system.

Operationalizing the Theory
Describes how the ADT is operationalized by modality specific models and IS support. These models are deliberately chosen to facilitate common understanding. Suggests a method, the *domain construction strategy,* for constructing the models and common understanding of them.

Figure 6. The stakeholder focus Section 4

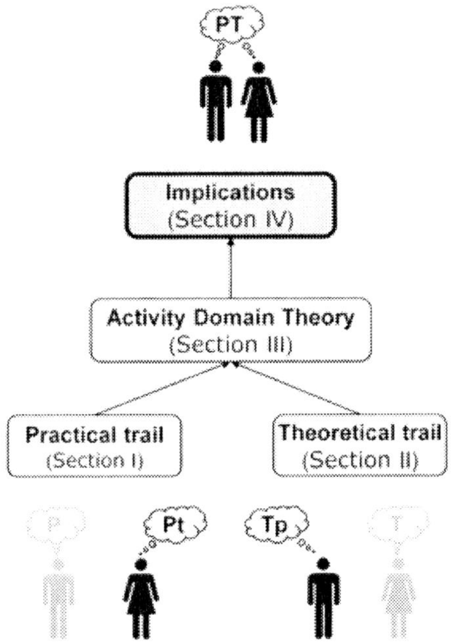

Positioning Against Other Theories
Positions the ADT against complexity theory and other akin, practice-based theories such as to Structuration Theory, Actor Network Theory, Organisational Semiotics, Work System Framework, and the Cultural-Historical Activity Theory.

The Practical and Theoretical Trails in Hindsight
Recapitulates the practical and theoretical trails from the ADT perspective; how are practical experience and theoretical roots reflected in ADT?

Section 4: Implications

Reflects on implications of the ADT in various areas. Since Section 4 represents a synthesis of practical and theoretical insights, stakeholders of the Pt and Tp types would be interested in this part. Moreover, it is my intention that Section 4 will provide arguments for the need of true border spanners; PT actors that combine deep practical and theoretical knowledge.

The chapters in Section 4 are as follows.

The Anatomy-Centric Approach for Coordinating Complex Projects
Reviews the anatomy-centric approach for coordinating complex projects. Elaborates the *anatomy* as a means to manage dependencies between capabilities in a system. Describes the anatomy-centric approach and discusses its implications. Suggests how the anatomy may be seen as an architecture, supporting agile development of software. Analyses the approach from the perspective of activity modalities. Demonstrates that the three images used – the anatomy, the increment plan and the integration plan –

each manifests a dominant modality: spatialization (the anatomy), transition (the increment plan) and temporalization (the integration plan). Discusses implications of this observation for the coordination of complex projects.

Enterprise Architectures: An Alternative View
Reconceptualizes enterprise architectures by using the activity domain as the basic architectural entity. Emphasizes the elements of communal meaning and transition between domains. Compares this view of the enterprise with other influential frameworks such as the one proposed by Zachman (1987). Discusses implications and how to operationalize the construction of enterprise architectures.

Product Lifecycle Management Revisited
Elaborates an alternative platform for Product Lifecycle Management (PLM). PLM is becoming increasingly important for product development companies as a means to manage products and product related information for a product during its entire lifecycle. The core of PLM is to implement large, often global information management system supporting PLM. Suggests how to operationalize PLM projects. Discusses implications of using the ADT as a theoretical guide for PLM projects.

Alignment: The Activity Domain in Focus
Proposes the activity domain as the central theme in aligning business strategies, knowledge strategies and IS/IT. The activity domain as an intermediate construct between the individual and the organization. Proposes an approach for how to operationalize alignment and discusses empirical evidence for the approach. Discusses implications of the ADT approach to alignment.

In Search for an Integrating Construct
Investigates the activity domain as an integrating construct. Questions prevailing conceptualizations of the business process as such a construct. Analyses different business process models from the activity modality point of view. Discusses IS development from the ADT perspective. Proposes to reconceptualize IS development as activity domain construction, where the IS and communal meaning are two designable elements.

In Conclusion
The concluding chapter. Describes the book as a parable from practice over theory development and back to practice with the theory as a new guiding lens. Summarizes the main messages of the book: the activity domain as an integrating construct, communal meaning to the forefront, the activity modalities – bridging individual cognition and social reality, paying heed to the unity of opposites. Concluding words.

REFERENCES

Giddens, A. (1984). *The Constitution of Society: Outline of the Theory of Structuration*. Los Angeles: University of California Press.

IS theories (2008). *Theories Used in IS Research*. Wiki, Retrieved Sept 21, 2008, from http://www.fsc.yorku.ca/york/istheory/wiki/index.php/Main_Page

Jones, M., & Karsten, H. (2008). Giddens's Structuration Theory and Information Systems Research. *MIS Quarterly, 32*(1), 127–157.

Kaptelinin, V., & Nardi, B. (2006). *Acting with Technology - Activity Theory and Interaction Design.* Cambridge, MA: The MIT Press.

Kraut, R., Streeter, L. (1995). Coordination in Software Development. *Communications of the ACM, 38*(3), 69–81.

Latour, B. (1992). Technology is society made durable. In J. Law (Ed.), *A Sociology of Monsters. Essays on Power, Technology and Domination* (pp. 103–130). London: Routledge.

Malone, T., Crowston, K. (1994). The Interdisciplinary Study of Coordination. *ACM Computing Services, 26*(1), 87–119.

MatrixOne. (2008). Retrieved Feb 2, 2005, from http://www.matrixone.com/index.html§

Orlikowski, W. (2000). Using Technology and Constituting Structures: A Practice Lens for Studying Technology in Organizations. *Organization Science, 11*(4), 404–428.

Parnas, D. L. (1995). On ICSE's 'most influential' papers. *ACM SIGSOFT, 20*(3), 29–33.

Schatzki, T. R., Knorr Cetina, K., & von Savigny, E. (Eds., 2001). *The practice turn in contemporary theory*. London: Routledge.

Standish Group, Inc. (2001). *Extreme Chaos*. February 2001, The Standish Group. Retrieved from http://www.standishgroup.com/

Taxén, L., Svensson, D. (2005). Towards an Alternative Foundation for Managing Product Life-Cycles in Turbulent Environments. *International Journal of Product Development (IJPD), 2*(1-2), 24-46.

Wartofsky, M. (1987). Epistemology Historicized. In A. Shimony & D. Nails (Eds.), *Naturalistic Epistemology* (pp. 357–374). Dordrecht: Reidel.

Zachman, J. A. (1987). A framework for information systems architecture. *IBM Systems Journal 26*(3), 276–292.

ENDNOTES

1. This is a common procedure in the IS discipline. For example, Structuration Theory (Giddens, 1984) has been used extensively in IS research (e.g. Jones & Karsten, 2008).
2. I will use two expressions to denote meanings that in some sense denote common stock in a society: "common understanding" as a practice oriented term and "communal meaning" as oriented towards theory.

Acknowledgment

The road to wisdom?—Well, it's plain and simple to express:
Err
and err
and err again
but less
and less
and less.

~Piet Hein, inventor, philosopher, poet

This book is the result of an enduring educational journey, which has been swaying between the attraction points of a deep interest in philosophy, my work at the Ericsson™ telecommunication company and my research at Linköping University. In a sense, I summarize my experience from a life-long career in industry and academia. Accordingly, I have allowed myself to be generous with what material I have included in the book. On the surface, issues like the philosophical question of what knowledge is, or how the cognitive system of humans conceives signs may seem rather loosely related to, say, requirement management. However, it is my firm conviction that it is not until we reach an integrated view on these seemingly non-related issues that we can truly address the extraordinary intricacies in managing complex systems in which humans are parts.

As is evident from the outline of the book, my figurative journey proceeded along several trails. The practical trail originates from 1968 when I started to work at Ericsson in Stockholm, Sweden. The experiences gained there form the empirical foundation of the book. I am deeply grateful to Ericsson for their overwhelming generosity in allowing me to use the detailed empirical material. Special thanks go to my contact person, colleague, and friend at Ericsson, Sören Ohlsson, whose continual interest in my results and life-long experience at Ericsson has been invaluable to me.

The theoretical trail originates from my personal interest in the praxis philosophy as interpreted by the young Marx and Engels in the nineteenth century. This perspective formed a kind of background fabric by which I reflected on what happened at Ericsson. This trail, no doubt, would have remained dormant if I had not met professor Sture Hägglund at Linköping University in 1997. He looked at my ideas over a cup of coffee, rubbed his chin and said: "Well, this might be something...". Sture put me in contact with my thesis supervisor Bengt Lennartson who has been my guide throughout my academic career. His patient and thorough criticism and encouragement have kept me on track whenever I was about to get lost at some interesting waterhole. Hail to you, Bengt!

The same goes for my other supervisors, Roland Ekinge at Whirlpool and professor Christian Berggren. Roland, who always asked those provocative questions, which (hopefully) taught me to challenge my own preconceptions and think instead of becoming defensive; Christian, with whom I often discussed shaky thoughts on shaky trains between Linköping and Stockholm.

Perhaps the trickiest part of the theoretical trail was to stay clear of philosophical pitfalls. Oh, how easy it is to get lost in a complete confusion here! "Philosophy is like trying to open a safe with a combination lock: each little adjustment of the dials seems to achieve nothing, only when everything is in place the door does open." (Ludwig Wittgenstein). For their help in break open this safe I am deeply indebted to two people: professor Göran Goldkuhl and Johan Schubert. Göran, whose graduate courses and seminars together with his students provided me with countless occasions for discussion and reflection. Johan, with his long experience as an opponent and supervisor, was always there to inspire, read and gently point out the flaws in my thinking. My special thanks also go to Joakim Lilliesköld for long, interesting, and truly rewarding research collaboration over many years.

As all authors know, it is impossible to write a book in solitude. I would like to honor the publishing team at IGI Global whose assistance throughout the writing process has been invaluable. Last, but not least, I am deeply grateful to Stefan Torneld for scrutinizing my English with his Argus eyes, always prepared to suggest corrections, alternatives, and improvements.

The practical and theoretical trails were eventually joined in the Activity Domain Theory, and from this perspective, possible, unbeaten trails have been outlined for the management of complex systems. With that, my final thoughts go to you who have been with me all the time: Tina for your ceaseless love and endurance, my beloved children Kerstin and Gustav for your everlasting encouragement and readings of the manuscript. And of course to Eva, my daughter-in-law and Anna and Erik, my grandchildren. Perhaps the essence of the book is best expressed by Anna. When three years old, looking at a colored version of the anatomy of the mobile switching center node in Figure 1, p. 13, she thought that the work packages looked like "square clouds over an island in the sea." Perhaps we should think about work packages as square clouds. Or as my favorite philosopher, Karel Kosík, expressed it: "Familiarity is an obstacle to knowledge."

Lars Taxén
Tullinge, April 2009

Copyright Acknowledgments

I am truly grateful to the following copyright holders for permission to reproduce their material in the book:

- **Inderscience Enterprises Limited** (www.inderscience.com): Figure 1, Figure 93, Figure 98, Figure 99, Figure 100, Figure 109, and Figure 110, which originally appeared in Taxén, L., Svensson, D. (2005). Towards an Alternative Foundation for Managing Product Life-Cycles in Turbulent Environments. *International Journal of Product Development (IJPD), 2*(1-2), 24-46.
- **PICMET** (http://www.picmet.org/main/): Figure 7, which originally appeared in Lilliesköld, J., Taxén, L., Karlsson, M., & Klasson, M. (2005). Managing complex development projects – using the system anatomy. In Proceedings *Portland International Conference on Management of Technology and Engineering, PICMET '05.* July 31st – Aug 4th, 2005, Portland, Oregon - USA.

- **Elsevier** (http://www.elsevier.com/wps/find/homepage.cws_home): Figure 12, Figure 13, and text extracts, which originally appeared in Taxén, L. (2006). An Integration Centric Approach for the Coordination of Distributed Software Development Projects. *Information and Software Technology, 48*(9), 767-780.
- **Wiley Inter Science** (http://www3.interscience.wiley.com/cgi-bin/home): Figure 15, Figure 20, Figure 38, Figure 105, and Figure 106, which originally appeared in Taxén, L. (2005). A Sociotechnical Approach Towards Alignment. *Software Process: Improvement and Practice, 10*(4), 427-439.
- **Ashgate Publishing, Ltd.** (http://www.ashgate.com/): Figure 78, which originally appeared in Remington, K., & Pollack, J. (2007). *Tools for Complex Projects.* Burlington, USA: Gower Publishing Company.
- **Elsevier** (http://www.elsevier.com/wps/find/homepage.cws_home): Figure 33, Figure 57, Figure 58, Figure 59, and Figure 71, which originally appeared in Taxén, L., & Lilliesköld, J. (2008). Images as action instruments in complex projects, *International Journal of Project Management, 26*(5), 527–536.
- **John Duncan Collection:** Figure 24, which originally appeared in Bryant, W. C., & Gay, S. H. (1883). *A Popular History of the United States.* Vol. I, New York: Charles Scribner's Sons. Original wood engraving by E Bayard.
- **Ericsson:** On several occasions, mainly in Section 1, I make use of empirical material that is the property of Ericsson. Moreover, Ericsson is a registered trademark. Whenever Ericsson occurs in the text, it should be read as Ericsson™.

Section 1
The Practical Trail

The dawn of the Activity Domain Theory: A pattern emerges (1990-1995). Coordinating incremental development (1996-1997). Termination and despair (1998). Renewal in conflicts – the pioneering S-domain (late 1998-1999). Other practices catch on – the C-domain (late 2000-2001). From federalism to monarchy (2002). Analysis of the Framework trajectory. The broader scope: Business processes. Overloading the process concept. Glitches between processes. Interdependencies. Information management projects. IS/IT architectures. Fragmented responsibility. Insights and needs: Common understanding. Integrating construct. Contextualization. Historical awareness. Recurrent patterns. Enactment.

INTRODUCTION

Section 1 is an historical account of experiences, observations and incidents from the Ericsson development practice that somehow have influenced the Activity Domain Theory (ADT). There are two main paths that contributed to this. The first one is the introduction in the early 1990s of the anatomy as a main instrument for managing the complexity of mobile telephone development projects at Ericsson. The second path is the evolution of coordination support based on a particular information system (IS), Matrix, in which I was personally involved. This path is called the domain construction strategy (DCS) in the book.

Figure 1. An example of an anatomy of a processor (Lilliesköld, Taxén, Karlsson, & Klasson, 2005. ©PICMET. Used with permission)

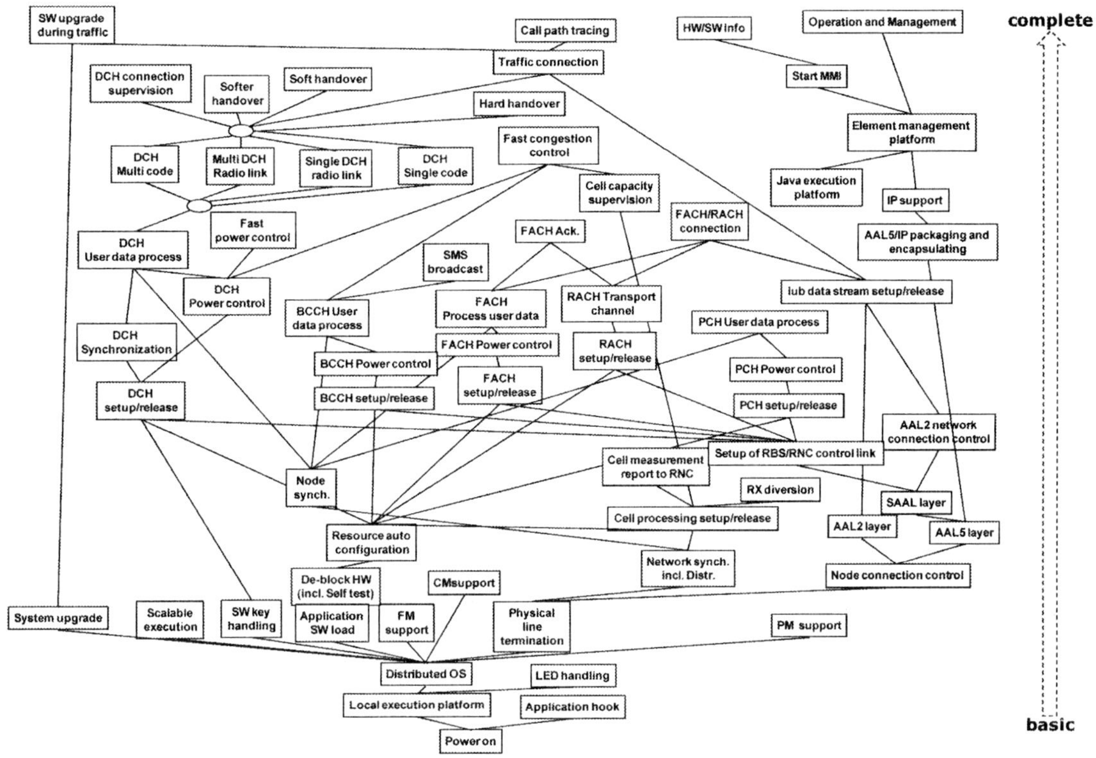

The Emergence of the Anatomy

The evolution of the mobile systems has been described as a sequence of generations (Lindmark, Andersson, Johansson, & Bohlin, 2004). The first generation, 1G, was developed in the 1980s until replaced by 2G generation in the 1990s. The main difference between 1G and 2G is that 1G radio signals are analogue, while 2G signals are digital and handed over between "cells" in the network as the mobile moves along. The architecture for the 2G systems is based on the GSM standard (Global System for Mobile communications; originally from Groupe Spécial Mobile).

By the early 1990s, the wireless industry was ready to go for digital cellular systems. On Jan 16th, 1992, Ericsson was awarded a contract with Tokyo Digital Phone for a mobile telephone system to be delivered on April 1st, 1994. At that time, the cellular mobile technology was in its infancy, the telecom market had been deregulated, and several new operators began to take advantage of the new technology. The project turned out to be a success, and by 1996, Ericsson passed the 1 million subscriber milestone for its CMS 30 system, making Japan one of the fastest growing markets for cellular telephony.

It was in the Japan project that Ericsson introduced the concept of the "anatomy" of a system for the first time. The anatomy is an architectural view, illustrating the dependencies between capabilities, from the most basic one to the complete capability of the system. An example is given in Figure 1. Each box represents a capability, and each line a dependency. The details are not essential here; however, note the "Power on" capability at the bottom of the anatomy! If the power cannot be switched on, all other capabilities will fail.

The anatomy differs from most other architectures in the sense that its purpose is to provide common understanding about the most critical dependencies in the system. The simple motivation for concentrating on this view is that if these dependencies are not clear to all project members, there is a great risk that the project will fail.

The anatomy was one of the reasons behind the successful outcome of the Japan project. This way of working represented a complete break with the previously dominant "waterfall" method. Due to various reasons, however, the lessons learnt were not transferred systematically to other projects. Rather, the experiences were, so to say, sedimented way back in "organizational memory" as a faint insight that complexity could be successfully managed by something called the anatomy.

In the late 1990s, the telecom industry was ready for the next, technical "quantum leap" – the transition to the 3rd generation of mobile systems. These systems were to be based on a new standard called UMTS (Universal Mobile Telecommunications System) that enabled substantially higher transmission rates needed to access the Internet, transfer video streams, data, etc. The challenges that Ericsson faced were expressed by the total project manager:

The total technical changes being implemented in this project are enormous. Such changes are needed in order for Ericsson to get a world-leading product first to market. Using traditional methods then the scope of change implemented in single steps will be too large and cannot be managed (Total project manager, Ericsson, Dec 1999).

The method chosen to implement the 3G systems was an elaborated anatomy-centric variant that had been worked out at the Ericsson development unit in Aachen, Germany. It was quite clear, however, that the coordination support used in the 2G projects was insufficient for the 3G development. It so happened that a new kind of support, the domain construction strategy, had been developed along a quite different path.

The Domain Construction Strategy

The second path can be traced back to another "quantum leap" that Ericsson tried to achieve during the 1990s. At about the same time as the 2nd generation of cellular systems was launched, Ericsson set out on a major endeavor to replace its AXE switching system. This system, which turned out to become a tremendous success for Ericsson, had been in development since 1958 and was formally launched in 1978 (Vedin, 1992).

By the late 1980s it was clear that the need to transfer data, video and other information over the telecom net was increasing. This need was to be met by a broadband successor of AXE – the AXE-N system where N stands for "network". This project ran between 1987 and 1995 without much interaction with the mobile systems that were developed simultaneously. However, the AXE-N project was terminated without having delivered a commercial product. The cost for the project, which was a major failure in the Ericsson history, has been estimated to a stunning 10 billion SEK, roughly about 1.1 billion Euros (Lindmark, Andersson, Johansson, & Bohlin, 2004). Even so, this money was not entirely wasted since many experiences made during this period could be reused in subsequent system development projects.

In 1990 I started to work in the AXE-N project in the Hardware Support System group. The purpose of this group was to provide support systems and processes for developing the hardware in the AXE-N system. For the evolution of the ADT, this period can be seen as the instigator and inspirer of many fragmented ideas that later were brought together.

When the AXE-N project was terminated in 1995, I was assigned to a group whose purpose was to consolidate various initiatives to develop systems in an incremental, or stepwise, way. This resulted in a method package, that is, a collection of instructions, templates and descriptions of how to perform incremental development. In order to support this kind of development, some kind of tool support was needed. In 1996, I came in contact with a system called Matrix (MatrixOne, 2008), which became the backbone of the coordination support for the 3G development projects.

During 1997, Matrix was tried out in a number of pilot projects that I was responsible for. Late 1997 these projects were finished, and the results achieved so far were encouraging. Nonetheless, due to internal conflicts about what IS to base the support on, the continued work based on Matrix was questioned by several influential persons. The entire initiative was about to be terminated early 1998, which coincided with the start of my Ph.D. studies. However, in May 1998 the project manager for a new 3G switching project was convinced by the good results achieved, and decided to use Matrix in his project. An extensive period of development took place. On May the 10th, 1999 the traditional way of managing engineering change orders and requirements was shut down, and the corresponding support based on Matrix "went alive" in a sharp production environment for the first time at Ericsson. The coordination environment that gradually evolved at the Stockholm site will be referred to as the S-domain in the book, where "S" stands for Stockholm.

The Anatomy-Centric Approach

Later in 1999, three more projects started to use the Matrix-based support in the S-domain. At the same time, project managers at other Ericsson sites were ramping up their 3G activities. Early 2000 it was decided to use the same support strategy also at these sites. This was a crucial point in the evolution of the coordination support in the sense that the anatomy and the DCS met for the first time. The result, i.e., a close combination of a method and a support strategy make up what I call the anatomy-centric approach in the book.

The overall coordination responsibility for the 3G projects was situated in Aachen, Germany (referred to as the A-domain in the book). Still another domain, the L-domain, was constructed at Linköping, Sweden during 2000. In April 2001, Ericsson signed a corporate agreement with MatrixOne for Matrix licenses. By this, the Matrix tool was accepted for the first time as a corporate matter of concern. Altogether, the coordination support achieved at the different domains was previously unmatched in the Ericsson history. Between approximately 1999 and 2003 around 140 main projects and subprojects were successfully coordinated using the anatomy-centric approach.

During the year 2000, the telecom crisis hit Ericsson (Lindmark, Andersson, Bohlin, & Johansson, 2006). In order to save money, investigations were conducted to consolidate the S-, A- and L domains into a single domain: the C-domain. This eventually resulted in the termination of the S- and L domains. The consolidation was, however, far from straightforward, and in the end the A-domain remained as an independent area. Finally, the responsibility for the C-domain was outsourced to another company, an event that efficiently shut the window to the innovative way of constructing coordination support at Ericsson.

Figure 2. The time line of the practical trail

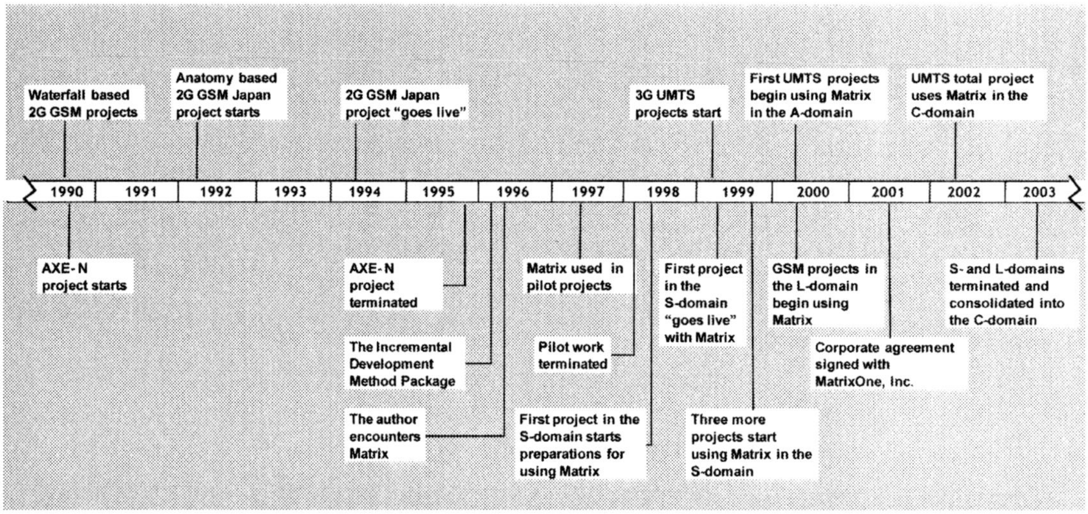

In Figure 2 the time line of this story is illustrated. In a way, this story can be seen as a cut through a large organization at a particular place and time as seen from my personal perspective. Others would most likely interpret the events that took place in a different way.

REFERENCES

Lillieskôld, J., Taxén, L., Karlsson, M., & Klasson, M. (2005). Managing complex development projects – using the system anatomy. In *Proceedings Portland International Conference on Management of Technology and Engineering, PICMET '05,* July 31st – Aug 4th, 2005, Portland, Oregon, USA.

Lindmark, S., Andersson, E., Johansson, M., & Bohlin, E. (2004). Telecom Dynamics, History and State of the Swedish Telecom Sector and its Innovation System 1970–2003. *VINNOVA Analysis VA 2004:04.* Stockholm: VINNOVA.

Lindmark, S., Andersson, E.J., Bohlin, E., & Johansson, M. (2006). Innovation system dynamics in the Swedish telecom sector. *Info: The journal of policy, regulation and strategy for telecommunications, 8*(4), 49–66.

MatrixOne (2008). Retrieved Feb 2, 2005, from http://www.matrixone.com/index.html

Vedin, B. (1992). *Teknisk revolt - Det svenska AXE-systemets brokiga framgångshistoria.* [Technical revolt – The diversified success story of the Swedish AXE system (in Swedish)]. Värnamo, Sweden: Atlantis.

Chapter 1
The Dawn of the Activity Domain Theory

In this chapter, the evolution of the domain construction strategy is recapitulated. This story is divided into a number of phases that can be seen as "life ages" of the strategy at Ericsson. It was born during particular circumstances in the late 1990s, it had its peak during the millennium shift and it died with the collapse of the telecom market around 2002-2003. Its remains still linger on at Ericsson, but in a completely transformed way where the essential elements of the strategy have evaporated into thin air.

However, the experiences gathered during this period became the fertile soil for the ADT and its operationalization. The story I will recapture in this chapter is not an account day by day. Rather I highlight such events that in hindsight come through as key insights but, when they happened, might not have been particularly noticed. As in many cases where theory and practice mutually influence each other, it is not possible to say that theory came before practice or the other way around. It was not until I could frame my practical experience in the theoretical perspective given in Section 2 (The Theoretical Trail) that the ADT began to materialize as a full-fledged theory.

In the midst of the maelstrom of events, the signs of the ADT came forth as certain elements that later was conceived as the operationalized form of the ADT, that is, the expression of ADT in elements that can be manipulated, measured or observed in a particular situation in order to influence this situation. The main elements that appear in the story – the Matrix tool, the information model, the process model and the transition model – gradually emerged as vital for supporting coordination. Rather early in the life of the domain construction strategy, around 1996, I began to gather these under the banner of a "Framework", a construct I will refer to occasionally in this chapter. Thus, a peculiar observation is that the ADT first appeared in the guise of its operationalization. So, let's go back to the dawn of the ADT in the early 1990s!

DOI: 10.4018/978-1-60566-192-6.ch001

A PATTERN EMERGES (1990-1995)

Many of the ideas that subsequently were included in the ADT came from my participation in the AXE-N project. I do not claim to be the first to come up with these ideas. On the contrary, taken one by one these are well known. My contribution, I believe, is to have selected and organized these ideas into a coherent theory.

Concept Elaboration

In the AXE-N project many new concepts were invented, and these concepts had to be relevant, unambiguous and understandable to the actors. To this end an ambitious subproject was launched to collect and define concepts. The project had special teams whose purpose it was to collect candidates for new concepts. Pending concepts were taken to a reference group where it was decided if a particular concept would be included in the AXE-N project encyclopedia.

In spite of this work, it was very difficult to get an overall picture of the project where concepts could be seen in their context (LTX-1994-08-15, ERI-1995-03-15)[1]. For example, in the new system development process to be used in the project, more than 120 new concepts were defined in a list without any conceptual map that could explain how they were related to each other (ERI-1993-09-23).

To make matters worse, directives were issued towards the end of the project to use the AXE-N system development process for all kinds of developments, for example, of educational material. This meant that concepts in the AXE-N system like "node", "logical reference model", "system entity", etc., had to be appropriated into the educational area. This created even more confusion: "Is a manual a node or system entity?" "Can a chapter be regarded as a logical reference node?" And so on.

Thus, in a few years (1990-1995) a completely new organizational language was to emerge, often in conflict with the traditional one. This turned out to be an overwhelming task. In parallel to this, a separate unrelated initiative was started outside the AXE-N project to define one hundred core concepts within Ericsson (ERI-1992-09-30). These experiences indicated that the effort to create common understanding in a work setting is in general underestimated, if paid attention to at all.

The System Design Environments

In the AXE-N project my main task was to contribute to the development of a Hardware Design Environment. The principles for this environment were laid down in a system description in 1991 (LTX-1991-10-09). The guiding principle was expressed in the following way:

The Hardware Design Environment shall combine a long-lived architecture with a flexible functionality based on purchase (LTX-1991-10-09, p. 4).

In this document a modular process architecture is described in which a *process core*, *process components* and *applied processes* are the main elements. The idea was to treat a process like any other product and configure tailor-cut processes from the process components. The purpose of the process core was to collect basic principles and guidelines that would apply to both process components and applied processes. This architecture was further refined in the following years (LTX-1994-07-06, LTX-1994-08-15).

A process component was a package of process descriptions, status checks, rules, tools, templates, etc., with the purpose of providing a designer with a complete set of utilities for a particular design task. The hardware development process was more heterogeneous than the software development process since different types of hardware were being developed such as printed circuit boards, ASIC's (Application Specific Integrated Circuits), multi-chip modules, etc. Thus, the contextual aspect was more salient in hardware design than in software design.

Early on it was recognized that the data used in the process components needed to be translated between the interior and exterior of the component. One purpose was to archive data in a format which was independent of the particular tools that might be used in a process component. To this end a mechanism to translate data between process components was developed. This mechanism, called the design information interchange model (DIM), was defined in the following way:

DIM is a kind of framework for how to handle interfaces between data. DIM sets up rules for which formats may be used for different views for hardware information, for example behavior descriptions, logical circuit descriptions, net lists, test data, layout. Moreover, DIM provides tools to simplify translations between formats. (ERI-1991-08-08, p. 2)

The work with DIM was one of the original indications of the transition modality in ADT. Another source of transition was the specification-based data model (SBDM) suggested by Gandhi & Robertson (1992; 1995). This model directly appealed to me as a potential model for handling the borderline between different contexts:

The specification-based data model (SBDM) is a unifying framework to model configurations of systems that contain components from differing engineering disciplines. (Gandhi & Robertson, 1995, p. 338)

My interest in modeling the separation of contexts originated in experiences from hardware design methods. For example, during 1990 I became a project manager for a study with the purpose of investigating the suitability for VHDL (Very high speed integrated circuit Hardware Description Language) as a design language. One of the key issues was studying *translators* (LTX-1990-06-18). The purpose of a translator is to make it possible to move between abstraction levels, for example, where a transition took place between "integer" and "byte" levels running at different clock speeds. The translators were early examples of the importance of attending the transition between contexts.

The basic construct of SBDM is a recursive structure of specifications, which are implemented by implementations and in turn need other specifications. This was utilized in an attempt to combine the process modules with an SBDM model of the system to be developed (LTX-1994-03-29b). In Figure 1 an example of this is shown where a certain specification can be implemented in either software or hardware using different process components for the design of the implementations.

Early in the AXE-N project, a project member suggested something called Information Interaction Models (IIM) as a comprehensive way of representing processes. A striking example is the IIM used for developing a multi-chip module (a complex integrated circuit) (Figure 2):

The horizontal lines represent information entities, and the vertical ones input and output to various activities lines up horizontally at the bottom of Figure 2. At the top, diamond shaped icons indicate progression control points. The detailed structure of IIMs is explained in detail in the Section *The process model*, p. 128; here it suffices to grasp the overall composition of the IIMs.

Figure 1. An example of the SBDM model

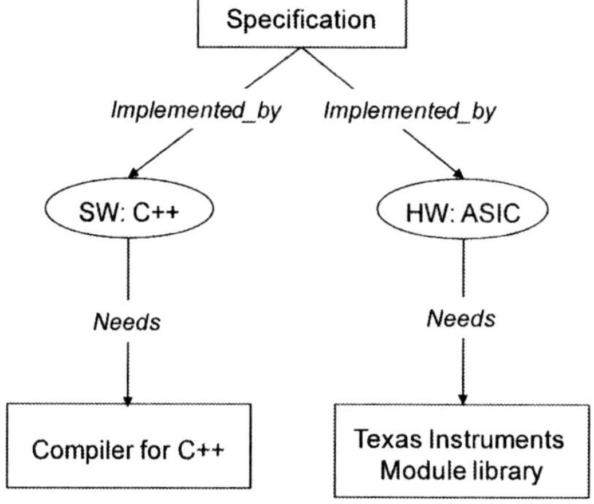

Figure 2. An IIM for the process component "Multi-Chip Design"

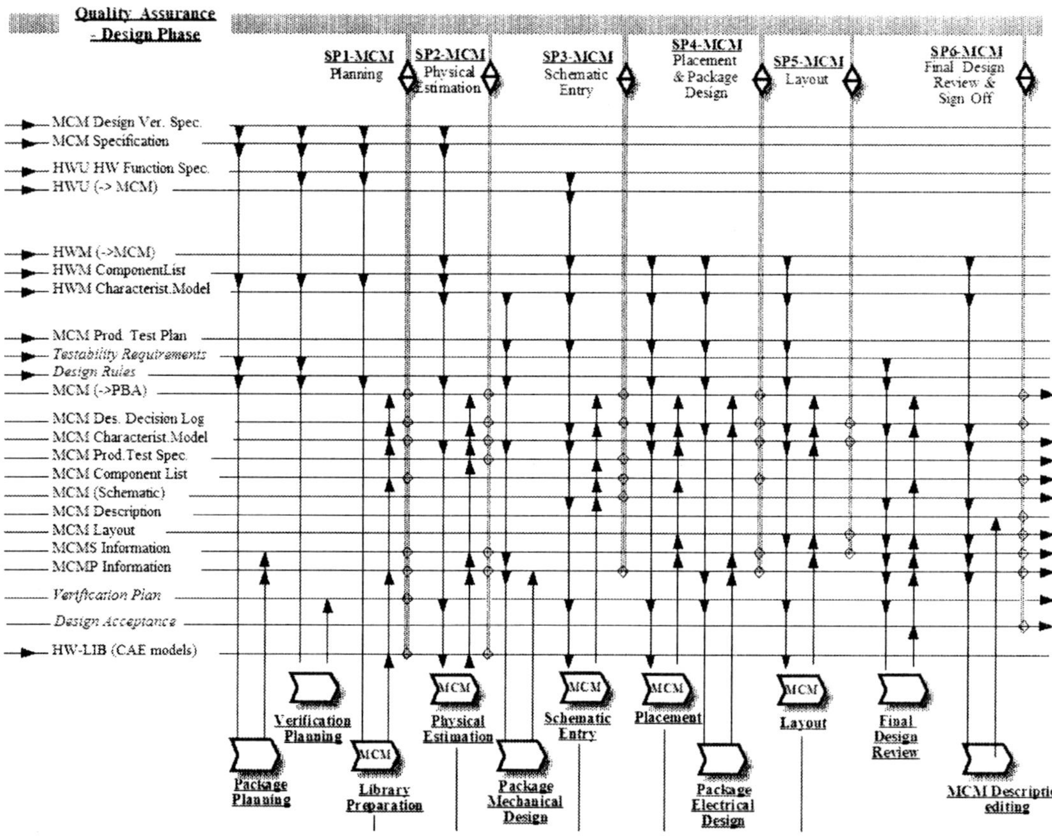

In the internal document LTX-1994-07-06 it is stated that IIMs have a number of advantages. For example, they have a core structure which can be applied to any process module and any layer in a process hierarchy. Another observation is that it is necessary to separate between common aspects of the process and those which are unique to a particular group. These two aspects are called the "administrator" and "entrepreneur" side of the process respectively. Methodology and support systems are considered commodities and, as such belonging to the entrepreneur side, while rules for identification, quality systems, etc., are more long-lived and belong to the administrator side. Thus, maintaining a balance between more stable and more agile areas seemed important.

The IIMs were later adopted as the standard way of representing processes for hardware design in the AXE-N project (LTX-1994-08-15, ERI-1995-03-18). One observation from the work with the multi-chip module was that an enlarged paper copy of this process model was hanging on the wall in the project room. The progress of the design was marked on the paper for everybody to see. Thus, the model served as a communication instrument in the project. During the AXE-N project the issue of automating process support was discussed a lot, but it occurred to me that the most important issue was the common view of the process. The automated support did not seem nearly as important, it was quite sufficient to see the paper on the wall.

During 1992, discussions started to coordinate the software and hardware development processes. The AXE-N system consisted of a large amount of hardware, yet the software process was constantly prioritized over the hardware process. In order to achieve a more even balance a strategy was defined as follows:

Starting from a common process architecture we first focus on each process area by itself and then we integrate them. (ERI-1992-12-21, p. 2)

The progress in a number of consolidated areas was summarized in a report (LTX-1994-04-01). Only in 7 out of 18 areas did some alignments exist. Most remarkable was that the foundation, the process architecture, was not aligned after more than a year's discussion. My recollection from participating in the consolidation team is that persons from the software and hardware communities could not agree on the basic perspective. The modular process architecture in the hardware process was never accepted in the software community. There were constant discussions about basic concepts and how to organize the alignment work. The effort to align the software and hardware processes was finally terminated when the hardware development was moved out of the AXE-N project in early 1995.

The experiences from this undertaking triggered many reflections which were later manifested in the ADT. One observation is that we tried to align too much. A common understanding cannot encompass everything. There must be a balance between what should be coordinated and what shouldn't. The separate practices of software and hardware design should be regarded as cooperating practices, not as one uniform, single practice. Again, this experience indicated that the effort to agree on the meaning of concepts is vastly underestimated.

Information Management

In the report ERI-1989-04-03 general principles for handling of the AXE-N system were defined. In this report fundamental characteristics regarding "surviving systems" are discussed, such as stability, flexibility, integration of new components, and autonomy of different system parts. The importance of stability is expressed in the following way:

A stable foundation is necessary. For humans this is provided by the stable structure of the DNA, which is then found at all levels in the human body. In an organization the business processes are adaptable while the type of information managed in general is indifferent... [the type of information] is structured in information models [...] which make up the stable structure. (ERI-1989-04-03, p. 3)

In another report, ERI-1991-04-03, principles for information management in the AXE-N project were defined. The report proposes a way of managing information which is a radical break with the traditional AXE world. For example, "documents" are to be replaced with "information elements" as management items. The report also includes a suggestion for an information model for the management. However, there is no discussion about how to arrive at a common understanding about this model. This was simply not an issue.

Outside the AXE-N project there was an uneven feeling that the importance of maintaining strict handling rules was downplayed. It was pointed out that the product handling rules in the traditional AXE system must be upheld also in the AXE-N system. This task was aggravated by the fact that functions in the system are realized by both hardware and software. These areas are more or less two different worlds alien to each other. The report ends with a strong statement regarding the importance of common understanding:

[The bottom line is that] we must in all of our operations have a common interpretation and understanding of what kind of product we are dealing with and what terms and prerequisites are valid for this. (ERI-1991-04-04, p. 17)

In the report ERI-1991-04-04 important principles concerning product handling in the traditional AXE world were treated. The report points out that these rules must be upheld vigorously also in the AXE-N system. There simply had to be some stable part of the system. These observations pointed towards the stabilization modality in ADT.

Information Systems

The IS in the AXE-N project had very ambitious goals from the start. The perspective was to develop a first class system which was superior to any commercial system available on the market. The main principle was that all information was to be managed in one IS only (ERI-1991-04-03).

A forerunner to the IS was being developed during the early years of the project. This development was however terminated in December 1992 on the very same day it was supposed to be released. The reason was that it did not match the requirements for sharp usage in connection with the development of early test nets of the AXE-N system. An interim IS solution was launched with the main purpose of securing the storage and retrieval of files. Gradually the ambition level was lowered and in December 1993, the internal development of the main IS was terminated. The strategy from that point on was to use a commercial system for software configuration management. This "devolution" was summarized in 1994 by one of the participants as follows:

The interim IS was and remains a quick and dirty solution. It cannot possibly evolve into an information management system worth the name. The interim IS was a conscious sub-optimization when it was introduced. It brought the information management ten years back in time. The interim IS is becoming a part of the problem instead of the solution [...]. Now a commercial system for configuration management is introduced. The intention is good. But I see an imminent risk that this will be a continued bottom up development which makes the "temporary" conceptual world of the interim solution permanent. (ERI-1994-04-26)

Another project member expressed this even more bluntly:

The current state of affairs concerning information management in AXE-N is a great scandal; the new clothes of the Emperor multiplied by two. In particular, I mean that the introduction of the interim IS as an information system borders to fraud.

During this period it gradually became evident that the original idea of managing every piece of information in one system was altogether unrealistic. The configuration needs for developing software were quite different from that of managing, for example, product structures. This observation indicated that there was a need to distinguish between different contexts. Different activities simply need different tools. This may sound very sensible and straight-forward, but during this period the opposite "truth" was common stock.

In 1993 I became responsible for the technical area called Hardware Support System in the AXE-N project. The need for various information systems tailored to specific tasks was obvious in the hardware community. Also, there were a number of commercial support systems available. This sparked off an investigation of commercial Product Data Management (PDM) systems as backbones tying all specific ISs together. The investigation led eventually to the establishment of a corporate standard PDM system in December 1994. I will refer to this system as C-PDM[2] in the following. Later on this decision became a source of conflicts, since the IS used in supporting incremental development in the AXE-N project – Matrix – was considered a violation of the Corporate policy concerning PDM systems.

The final step in the saga of the ISs in the AXE-N project was taken in July 1995 (about half a year before the project was terminated) when the basic principle concerning the IS was redefined as "The information will always be managed in several systems" (ERI-1995-07-06, p. 5). Thus, the original principle was completely reversed. No strict analysis was ever made of this calamity. However, a number of issues were brought up:

- The conceptual world of the main IS was never demonstrated and tested. Neither were the theories behind the system (ERI-1994-04-26).
- It was not possible to understand the system without confirmation in practice. Therefore, the development must be made in steps. (ERI-1994-04-26). Prototyping should have been included in the project from the beginning (ERI-1994-04-26) (this experience points towards the importance of enactment of the technology at hand).
- The long term solution was too long term and the short term solution too short. The world of the interim IS become more or less permanent (ERI-1994-04-26).
- One IS cannot be used for wildly different activities (ERI-1994-06-02). The intention to create one support environment that everybody could use was futile (ERI-1995-12-12). Each activity should have ISs which suit the tasks of that activity.

- The management of structures is different from the management of files during software development (ERI-1994-06-02).
- The management issues were largely neglected by the project management (ERI-1994-08-15b)
- Lack of a common language (ERI-1994-08-15b).
- Difficult to see the whole picture, nobody had control of this, deep isolated islands of knowledge without communication in between (ERI-1994-08-15b).
- Requirements were unclear and a dedicated customer did not exist (ERI-1994-08-15b).
- There was a lack of governing directives. Standards were not followed. We invented our own standards (ERI-1995-12-12).
- The bureaucracy in the project was terrible. The development should have been headed by experienced senior designers rather than line managers (ERI-1994-04-26).

Towards the ADT

In 1995, about half a year before the AXE-N project was terminated, I wrote a report which summarized the experiences from working with the design environment for hardware in the AXE-N project (LTX-1995-04-26). This report contains several examples and suggestions which later became vital elements in the ADT. The first attempts to theoretically ground ADT in praxis and dialectics were reported at a conference in Austin in December 1995 (Taxén, 1995). It so happened that the AXE-N project was terminated while I was there. Much of the continued work towards the ADT and its operationalization is grounded in these two documents and in an internal report: "A Coherent Framework for Development" (LTX-1996-02-23). The latter may be seen as the first sketch of the ADT.

COORDINATING INCREMENTAL DEVELOPMENT (1996-1997)

The experiences I got from working in the AXE-N project provided a kind of mind-setting for the next phase towards the formation of ADT. As described earlier, Ericsson had for some years started to change their way of working towards an incremental model for developing large software systems. The evolution of ADT between the years 1996-1997 was mainly related to my work with providing support for the coordination of incremental development projects.

The Incremental Development Method Package

It turned out that many projects at Ericsson were using different variants of incremental development in the early 1990s. Such variants concerned the definition of "increment" ("feature increments"; "design increments"; "integration increments"), whether increments should be considered as tasks or products, what types of increments should be testable or not, and a number of other aspects[3]. Discussions had been going on for quite some time about what incremental development was all about, and it was difficult to agree on a common understanding of this concept. There was a need to consolidate the different experiences learnt.

Early 1996 I joined a project which should compile best practices into an incremental development method package (IDMP). In February 1996, I suggested to use a conceptual model in order to make our discussions more efficient. The intention was merely to have a picture to which you could point and say

Figure 3. The conceptual model for incremental development (1996)

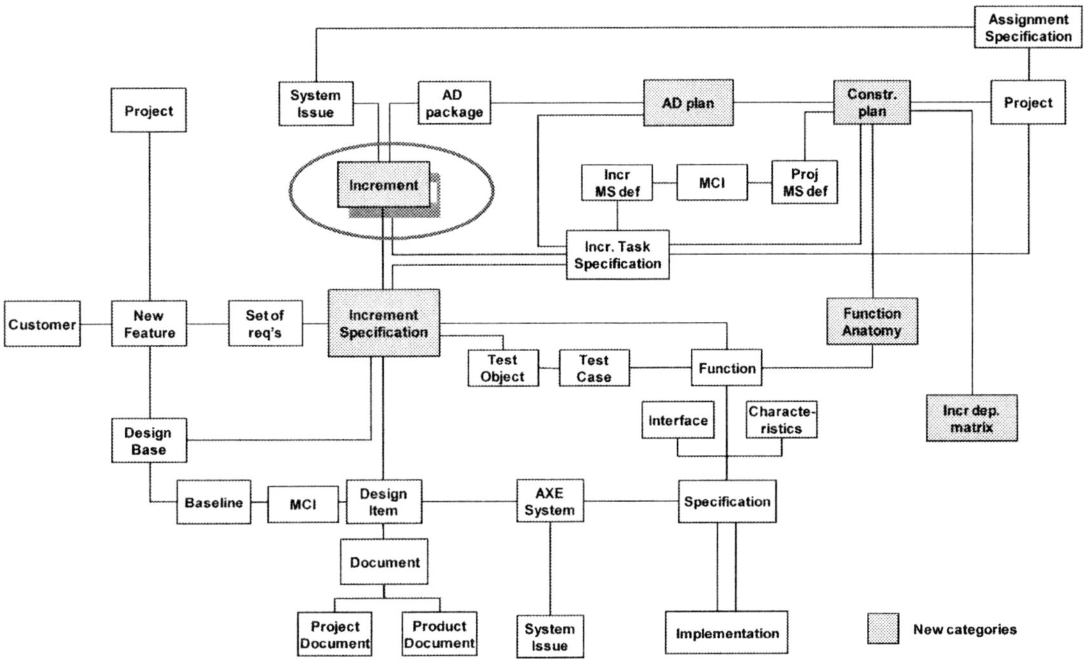

something like: "I want this relation to go from here to here instead" or "This object is not important to incremental development", and so on.

With the help of the model the discussions slowly did begin to converge. In March 1996 the first version of the model was ready. This was further refined into the one illustrated in Figure 3. The boxes signify things that were considered important to incremental development and the lines signify relations between them. Some parts of the model were already existing in the traditional way of working (the waterfall oriented model), but some parts were new. Since the coordination items at this time were mainly documents, most boxes represent different document types. The focal item, the "Increment" is indicated by the oval. It can also be noted that the SBDM is included (the "Specification" – "Implementation" part in the lower right hand corner).

The IDMP was refined into an Ericsson product containing guidelines, method descriptions, document templates, etc. Several Ericsson internal seminaries were held in Sweden and Holland during 1996 and 1997. In January 1997 the first product release of the IDMP was made (ERI-1996-12-02).

All throughout this period, discussions continued about the meaning and interpretation of the different revisions of the conceptual model. One of these discussions concerned how to modify the standard Ericsson project management method PROPS[4] for incremental development. Here is an excerpt from a discussion in November 1996:

During our work with incremental development and PROPS, the generic version, we have come across ambiguities in the definitions in the UAB[5] method package. The basic inconsequence lies in the fact the activity and the result of an activity are confused by giving them a common name. (PROPS developer, ERI-1996-11-05)

The reply to this statement was as follows:

To me it is obvious that "project" stands for both result and activity. How we separate between these things can best be clarified in a conceptual model. All in all, the model shows how we thought when we integrated incremental development into existing methods. We can't describe the world any better than anyone else can. But we are pretty good at describing how we think about it. (IDMP project manager, ERI-1996-11-05)

The method package was used in a project called CMS-30 phase 7, which had a Japanese operator as a customer. Some conclusions from the evaluation of this project were (ERI-1997-10-23):

- Incremental Design is a possibility, but needs some improvements.
- It must be agreed that we need several TG2's for one project[6].
- Milestones are to be planned per increment (a good tool is needed).

It so happened that a candidate for a tool to support incremental development was making its way into Ericsson.

Matrix Enters the Stage

Already in January 1996 the tool issue was discussed and a requirement specification was written in March 1996. It was clear that none of the existing tools did fulfill the specification. In May 1996 I was visiting a construction fair in Gothenburg when, by coincidence, I came across a lonely looking salesman at a terminal. Roughly the following conversation took place:

- Lars: What have you got here?
- Salesman: It's a tool called Matrix, I'll show you.
- Lars: Can you implement this model? (I drew a sketch of the SBDM).
- Salesman: Sure, no problem (a couple of minutes later he had implemented the model and instantiated some objects from it).
- Lars: Wow! That was impressing!

This unlikely event led to further contacts which eventually established Matrix as the IS in the Framework. Matrix was originally a PDM system aimed at industrial management of globally distributed, large quantities of product data and with many thousands of users (MatrixOne, 2008). In September 1996 Matrix was suggested as the tool for supporting incremental development and in October 1996 a demonstration license was bought by Ericsson. The first installation of Matrix was done in April 1997. The next step was to start up some pilot projects.

The Pilot Projects

In May 1997 a consultancy company was engaged to contact potential pilot projects within Ericsson. A tool prototype called the construction planning tool (CPLtool) based on Matrix was developed. The first demonstration was held on June the 9th, 1997. Later that month contracts were written with pilots in the

Netherlands, Germany and Karlskrona. I became the technical project manager for the development of the CPLtool. During this period the information model[7] evolved into the one shown in Figure 4.

Each box in the model was directly implemented as a type in Matrix and each line as a relation. In addition to this, state chains, attributes, cardinalities and revision stepping rules were implemented. The implementation was tried out in practice and modified whenever needed. In comparison with the model in Figure 3, the tool supported model shows less document oriented types. For example, the so called master configuration index (MCI) visible in the first model is now replaced by relations. The boxes at the bottom (ANT, CNT, etc.,) denote Ericsson specific product and document types.

The model also contains types that are related to project management, for example Resources, Increment Task Specification, etc. The box "Impact" is in fact an attribute on the relation between "Resource" and "Feature Increment" which holds estimates for the effort to develop a certain increment. In Figure 5, an example of a Matrix view is shown, where instances of the types defined in the models can be seen along with their relations. Basically, the figure shows the tracing from customer needs all the way down to various Ericsson-specific products and document implementing these needs.

Thus, it can be seen that the introduction of the tool opens up new possibilities to construct the coordination practice. Furthermore, the information model and the implementation in Matrix were continuously changed as the construction of the domain evolved. The models and the implementation were however different in all pilot projects, indicating the need to adapt these to local circumstances.

Figure 4. The information model after the introduction of the Matrix tool (1997) (Taxén, 2006. ©Elsevier. Used with permission)

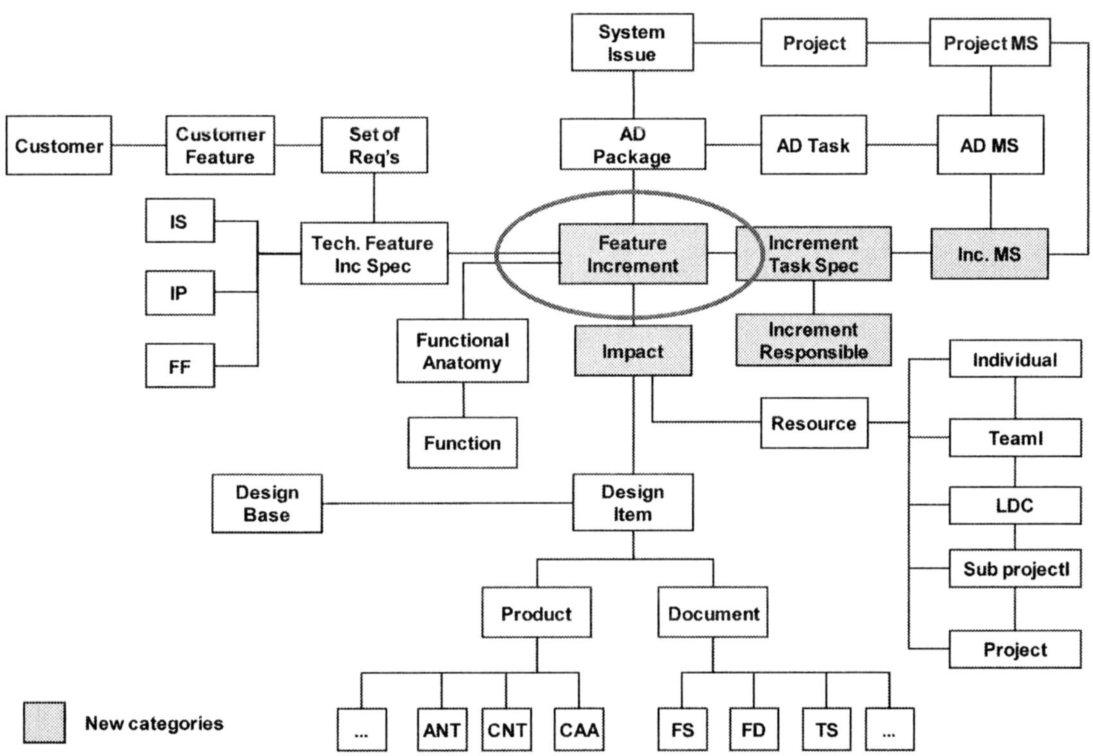

Figure 5. A screen dump of the Matrix tool (Taxén, 2006. ©Elsevier. Used with permission)

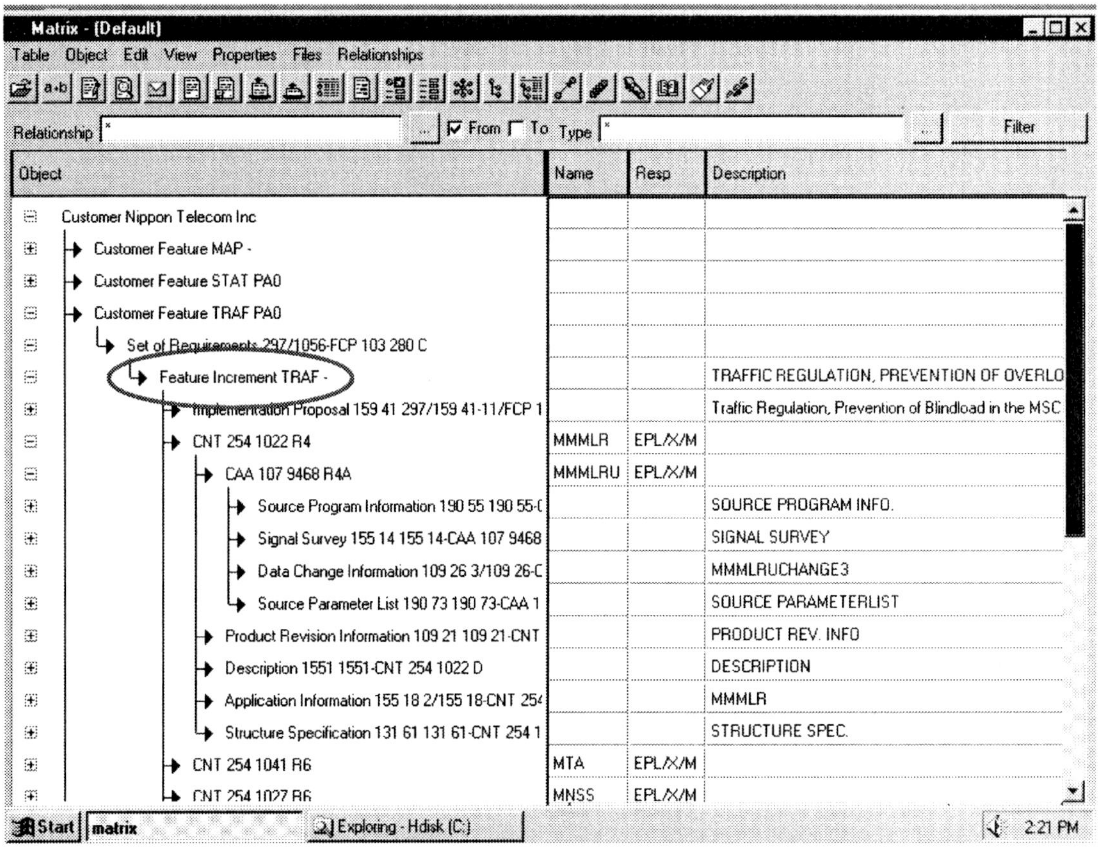

The results of the pilot project were in general positive. Below is a statement from the pilot in the Netherlands:

To summarize the reactions: very positive! General comments were: "It is very flexible", "That tool would be very helpful", "I am really impressed and had not expected such an advanced tool". (ERI-1997-10-20)

The pilot team anticipated that the procedure for checking statuses at milestones in projects could be made more efficient. The traditional procedure was as follows: a configuration manager accessed a separate document library where the documents to be checked in a milestone survey were stored. From the status of each document he/she made a list containing all impacted documents and their status; a list that was subsequently used at project meetings.

With the tool at hand, this procedure could be made fully automatic and even run in real-time at project meetings. This feature was later realized in the Matrix application in the S-domain. This anticipation of future enhancements during the actual work with the tool is a nice example of enactment, that is, how emergent structures are realized by recurrent interaction with the technology at hand.

Positive reactions also came from the pilot in Karlskrona from one of the consultants:

The demo this time went excellent! What the participants mentioned and I also noticed was that implementing their own project into the prototype made a BIG difference. They were able to see ALL of their design base, ALL of their increments, ALL of their impacts for each of their increments, a very detail and accurate Increment Impact Matrix, some Milestones definitions, and more... (AND ALL THESE WITH ONE TO TWO DAYS OF WORK). (ERI-1997-11-05)

During the autumn of 1997 the existence of the CPLtool became more widespread at Ericsson. In October a demonstration was held for a Corporate IT group working with IS architectures. The group showed interest in the tool. However, Corporate IT had already decided to recommend the C-PDM system. In November 1997 the system manager for the Methods & Tools unit at UAB decided that the C-PDM system should be used as the tool supporting incremental development (ERI-1997-11-10). Since it was much harder to change the implementation in the C-PDM system this decision, if enforced, would have made the iterative way of working utterly awkward if not point-blank impossible[8].

There were also other voices raising concerns about moving towards a centralized solution based on a single application:

The greatest risk as I see it [...] is the role that the C-PDM will get as a central system for the entire enterprise [...]. It will be extremely arduous to coordinate, discuss and prioritize all requirements that will be put on C-PDM. Believe me, I know. We had a similar arrangement on the hardware side many years ago: a central system that would serve all hardware designers. Nothing happened, lots of "revolt outbreaks" all around, an eternal compromising without end. The functionality became a kind of least common denominator, and no one was happy with it. (Hardware support architect, ERI-1997-11-12)

I argued against the decision to replace Matrix with the C-PDM system. It was decided that further investigations were needed in order to clarify the requirements on the production version of the tool, particularly the need for flexibility and adaptability. In December 1997 the steering group for the Methods & Tools unit decided to put the work with CPLtool in a pending state until more information had been collected from the users. In December 1997 I sent the following e-mail to the participants in the CPLtool project:

Finally, I would like to thank you all for your outstanding work this year. I firmly believe that this is the forerunner of the next leap in the methods and support area for AXE development, and it is not easy to be up front. (ERI-1997-12-17)

However, at the end of 1997 the demonstration license that the pilot projects had been running on would cease to be valid. This meant that the pilots would have to be stopped immediately. Since the steering group was reluctant to enforce this, a decision was taken on the 8th of January 1998 to buy the first commercial set of licenses for Matrix. This event became one of the turning points in the evolution of the Framework at Ericsson, an event that was disclosed to the different team members in an e-mail:

Hi all, the steering group has just approved the cost for the Matrix-installation! This means that we can go ahead with the pilots. This also means that we have to detail plans for installation and support as soon as possible. Now I'm going to get that beer... /Lars

However, it would become distressingly clear in the period that followed that the road ahead was paved with impeding stone blocks.

TERMINATION AND DESPAIR (1998)

For some time Ericsson had been negotiating with a software company called Rational™ (Rational™, 2008) to take the full responsibility for software development methods and tools at Ericsson. An agreement about this was signed in December 1997. The intention was that Ericsson should concentrate on its "core business". According to the prevailing management opinion at this time, internal development of methods and tools was not considered core business. Consequently, the Methods & Tools unit at UAB was shut down in late 1998.

Rational™ claimed that incremental development was included in their portfolio; thus, the IDMP was not needed. No analysis, however, was made of the similarities and differences between the approaches. The maintenance of the IDMP was transferred to the Netherlands in January 1998 where it eventually ceased to play a role in the organization.

Moreover, Rational™ stated that Matrix could be replaced with their tools. This meant that it became virtually impossible to get funding for the continued work with the pilots. During the first half of 1998 I made a number of demonstrations of the CPLtool in order to engage allies from wherever. This period can most properly be characterized as a "peddling" one of utmost despair mixed with a conviction that the Framework initiative was on the brink of being terminated. One consequence of this was that I started my Ph.D. studies in February 1998 in order to "save" the Framework ideas and results in some way[9]. This event can be seen as another turning point in the development of the ADT, since it was the starting point for elaborating the theoretical foundation of ADT at the university.

In the meantime a major project called ICH[10] had been launched at Ericsson Canada to appropriate the methods and tools of Rational™ for the traditional AXE development. All of a sudden, internal development of methods and tools were back as core business again. By coincidence, I happened to meet one of the project managers of ICH in the printer room and I showed him some tool snapshots. He became very excited. "This is what we need!", "Has Rational seen this?", "Why haven't they told us about this?", "Can you do this and that?" and so on. Then he asked me if I could go to Canada the following week...

This unlikely event resulted in a workshop in Canada in June 1998, the objective of which was to determine the suitability of Matrix in the ICH project. Two of the consultants working with the CPLtool participated in the workshop. However, the result was a complete breakdown. The expectations were quite different, as somewhat sarcastically related in the consultants' report:

The purpose of the workshop was reviewed together with Mr. S and Mr. R. Our understanding of the task was to make adjustments of an existing Ericsson Matrix model to make it fit the ICH environment and thereby create a clearer understanding, from a technical standpoint, whether Matrix can be used as a complement to the already defined ICH tools: a proof of concept exercise. Mr. S's understanding was that we during the 3 days should produce a system ready for use in production. Mr. R wanted to discuss licensing policies and prizing. (ERI-1998-06-26, p. 1)

Meanwhile, the pilots managed to continue by allocating resources on their own. The following statement came from the pilot in Karlskrona:

By the end of this week we'll have all their IP's [Implementation Proposal] entered and the tool will be ready for real use in the project! They are very excited and they are really committed to the project. Jorgen already entered the increments and allocated them into ADs [Amendment Directive]. The more we use it the more we come with innovative ideas of how it helps the project. Mailing a whole increment to a team who then expands the increment to view their assigned items is a really COOL and HI-TECH feature. (E-mail from consultant, 1998-02-02)[11]

This statement is nice example of enactment showing that new insights are gained in actual use. The remark about mailing increments refers to the fact that objects in Matrix can be mailed to a receiver like an ordinary e-mail. Thus a central increment planning could be combined with decentralized increment development by a team.

In July 1998 the Karlskrona site decided to continue with the CPLtool. However, this initiative was never firmly established in the organization and was finally terminated when the product line in Karlskrona was closed.

In early 1998 a large and important project in Germany, the so called AXE Mobile Core project (AMC), showed interest in CPLtool. After a long evaluation period AMC decided in October 1998 to become a pilot. However, this initiative went into a hibernating phase during 1999 and was not revived until early 2000 when the UMTS project decided to go for Matrix.

In a report in April 1998 the experiences from the CPLtool so far were summarized by the consultant working in Germany (ERI-1998-04-20):

- **Prototyping in real projects***:* It is essential for getting the right requirements to use the prototype in a real development project. It is possible to get good requirements and suggestions by just giving demos and having interviews with projects, but to evaluate the requirements regarding usefulness and feasibility they have to be tested in a real project under normal circumstances. When the people start to work with the prototype they come up with numerous ideas how to further improve and extend the functionality of the prototype. Only the people later working with the tool can find out the real needs (requirements) and identify potential problems.
- **Commitment from pilot:** It is absolutely necessary that the pilot project commits to the pilot work. They have to invest resources, time and money. They have to provide all project related information, come up with requirements, use, test, and evaluate the prototype. They have to take an active role and should have a real interest in the tool. It is vital that the pilots come up with the functionality they need and take part in the implementation of the adaptations. If you have to fight for every meeting and they have no money to spend this will block the work significantly.
- **Identify crucial prototyping parts***:* It is necessary to identify the really crucial parts of the work with the prototype. If these parts are not found or not considered they can kill the whole pilot work. In our example, the crucial part is to enter the IP/FF data[12] and always to keep it updated. For big projects it is impossible for a single person not directly involved in the project to get all the data and to enter it. Even for the people in the project it is very hard and time consuming to keep track of all changes in the data. If the IP/FF data in the prototype is not complete or out of date, then the prototype will produce wrong results. Before starting the pilot work, these killing factors have to be identified and solved; otherwise they will threaten the pilot success.

- **Flexibility***:* The platform of the prototype has to be extremely flexible, i.e., numerous adaptations of the prototype must be possible (so that they can be implemented, demonstrated, used and evaluated) and it must be possible to incorporate the changes and new features very fast (time is always crucial in pilot projects). Working with the pilot is not possible when new requirements from the pilot will take 3 months to implement due to time and cost problems. The flexibility is especially important when several pilot projects are started because every project (although in the same corporation) is quite different from every other and needs various adaptations (different usage scenarios, different documents, working procedures, different sizes of the projects,...). It is impossible to develop a nice tool, which is suitable to all kinds of projects without adaptations. The Matrix platform has the needed flexibility.
- **Consider different views***:* It is important to talk to numerous different persons of the pilot project because everybody has a different view on the prototype according to his/her role in the project. Very important is the opinion of the people who will later work with the tool (e.g., PM, IP writer, CM)[13]. Involve as many different people in the prototype work as possible to minimize the possibility of missing important requirements for the tool or to miss blocking factors of the work with the prototype.
- **Different needs:** When selecting pilots try to get different ones (small and large, AXE and non-AXE, total projects and subprojects,...) because they have different requirements for the tool.

In these statements many of the constituents of the ADT can be found: contextualization, enactment, stabilization, and transition.

So, the first half of 1998 was a schizophrenic period. On the one hand, very promising results were achieved in the pilot projects. On the other hand, the whole Framework endeavor was on the brink of termination. This was however soon to change.

RENEWAL IN CONFLICTS: THE PIONEERING S-DOMAIN (LATE 1998-1999)

One day in May 1998 a person asked me to demonstrate the CPLtool to him. It turned out that he was the project manager for a project called Beamon that was developing new switching equipment for the AXE platform. This project manager had tried out several management tools in previous projects but found them inadequate. After this demonstration he decided to try out the CPLtool in Beamon. This was to become the first, sharp project using the Matrix at Ericsson[14].

The Beamon Project

Beamon was a large, globally distributed project with the main development sites in Stockholm, Italy and Australia. The work to adapt the CPLtool to the needs of Beamon started during the summer of 1998. One focus was to manage engineering change orders in a better way; that is, requests for making controlled changes in the project. At Ericsson these are called change requests (CR). The CRs are closely associated with baselines in the project. When a project has reached a certain stage, the state of the product and its describing documents and various other items are frozen and included in one or several baselines. A change of any of the baselined items must be proposed in a CR. This is processed according to an established procedure where the CR is analyzed in detail and approved or rejected by a specific group called the change control board (CCB).

Another focus was requirement management. The traditional way of managing requirements at Ericsson was to write them down in a requirement specification (RS) and make new revisions of the RS when the requirements were changed. The project manager wanted to have better control of individual requirements, something that was not possible with the RS document.

This meant that Beamon had quite a different scope than the pilot projects where the support of incremental development was in focus. This was still interesting in Beamon but the management of CRs and requirements were prioritized. In July 1998, I started as project manager for the adaptation of Matrix for the Beamon project. During the autumn 1998 extensive prototyping was carried out together with configuration managers, requirement coordinators, the project manager, and consultants from the vendor of Matrix.

On October the 30th 1998 a decision was taken to go ahead with the use of the Matrix in Beamon. This was however not uncontroversial. The initiative came from a single project manager, that is, it was not endorsed by top management. Moreover, it was against Corporate policy and the agreement with Rational™. This is reflected in the protocol from the decision meeting:

Decision: to use Matrix in one project, Beamon, and if that is successful to transfer it to other projects. It is important to remember that Matrix is a "gap filler" until Rational has a corresponding product. (ERI-1998-11-03)

This decision was followed by an extensive preparation and testing period which included issues like

- Elaborating the information model and its implementation,
- importing product data from the Ericsson product archive,
- defining users and access rights in Matrix,
- checking the infrastructure, network, etc.,
- developing integrations to document archive systems,
- installing client systems on user terminals,
- training users,
- and providing documentation, etc.

On May the 10th, 1999, the traditional way of managing CRs and requirements was shut down, and Matrix was in use in a sharp production environment for the first time at Ericsson. On May the 19th the first CR was issued in the new system:

Yesterday you all received a mail from me (automatically generated from the MATRIX tool) about Change request 109 020 CR-101 PA1. This CR is the first to be handled by the MATRIX tool. (E-mail from the Configuration Manager in Beamon, 1999-05-10)

In June 1999, the specification of what the Matrix should achieve in Beamon was fulfilled. This work was carried out with quite a few resources: one assistant, two consultants from the vendor of Matrix

Figure 6. The information model late 1999

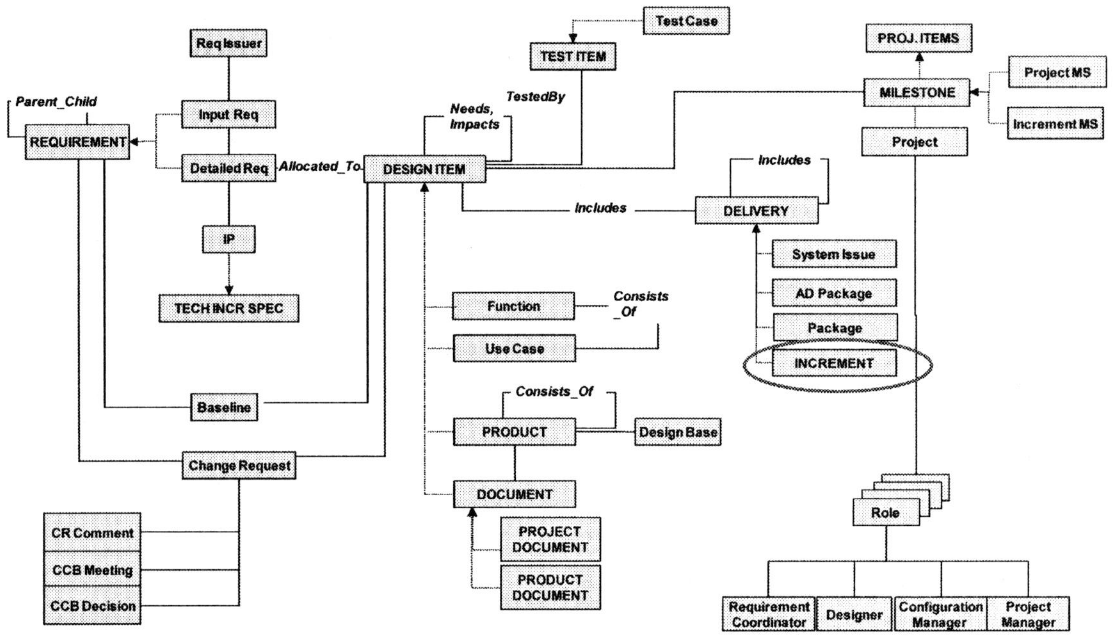

and me. This effort is surprisingly small, something that Larry Bowen, an authority on Configuration Management lecturing at Ericsson, expressed in a mail:

Lars, what a Herculean[15] job you have done with very limited resources! It's going to really be difficult for you to do too much with the system with your manpower level. It amazes me that you have done so much with so little so far. Keep me posted on where Ericsson is going with the system (E-mail conversation, 1999-07-07, Larry Bowen, Jet Propulsion Laboratory).

Later during 1999 the application in Beamon was amended with support for test management. This was not in the original specification. All in one, the information model in late 1999 had evolved into the one in Figure 6. As compared to the models in Figure 3 and Figure 4, this model looks quite different, although some reminiscences from the earlier models can still be traced. This indicates that actors in different areas of Ericsson have different meanings about the coordination context. Moreover, the needs are different in some aspects and common in other.

Fighting Problems

During the deployment of Matrix in the Beamon project several problems were encountered. The set-up of the Matrix system was based on one server in Stockholm serving various types of clients[16]. This meant that the set-up became very dependent on efficient web-technology and a fast network. Moreover, the clients should work in a heterogeneous environment both on PCs and work stations.

Already in February 1999 it was noticed that the web-client in Matrix was unstable. The so called full client could not be used at all in Australia because of very long response times. These problems became a risk in the project as stated by the system manager for Beamon in Italy:

Hello [...] yesterday we raised some issues [...] concerning matrix tool. I started making some sort of checks towards especially the technical coordinator about the feeling in using this tool: the answer something like: - almost nobody is using the tool. -It seems quite difficult to start using it. - For EPA and TEI there are additional problems due to the remote access to the matrix server, usually quite slow. - Moreover, there are problems using the tool from the web client interface, since some functionalities (like get a report from an item, to printout or to save) are not working properly. I had also the confirmation from our LSO in TEI that to work properly, it should be used from PC.[17] *(E-mail, 1999-07-09)*

The problems could be controlled by various workarounds, but the attitude of the users towards the tool in Italy and Australia was quickly becoming hostile. These problems were further escalated in July when the situation in Australia jeopardized the continued use of Matrix in Beamon:

It is very important that we can fix this. Otherwise we may have to stop using Matrix in Beamon and that would be a dire fiasco. (Project manager, e-mail, 1999-08-06)

One consequence of this was that a Matrix consultant was sent to Australia in August to try to fix these problems. On site he discovered an inconsistent picture; some clients worked fine and others were very slow. Also, the proxy server to the local Australian network was not working properly. Another issue was that the support on-site at times felt abandoned by the support group in Stockholm:

We at EPA have been having very serious reliability/stability problems with MATRIX. I have sent a number of mails to various people to follow up but the response has been slow and not helped when it does come. I know it is summer up north and this means many vacations but we are quite busy here and need better and faster support. [...] There needs to be some improvement and coordination in available MATRIX expertise. As I said I have sent various mails with MATRIX error messages and problems. Now I need some answers. (Local Matrix support in Australia, e-mail 1999-08-04)

The situation was however stabilized and remained so until October 1999 when an error in the web-client was discovered. This error occurred only in the specific combination of Windows 95 and the current Matrix-release installed at UAB. The error was found also in a new release of Matrix which caused a five month delay in the transition to a newer release. In February 2000 the vice president of MatrixOne was visiting UAB for "serious and frank" discussions of how to remedy this situation.

The problems with long response times and unstable performance nearly caused a user mutiny in Australia. Here are some examples:

You are probably aware of this, but there is a large number of CRs that should be written but have not as yet. There are mainly related to changes made to FDs[18] *after they have been reviewed but there are more. The main reason for this is the MATRIX tool. I know that you at UAB have no problem but we at EPA (and I think TEI as well) have nothing but trouble. It is the exception rather than the rule that the tool actually works here at EPA, which causes everyone frustration and hence a backlog of unwritten CRs. (Subproject manager, Australian site, e-mail 2000-02-18)*

As you most undoubtedly heard, MATRIX at EPA is a nightmare to use. I couldn't get it to work at all today and was wondering if you could help me out. (Design engineer, e-mail 2000-02-18)

During this period, one of the consultants had begun to investigate a new web-client based on a Java Servlet technique. This turned out to be much faster than the Java-based standard web client. In March 2000, the Servlet client could be released together with a new release of the Matrix tool. This finally solved the problems with long response times and unstable performance:

Hi, I used this today to enter our comments on the current CRs. It turned what would have taken me more than an hour using the web client into a 20 minute job. Fantastic! (Subproject manager, Australian site, e-mail 2000-05-24)

The HTML applications/wizards technology has shown drastically improvements in speed and could work acceptable for Australia and Italy. (Beamon meeting, minutes, 2000-04-20)

The nightmare was over. From that point on, the performance of Matrix was not an issue.

Escalation

During 1999, two more projects became interested in using Matrix. One project, Uranus, developed a new version of the central processor in the AXE system. The other project, Hercules, developed an exchange terminal to the central switching equipment. Both these projects consisted of both hardware and software. In June 1999, an installation of clients was made in Norway for the exchange terminal project and later on Croatia joined in as a subproject. In August representatives for these two projects were included in the Matrix project meetings in Beamon.

At this time the support environment was not fully developed in Beamon. Moreover, the two new projects worked differently from each other and from Beamon, which created lengthy discussion about the information model and its implementation in the Matrix tool. For example, Beamon and Hercules could never agree on the state chain of a baseline which resulted in two different chains being maintained in the tool.

On February the 15th, 2000, Hercules project started to use Matrix. Thus, it became the second project to rely on the Matrix for its work. It was followed by Uranus and still another project called HANSI. However, all throughout this period the distrust from the line organizations prevailed. The resource allocation was constantly questioned, which made it difficult to hire the necessary consultants.

Conflicts

At this time, the decentralized organization of Ericsson had resulted in various support tools being developed at different sites around the world. These tools were in general specific for each site and not coordinated. Although they fulfilled certain needs, the downside was high costs for development and maintenance, overlapping applications, problems with data transfer between tools, etc. For Ericsson it was of utmost importance to reduce these costs by reducing the number of tools. Part of this strategy was also that only one type of PDM systems should be used throughout the company.

When the first prestigious project started to use Matrix, the conflicts with Corporate IS/IT intensified. From the Corporate point of view Matrix had no place in this overarching strategy. However, since some projects had started to use Matrix, these could not be jeopardized. Instead, Corporate sought to contain the further expansion of Matrix.

One way of doing this was to obstruct the access to the IS/IT support group at UAB. Without the assistance of this group, new releases of Matrix could not be installed, network problems could not be investigated, backups could not be taken, etc. The IS/IT group was controlled by a steering board. One of the members in this group was K, the program manager of Rational™ program at Ericsson[19].

At a steering board meeting in February 1999 it was decided to support Matrix only as a prototype tool:

The introduction of the Matrix application: There is no decision or forecast of broad introduction. K remarked that Matrix is directly divergent from the upcoming [Corporate] standard. Decision: Matrix shall be regarded as a pilot. If a broader usage is anticipated this must be forwarded to the Steering Board. (ERI-1999-02-15)

Moreover, it was decided that the new projects (Uranus and Hercules) had to await further decisions before they could start using Matrix (ERI-1999-05-31). The concern from the Corporate point of view is clearly expressed by K in the following e-mail:

Hi, a clarification. I have no objections against letting the projects that chose Matrix half a year ago get the support they need. As I remember it was Beamon that was discussed, nothing else. They should pilot the usage and someone should support that. That's between them and the IS/IT group. However, I will never support a broad deployment since this is against all initiatives and directions that Ericsson as a whole (corporate) has. I think it is bad to talk about spreading this to Uranus etc. UAB is heading towards an own, internal solution now and I believe that will cost. (E-mail, 1999-06-08)

According to Corporate, the future solutions should be built on the existing and upcoming corporate standards. Matrix was not part of this since it had emerged from "the floor". This principle blocked Corporate from taking part in any analysis about pros and cons concerning the use of Matrix.

However, this picture was not unambiguous. A member of the Corporate IT group supported the incremental development method in IDMP and allocated resources to continue the pilot projects, although he did not want to see Matrix as the tool in the Framework. He also wanted me to contribute to Corporate conceptual modeling initiatives (E-mail, 1999-10-04). This ambiguous stance from Corporate is reflected in the following decision:

METHOD AND TOOL FOR SUPPORT OF INTEGRATIONDRIVEN DEVELOPMENT AT UAB. This is handled by Lars Taxén at UAB and [Corporate] has given some support to Lars awaiting the start of DTL[20] where the question should reside and be considered. DTL declared that the use and support of this method and tool is a UAB question. DTL do not see that supporting the method and the tool is a corporate question. (ERI-1999-10-21, p. 2)

Thus, in 1999 the prime concern of Corporate was to let the current projects continue by themselves, and arrest the further spread of Matrix within Ericsson. However, this was in fact already happening.

OTHER PRACTICES CATCH ON (2000)

During 2000 two more development sites began using Matrix. The domains being constructed at these sites were independent of each other, which meant that the information models (and the implementation in Matrix) evolved separately[21]. Thus, in 2000 three independent domains existed at Ericsson (the S-domain in Stockholm, the L-domain in Linköping and the A-domain in Aachen).

The L-Domain in Linköping

In October 1999 the Base Station System (BSS) unit at Ericsson became interested in Matrix. This unit was responsible for developing base stations and base station controllers in the mobile network. At a meeting in November I gave a demonstration of the Matrix application in the S-domain. After some discussions, BSS decided to investigate the consequences of using Matrix for their purpose. This report stated that a combination of Matrix and the Rational™ tool suite was the preferred alternative (ERI-2000-02-10). The Matrix implementation in the S-domain was copied to the L-domain in Linköping, and in May 2000 the work to adapt this implementation to the way of working at BSS began.

The A-Domain in Aachen

During 1999, the development of the 3rd generation of mobile systems, the so called UMTS[22] system, was intensified. Some nodes in the UMTS network are very complex and it was recognized early on by the project management that the traditional methods were inadequate. The approach was to divide the system into small parts, and build it up from basic to total functionality in a step-by-step manner, verifying system behavior as each step evolves. This was basically the same approach as in the IDMP. The terminology was changed and some new ideas introduced. For example, the Feature Increments in the IDMP was now called Work Packages and the Amendment Directives were called Latest System Version (LSV).

The UMTS projects were the first ones to use the software configuration management tool ClearCase from Rational™ in order to coordinate the work of about 500 software developers working all over the world. At first, the intention was to manage the integrations of the LSV's in ClearCase as well, but this soon turned out to be a bad solution:

[The] idea in the very beginning was to do this as well in ClearCase by some sort of browser or viewer on top of it. But what we have seen is that would take a hell of a lot of design work and that there are other tools which can manage this far better. What we said is that we should separate the revision handling part in the Configuration Management from the content handling. That's basically why we introduced Matrix, more planning aid. (Project manager, A-domain)

Figure 7. The information model in the A-domain (early 2000) (Taxén, 2005. ©Wiley Inter Science. Used with permission)

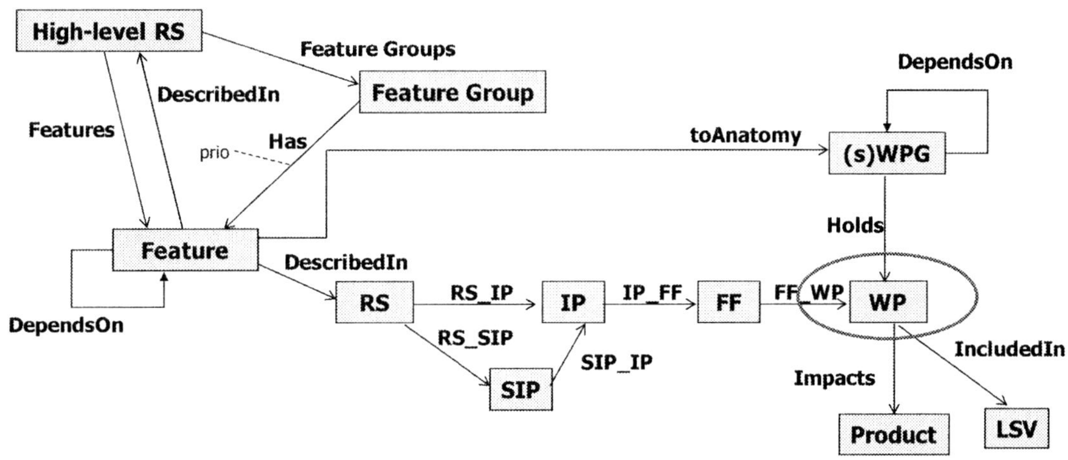

After the shut-down of the activities around CPLtool in Aachen, the contacts were taken up again in October 1999 when I gave a demonstration of the S-domain application. In December 1999, the project manager for the UMTS project ordered the use of Matrix in the project (ERI-1999-12-01). In January 2000, the adaptation work began and an extensive prototyping period followed. In May the same year, the UMTS started to use Matrix for controlling the status and tracking of Work Packages, as well as managing the integration of the LSV's.

The information model for the A-domain is illustrated in Figure 7.

As can be seen, this model hardly bears any resemblance to the previous models in Figure 4 and Figure 6. This reflects both a different focus and a different way of working.

The A-domain was to become the most successful at Ericsson. One of the project managers, PM4, expressed this in the following way:

Especially for the execution part I think we would not have been able to run this project without the tool. I think if you simply look at the number of work packages, the number of products that we have delivered, the number of deliveries that we have had, if we would have to maintain that manually, that would have been a sheer disaster. (Project manager, A-domain)

Thus, as I summarized in an e-mail on Dec 15th, 1999, a full circle was completed:

- The IDMP is developed in blood, sweat and tears at UAB during 1996-1997, including the Matrix support.
- The Rational™ agreement almost kills this initiative in 1998.
- The Work Package concept (in principle the same as IDMP) is developed in Aachen in 1999.
- UMTS decides to use the Work Package concept and the Matrix to support in 1999.

Thus, astounding results were achieved in the years around the millennium shift with the combination of the Work Package concept, and the way the Matrix based coordination support was implemented (later to become the domain construction strategy in ADT). Together, these two straits forged the anatomy-centric development method. However, at the peak of its performance, the dismantling of this highly successful method at Ericsson began.

A CENTRAL CONCERN: THE C-DOMAIN (LATE 2000-2001)

During the period between late 2000 and 2001, Matrix became increasingly a central concern for Ericsson. Voices were raised to make Matrix a tool to be used enterprise-wide at Ericsson. A number of meetings were held regarding the future evolution of Matrix. However, Corporate IT still regarded Matrix as a local initiative. For example, in a report early 2000 about the future positioning of ISs within Ericsson, Matrix is not mentioned (ERI-2000-01-11b). Still, with three different sites running Matrix in large development projects, and in particular the UMTS project, the need for a central initiative became imminent.

The Reference System

In January 2000 a new unit, SSES[23], was established at Ericsson. In SSES a number of SPMs[24] were each responsible for a specific method & tools area. One of the responsibilities for SSES was to define and manage a reference system which would be the coordination point for the different domains. A principle called "coordinated flexibility" for the reference system was outlined already in 1998 (LTX-1998-10-12). This principle was based on two assumptions:

- The autonomy of domains to evolve to their specific needs must be maintained.
- The items needed to coordinate different domains must be mandatory for all domains.

This meant that a federated architecture of information systems was proposed (Sage, 2001). The idea was that all mandatory and common parts of all the local domains would be implemented in the reference system. This system, which was called MARS (MAtrix Reference System), would evolve at a far lower pace than the applications in the local domains. New organizations would be able to use the reference system as a starting point for their own, local adaptations. No sharp project would be allowed to run in the reference system in order not to bias it towards the preferences of a particular project.

However, the federated architecture was soon questioned by influential actors in SSES in favor of a single, centralized domain in which all projects would run. This would mean that the three existing domains would be merged into one:

We think that the effort for reconcile and coordinate the existing applications is not that high. It will be possible to have several customers in the same environment. We see an environment or a database more as a building which has different lodgers. Some services they will use together some services are individual and we see it more as a challenge to find a solution for this. (Methods & Tools coordinator, e-mail, 2001-05-01)

A detailed business case was also made which showed that a centralized solution was the cheapest:

The most economical solution is the centralized Matrix solution with a common shared business model and one database. This is also the most common alternative for Matrix implementations in other companies. By using one centralized solution a company can harmonize its functions and create a common way of handling information. (ERI-2001-04-20, p. 20)

However, no effort was made to estimate the cost of achieving common understanding about the information model and its implementation in the Matrix system. I claimed that this cost would reverse the conclusions in favor of a federated architecture, although I had no compelling arguments for that claim. This discussion about a central architecture versus a federated one became a recurrent theme in SSES until the central solution finally took precedence over the federated one in 2002.

In line with this, SSES also started a number of other activities, one of which was to define a "common information model" for Ericsson. The starting point was the information model which had evolved over several years in the S-domain. The first meeting was held in February 2001, which was followed by a series of meetings during the spring of 2001. The premises for these discussions were that the model should be tool independent. This model, which was called SCIM (SSES Common Information Model), evolved only through lengthy discussions among the participants and was never tried out in practice. Eventually, SCIM was productified and managed as an individual artifact. It was also suggested that the steering of SCIM would be done by the SPMs:

I discussed how [SCIM] should be managed and controlled within SSES with N. His proposal was that the SCIM product should be controlled by all SPMs in SSES. (E-mail, 2001-05-28)

This meant the information model was detached from its implementation in Matrix, and that the control of the evolution of the model was moved from the actors in the domains to managers who had never worked with the model or Matrix. Thus, the enactment of the model was completely ignored, which meant that the model quickly ceased to be operational; it became just an image drained of all interventional capacity. This trend was to become more proliferated during 2002.

The Core Network Starts Using Matrix

In January 2001, I started to work at the unit in Stockholm which was responsible for the core network part in UMTS. The core network consists of a number of nodes which perform various tasks in the network. My responsibility was to introduce Matrix on the core network level. The strategy was to implement the requirements from the network level in the A-domain. In order to prioritize the requirements on that domain, a CCB for Matrix was initiated with me as chairman. The first meeting was held in March, 2001. All through the summer 2001 discussions continued about what requirements to implement. The prevailing mind-set was that detailed and proper requirements should be provided in advance:

To have a successful implementation of MSC[25] requirements within MATRIX it is of upmost importance that J and your guys provide proper requirement specifications!!! That was the major showstopper in the past. If you cannot get the required actions in place within YOUR project I see no chance to get the needed feature within to be available within the tool. So - detailed and complete requirement specifications are the key to get something working for MSC 2.0! (Project manager)

Although all parties had the ambition to compromise, it turned out that the requirements from the network level were quite disparate from the requirements on the nodes in the network. After five months, virtually none of the requirements from the network level had been implemented. As a consequence, the UMTS network level decided to move to the MARS C-domain. The CCB was closed, and the responsibility for the A-domain was taken over by the Aachen unit. This experience was yet another indication of the prohibitive efforts involved in achieving common understanding if the domains are too disparate.

Corporate on Stage

During 2000 and 2001 the contacts with Corporate intensified. A large investigation about the future PDM solutions at Ericsson was carried out during the autumn of 2000, where the Matrix experience was one input. I presented the Framework at an Ericsson meeting in Vienna in January, and the future of Matrix was discussed at a large meeting in March 2001. In the same year a project was started to evaluate Matrix and the C-PDM system. This was the first time these systems were officially recognized as "equals" by Corporate. However, no decisive recommendation resulted from this investigation.

Eventually, the budget responsibility for Matrix was transferred to Corporate IT, which meant that corporate representatives were participating in the steering group of MARS for the first time. Corporate IT was now in control of the further evolution of Matrix at Ericsson. However, suspicions about Corporate intentions were aired:

In the discussion I could read pretty clearly that Matrix will be the first target for budget cuts. It is evident that they have taken over the responsibility of Matrix just to kill it. It is quite clear that we within the R&D community are the only guarantees for Matrix at Ericsson. (e-mail, Dec 2001)

These intentions were however never attained. Discussions with MatrixOne Inc. concerning licenses had been initiated early 2000. On the 1st of May, 2001, a corporate agreement was signed and in December 2001, a joint press release from MatrixOne and Ericsson concerning this agreement was issued. This was still another turning point in the establishment of Matrix at Ericsson.

FROM FEDERALISM TO MONARCHY (2002)

During 2001 and 2002 the telecom crisis hit Ericsson hard. The staff was reduced drastically from 107,000 to less than 60,000 in a couple of years. The organization at the core network unit was shut down. I moved back to UAB in November 2001, where a new unit for methods & tools was established with the main purpose of reducing costs. In February 2002, the radio network development unit at Ericsson started to use MARS. In March 2002, the S-domain was moved to the central MARS domain. The A-domain in Aachen was put on hold, and the L-domain planned to move to MARS.

During 2002, the trends from the previous year accelerated. In the new method & tools unit, the division between the Ericsson common information model, the work package-based methods and the

Matrix tool were institutionalized. For each one of these areas, a technical manager was appointed with full control of their further evolution.

The work to define a common information model was organized as a regular project with four different tracks, a corporate steering group, a reference group consisting of 10 members from 8 different organizations and a coordination group consisting of 13 members from 7 different organizations (ERI-2002-07-31). Thus, instead of continuing with the "coordinated flexibility" strategy that had been proven successful at the A-, S-, and L domains, the course was steadily set for a "one-for-all" centralized domain. In the words of Davenport, Eccles, & Prusak (1992), the path went from "federalism" to "monarchy":

[Federalism] has a number of desirable features, and in today's business environment, it is the preferred model in most circumstances. Its distinguishing feature is the use of negotiation to bring potentially competing and non-cooperating parties together. [...] Each realm contracts with [...] other realms to cede some of its information assets in return for helping to create a greater whole. (Davenport, Eccles, & Prusak, 1992, pp. 58-59)

[Monarchy is when the] CEO, or someone empowered by the chief executive, dictates the rules for how information will be managed. Power is centralized, and departments and divisions have substantially less autonomy regarding information policies. [...]. As a model for information management, this means that dominion is established over what information is collected, in what form, by whom, and for what ends. (ibid, p. 58)

For my part, the story of the Framework evolution at Ericsson ended in June 2002 when I had to leave the company. In summary, the achievements were substantial. Between 1999 and 2002, the coordination of more than a hundred main projects and subprojects were supported by various applications built on Matrix (ERI-2002-06-06). These projects were distributed over more than twenty different development units around the world, and were carried out in a fiercely turbulent environment. Less than six years had passed from my first contact with the lonely salesman in 1996, to a corporate strategic tool impacting key development projects at Ericsson in 2002.

REFERENCES

Axelsson, K. (1998). *Metodisk systemstrukturering - att skapa samstämmighet mellan informations-systemarkitektur och verksamhet* [Methodological Structuration of Systems – to create concordance between information system architecture and business (in Swedish)]. Dissertation No. 1, Department of Computer and Information Science, Linköping University, Linköping, Sweden.

Davenport, T., Eccles, R., & Prusak, L. (1992). Information Politics. *Sloan Management Review*, *34*(1), 53–65.

Gandhi, M., & Robertson, E. L. (1992). A Specification-based Data Model. *Indiana University Computer Science Department Technical Report TR344*. Bloomington, IN: Indiana University. Retrieved Sept 30, 2008, from http://www.cs.indiana.edu/ftp/techreports/index.html

Gandhi, M., & Robertson, E. L. (1995). SDBM as a Model for Codesign Data. In J. Rozenblit & K. Buchenrieder (Eds.), *Computer Aided Software/Hardware Engineering* (pp. 338–355). Piscataway, NJ: IEEE Press.

Karlsson, E. A. (2002). Incremental Development- Terminology and Guidelines. In *Handbook of Software Engineering and Knowledge Engineering,* (Vol. 1, pp. 381–401). Singapore: World Scientific.

MatrixOne (2008). Retrieved Feb 2, 2005, from http://www.matrixone.com/index.html

Rational™ (2008). Retrieved Oct 1, 2008, from http://www-01.ibm.com/software/se/rational/

Sage, A. P. (2001). Systems Integration and Architecting: Contemporary Issues with Abundant Complexity. *Managing Complexity in Product Management.* Web-conference, hosted by Wade H. Shaw, April 9th, 2001.

Taxén, L. (1995). The Dialectical Approach to System Design. In *Proceedings of Integrated Design and Process Technology* (pp. 147–152), December 6–9, Austin, Texas.

Taxén, L. (2005). A Socio-technical Approach Towards Alignment. *Software Process Improvement and Practice*, *10*(4), 427–439. doi:10.1002/spip.245

Taxén, L. (2006). An Integration Centric Approach for the Coordination of Distributed Software Development Projects. *Information and Software Technology*, *48*(9), 767–780. doi:10.1016/j.infsof.2006.01.007

ENDNOTES

1 References beginning with "ERI" or "LTX" are Ericsson internal documents and as such not available outside Ericsson. "LTX" stands for documents written by myself, and "ERI" for documents written by other Ericsson employees.

2 This PDM system should not be mixed up with the legacy PDM systems that have been used for a long time at Ericsson for archiving products and documents (called PRIM and GASK within Ericsson).

3 A comprehensive overview of the state of incremental development at this time is given by Even-Andre Karlsson (Karlsson, 2002), who was partly working as a consultant at Ericsson.

4 There seems to be no official explanation what this acronym stands for. The terminology guide to PROPS merely states that "PROPS is the name of the Ericsson project management method." (ERI-1994-01-01)

5 UAB: Ericsson Utvecklings AB, the Ericsson unit where I worked at this time.

6 TG2: Tollgate 2, a decision point in PROPS where a decision is taken to go ahead or not with the execution of the project.

7 The conceptual model described all relevant concepts, while the information model signified item that were to be managed in Matrix.

8 For example, adding a new attribute in C-PDM implied closing the data base and re-compiling the system, something which was not needed in Matrix. What was done in a matter of minutes in Matrix was a matter of hours in the C-PDM system.

9 This research task was made worse by a decision in March 1998 to purge the entire document archive from the AXE-N project.

10 This is a pseudonym.

11 IP: Implementation Proposal, a document specifying a feature increment. AD: Amendment Directive: a document specifying the contents of a software build.

12 IP/FF data: A matrix showing the estimated effort for developing a particular increment.

13 PM: Project Manager, IP writer: Implementation Proposal writer, CM: Configuration Manager.

14 Much of the continued discussion is focused on Matrix as the tool in the Framework since this was the most salient and controversial part of the Framework. However, the other parts of the Framework (in particular the information model) were always involved.

15 It so happened that the next project to start using Matrix was called Hercules...

16 An alternative set-up would have been to install servers in Italy and Australia and replicate the data between them. This was however not realistic with the resources and budget that was allocated to the Matrix project in Beamon. It can be noted that the architecture chosen is in fact a return to the Information Resource Management (IRM) - architecture (Axelsson, 1998).

17 EPA, TEI: Ericsson designations of the Australian and Italian site. LSO: Local Support Organization.

18 FD: Functional Description, a document specifying how to implement a function.

19 The program manager was responsible for the contacts with Rational™ and the deployment of the Rational™ products within Ericsson.

20 DTL: a corporate decision board.

21 Technically, separate data bases for Matrix were used.

22 UMTS: Universal Mobile Telecommunications System

23 SSES: System and Software Engineering Support

24 SPM: Strategic Product Manager.

25 Mobile Switching Center, the central node in the 3G networks

Chapter 2
Reflections

In this chapter, I take a step back from the whirling flow of events related in the previous chapter, and try to make sense of what happened during these years. In the first section, I analyze the evolution of the Framework using the vocabulary of the Actor Network Theory (ANT) (Latour, 1992). Next, I relate some observations from Ericsson outside the activities I was directly involved in; observations that somehow seemed to be in line with the pattern that began to emerge. Finally, I summarize the insights and needs from the practical trail.

ANALYSIS OF THE FRAMEWORK TRAJECTORY

In ANT, the social and technological aspects of human activity are seen as a unity, which is well in line with the praxis perspective in this book. Humans and artifacts are both considered actors. For example, an IS may have interests of its own like any human actor. In order to underline this somewhat provocative position, the term *actands* is sometimes used instead of actors[1].

An innovation is considered in ANT to follow a certain trajectory in a society. This trajectory is called a *program*. The innovator *inscribes* certain intentions in the innovation. The inscription will change as

DOI: 10.4018/978-1-60566-192-6.ch002

the innovation becomes engaged in a heterogeneous *actor-network* of people, organizations, standards and artifacts. Eventually the innovation and its inscription may solidify into *irreversible* network elements called *black boxes*, which indicates that the innovation has become an institution. At that point it is no longer possible to go back to an earlier stage.

The program is opposed by an *anti-program,* which represents interest groups and existing black boxes that might clash with the program. For example, a legacy IS may be part of an anti-program opposing the introduction of a new IS. In order to engage allies who may adopt the innovation, a series of *translations* of the innovation must take place. The purpose of a translation is to align the interests of other actands with the interest of the innovation. The inscribed pattern may not survive in the translation process since the anti-program may inscribe other intentions in the innovation. For example, the innovation may be used in ways not intended.

The program trajectory can be illustrated in a diagram such as in Figure 1. The horizontal dimension, called *syntagm* in ANT, corresponds to actors who have adopted the innovation. The vertical dimension, called *paradigm* in ANT, indicates what translations have taken place. An expansion in the syntagm dimension to the right must always be paid for by going down in the paradigm dimension.

The Framework may be regarded as a program with me as the innovator. The intention inscribed in the Framework was to pursue a certain coordination practice at Ericsson that eventually would become a set of black boxes with this intention inscribed. How it in fact turned out is indicated in Figure 1. The Information Model (I), Process Model (P) and Transition Model (T) are the generic model names I use in the operationalization of the ADT (see Section 3, *OPERATIONALIZING THE THEORY*, p. 126).

Figure 1. The trajectory of the Framework during 1990-2002

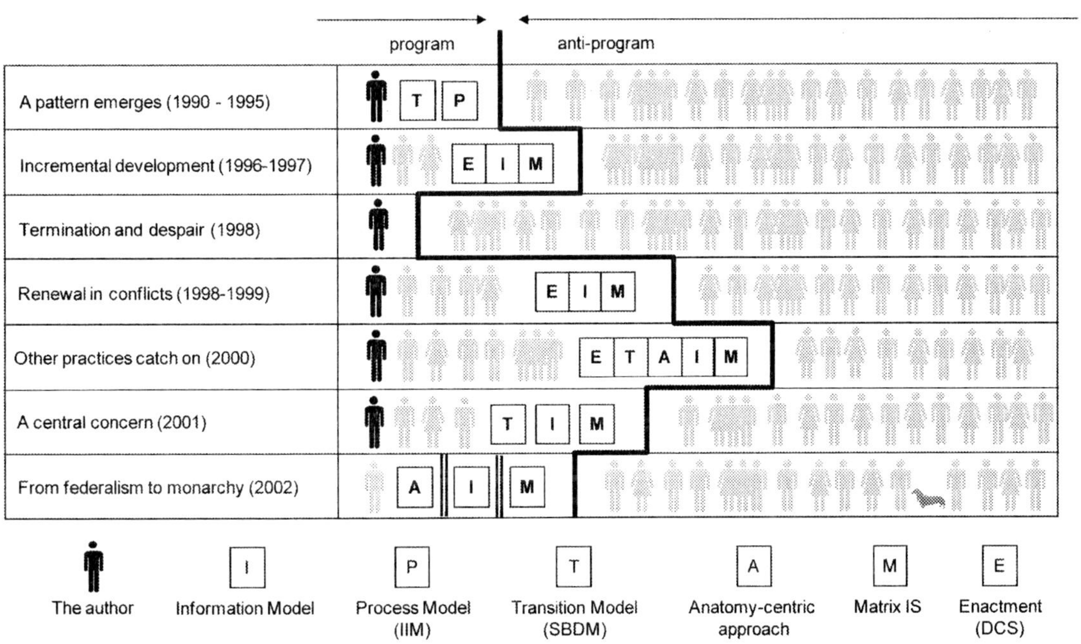

A Pattern Emerges (1990-1995)

In this phase, the ideas behind the Framework were articulated. The information interaction model (IIM) and the specification-based data model (SBDM) entered the program. At this point I was the only actor associated with the Framework. An anti-program could hardly be noticed.

Incremental Development (1996-1997)

This phase engaged more actands with the introduction of the incremental development method package in the CMS 30 Phase 7 project. The experiences indicated that a better tool support was needed in order to manage the incremental way of working. I came across the Matrix IS, and eventually this system became the IS actand in the Framework. Together, the IS and the information model were enacted into providing coordination support for a number of pilot projects with good results. During this period, my original intentions remained inscribed in the Framework and, in addition, became more elaborated and articulated. However, several actands in the anti-program began to influence the trajectory towards the end of this phase:

- The C-PDM system – this was the only allowed PDM system at Ericsson, and since Matrix was regarded as a PDM system, it was against corporate policies.
- The organizational unit where I worked – this unit did not want a separate solution outside the corporate strategy.
- The Rational initiative at Ericsson – Rational™ became responsible for the software development at Ericsson in December 1997, which resulted in dire consequences for the Framework program.

Termination and Despair (1998)

During this phase, the program was on the brink of extinction. Although the first set of licenses for the Matrix was bought, the budget for supporting the pilots was gradually squeezed. This meant that the pilots became isolated, and eventually ceased to be part of the program. The methods & tools unit in my home organization was closed down. For a period I kept the program alive by giving demonstrations and trying to get funding from wherever. One important actand in this phase was an insightful person at Corporate IT. He supported the program and wanted it to continue in the organization, however with Matrix replaced with the C-PDM system. This initiative was an indirect criticism of the Rational™ actand.

Even so, the anti-program had at this time more or less overhauled the program. At this point I was about to give up. However, more or less by coincidence the project manager for a new project named Beamon became interested in the Framework results. This was the turning point for the program.

Renewal in Conflicts (1998-1999)

During the summer 1998 a prototype for configuration and requirement management was developed. The experiences from this work gradually matured into a conclusion to use Matrix in the Beamon project. The most outspokenly actand in the anti-program most was Rational™. The peak of the clash between the program and the anti-program was the meeting on November 3rd, 1998. After some turmoil, a decision was taken to use the Information System as a "gap-filler" until Rational™ could provide an equivalent system, something that however never happened. In May 1999 Beamon started to use Matrix, which meant that the S-domain was established at Ericson.

Soon thereafter, three more projects joined the program. Again, the original intention of the Framework remained inscribed and became more elaborated and articulated during this period. Nevertheless, the severe stability and performance problems that were experienced with Matrix made it temporarily part of the anti-program. If these problems had not been fixed, the program might well have ended here.

Other Practices Catch On (2000)

In 2000, two more domains were established, the A-domain in Aachen and the L-domain in Linköping. At this point the anatomy-centric approach was at its peak in coordinating exceptionally complex projects, above all in the A-domain. The Transition Model, although never articulated, was apparent in the work to interface the Matrix to a number of other information systems needed. The iterative way of enacting coordination support was heavily employed.

In this phase, the anti-program was forced to retreat further since the Framework was backed by project managers in central and prestigious projects. The original inscription in the Framework was by now fully realized. It must however be remembered that the domains were only bridge-heads, albeit very important ones, in the overall Ericsson enterprise.

A Central Concern (2001)

A corporate agreement with the vendor of the Matrix system was signed in May 2001. This turned the Matrix IS into an irreversible black box in the Ericsson practice. Also during 2001, the need for coordination between the L-, A-, and S-domains became apparent. As expected, the information model differed markedly between the domains. This created foreseen problems since some 3G projects crossed several of these domains. Discussions were started to set up a federated structure of cooperating domains. The idea was to establish a central reference system called MARS, which would manage a core of mandatory elements and capabilities in the Framework. A product management function was established for MARS, and work was initiated to interface the reference system to other systems like software configuration systems.

However, MARS soon turned into still another domain for ongoing development projects, the C-domain. The original inscription of a reference system gradually changed into an inscription of being *the* one system backed by Corporate. Workshops were held to decide on a common information model and the numbers of actors around MARS increased significantly.

By this translation the original program was completely overhauled by the anti-program. The most apparent consequence was that the iterative way of enactment was replaced by a traditional, phased development strategy. The close interaction between developers and users was replaced with reference and steering groups that stated and prioritized requirements from all domains. This immediately slowed down the progress and increased the costs:

If you look at the functionality which has come out of the system it is incredibly little functionality during a very long time compared to the functionality we produced with two persons in half a year. (Matrix consultant)

At the end of this phase it was clear that the Framework would never reach the state of an irreversible black box with some of the original inscriptions intact. The close integration between actands in the Framework – the information model, the Matrix tool, the iterative enactment, and the anatomy-centric approach – was broken. The Framework as a coherent actand was gone. Instead the separate parts of the Framework began to appear as actands of their own.

From Federalism to Monarchy (2002)

In the last phase the development towards commonality accelerated. The upheaval in the telecommunication business during 2002 resulted in a fierce hunt to reduce costs. Corporate Ericsson control of Matrix, the information model and the anatomy-centric way of working was established. Projects were transferred to the MARS system, the S-domain was closed, the L-domain planned to move to MARS, and the A-domain was put on hold.

In 2002, Matrix was firmly established as a black box in the organization and used by a large number of projects and actors. The inscription associated with it however turned into "a requirement management system for large projects". This was a significant retreat from the original intentions with the Framework, where requirements were but one type of coordination items among others.

The evolution of the information model and the anatomy-centric approach was split in two branches. One branch was tightly coupled to the continued evolution of the C-domain. In the other branch, these actands were developed and maintained as separate artifacts that solidified quickly into relics without any practical significance. By that, the cardinal feature of the Framework; the interaction between the information model, the anatomy-centric approach and the Matrix IS "on the battlefield", was gone. The actands in the original program, including myself, had left the scene. Consequently, the Framework program was virtually extinguished by 2003. The original inscription could no more be discerned.

The Aftermath

In 2003 the development of Matrix was transferred to another company outside Ericsson, and handled through a strict order-delivery process. The result was a slow, expensive and bureaucratic process, where all the anxieties expressed by me in an internal discussion in 2001 had become reality. I argued that one, centralized database would imply that:

- All current and prospective users at the whole of Ericson shall utilize the same information model in spite of significant differences in their way of working.
- Alternatively, all the differences must be housed in the same information model and managed by one and the same group of people.
- All existing sites must be convinced to abandon their local control of its application and submit to the central control.
- Every disagreement about the contents of the application must be discussed, prioritized and decided upon by some steering group.
- In the S-domain we did several hundreds of changes to the conceptual model in one year. This was at one site. Still we had to manage two different ways of managing baselines and three different ways of managing Statements of Compliance[2] just because the projects could not agree upon common understanding of these items.

The main conflict was amply summarized by me in the following statement, also in 2001:

I mean that this points to the fundamental difficulty: to agree upon what should be implemented. The cost for achieving this will be prohibitive (all the time spent on meetings, discussions, solving conflicts, etc.) My main point about the central server concept is that this has very little to do with technology. I think maybe this is where we differ most. In my view, this is about the effort it takes for people to agree on something and the need for flexibility in organizations today. The effort to agree should be obvious to anyone working with the Matrix applications.

The practical experiences that molded the construction of the ADT took place roughly between 1990 and 2003. The coordination support achieved at the sites in Stockholm, Aachen and Linköping was outstanding in the history of Ericsson, both in terms of the functionality achieved, and comparatively small costs for the development. In hindsight, it can be safely said that a historical opportunity to continuously improve the coordination of complex systems at Ericsson was missed.

THE BROADER SCOPE

This chapter provides an account of additional observations collected over many years that somehow influenced the construction of ADT.

Business Processes

The concept of "business process" has a long history in Ericsson (as for that matter in other organizations as well). A business process can be defined as: "A collection of activities that takes one or more inputs and creates an output that is of value to the customer" (Hammer & Champy, 1993, p. 35). Typically, business processes are illustrated by "shoal of fish" diagrams such as the one from Ericsson in Figure 2.

Overloading the Processes Concept

In essence, a business process signifies a temporal dimension: the dependencies between a set of activities to be performed in a certain order. However, it is common to include other things besides activities in a process diagram. A typical example from Ericsson is shown in Figure 3. This diagram includes business units and views corresponding to commercial, design and supply perspectives.

A consequence of the tendency to include non-temporal aspects in the processes is the disintegration of the process concept; it becomes more difficult to grasp the essence of a process. In addition, the process diagram becomes more cluttered and hence more difficult to understand.

Glitches Between Processes

In general, the business process concerns the enterprise level of the organization. Each activity at that level may in turn be further split into sub-processes for software and hardware development, marketing, supply, service and support, and so on. Progress is usually signified by status chains, illustrating the evolution of a particular design item.

Figure 2. A business process illustration from Ericsson

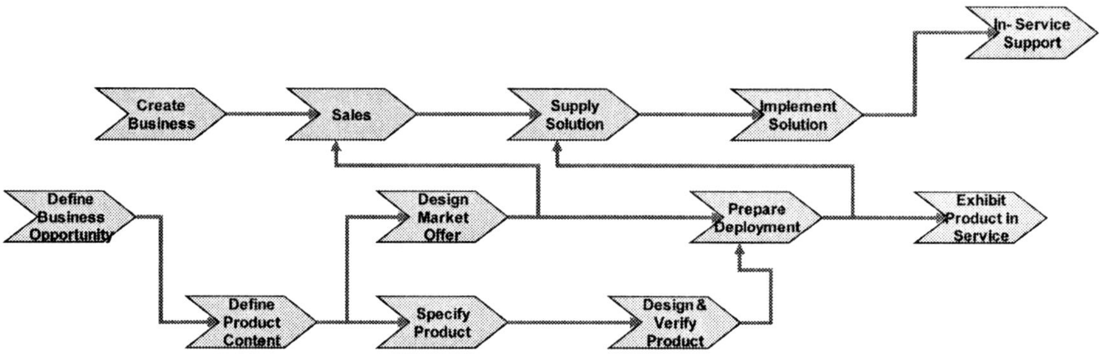

Figure 3. A business process including organizational units and different views

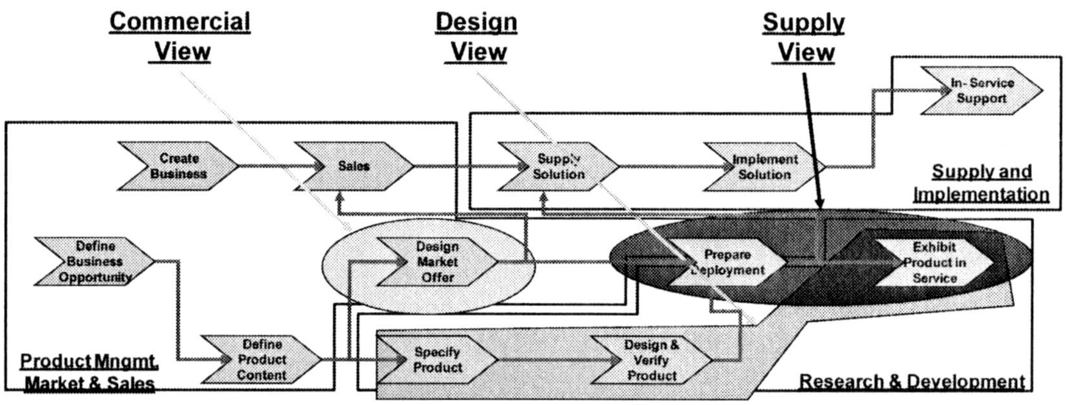

In each process, progression may be signified by entity states as illustrated by the Ericsson example in Figure 4. At the enterprise level, the statuses PC0 to PC6 and SC1 to SC8 are used. In the sub-processes for software (SW) and hardware (HW) design, other sets are used as indicated in the figure.

In each process, the set of states are defined from the specific needs of that process. The coordination between processes requires that states in different processes can be mapped into each other. However, analyses of the Ericsson processes showed that this issue was inadequately addressed. For example, there were no rules for how to map the HW statuses DS-, DS1 and PR- in the HW process into the statuses SC3 and SC4 in the business process. In other words, there were glitches in the transition between the processes. Quite naturally, this caused problems. The implication of this observation is that the transition between different areas of concerns should be more attended to when defining business processes and sub-processes.

Figure 4. State sets used in different sub-processes at Ericsson

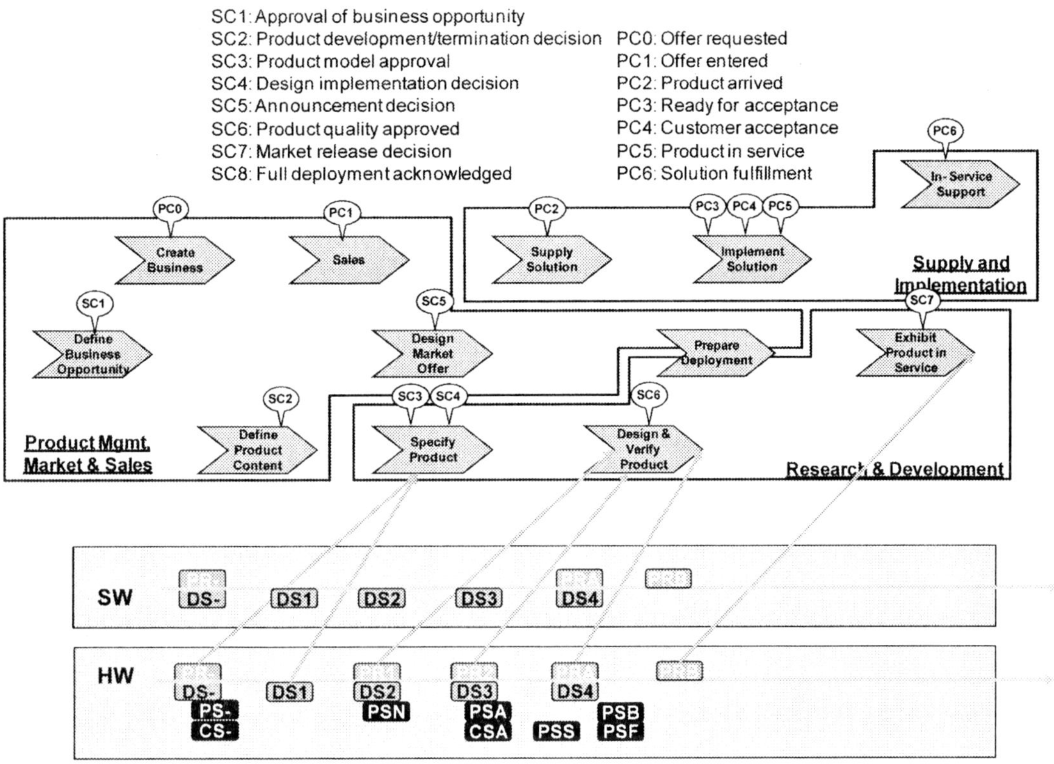

Interdependencies

The traditional way of illustrating a business process is to show how the various activities are interrelated (see the example in Figure 2). However, there is something unsatisfactory about this type of model in the sense that it does not show the information the activities are operating on other than as text in activity names, for example, "Specify Product". After all, the progression of information associated with the product should be the main concern when developing something. Just knowing that a particular activity has been carried out, or how the activities are interrelated is a secondary issue. Moreover, the order between activities cannot be laid out in an arbitrary way; this depends on how information entities are related to each other.

As alternatives to the activity based process models, entity based process models have been suggested (e.g. Humphrey, 1988; Humphrey & Kellner, 1989). An example of an entity based process model is the Information Interaction Model (IIM) described earlier. In Figure 5 the IIM corresponding to the model in Figure 2 is given.

The most conspicuous difference is that information entities are clearly visible as horizontal lines impacted by the activities outlined at the bottom. Although seemingly more complex, it is easier to comprehend how entities and activities are interrelated in IIM images that in traditional, activity-based images of processes. It is also clear from the IIM image how the state of each entity progresses through the process.

The significance of this analysis is that entities, activities and their interdependencies must be brought to the fore when conceptualizing how to coordinate product development. Just modeling activity de-

Figure 5. The business process as an IIM (Taxén, 2005. ©Wiley Inter Science. Used with permission)

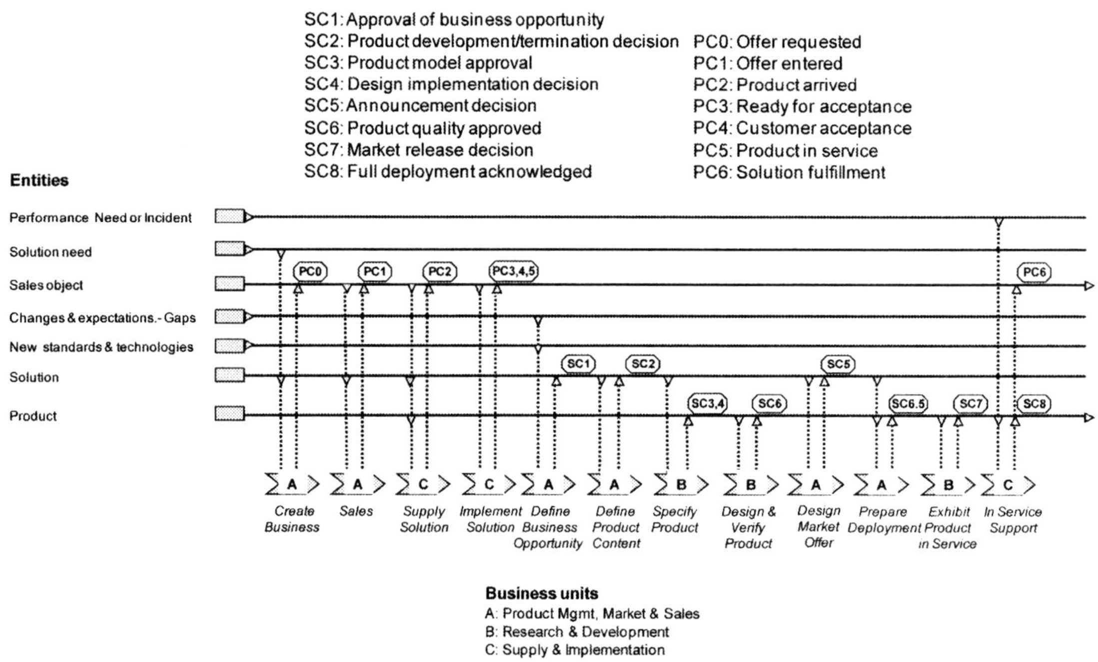

pendencies is not enough. Further, images that alleviate common understanding should be used. If the most important aspect – how entities evolve – is obscured, as in traditional business models, it can be expected that the effort to agree on the process is higher. Consequently, the risk of defining dysfunctional business processes is increased.

In summary, there is a tendency to include other elements than activities in business processes. This indicates that understanding coordination as "managing the dependencies between activities" is insufficient. Coordination is a wider concept. In particular, business units, information entities, activities, and dependencies between these elements must be clearly recognized.

Information Management Projects

Information is a key asset in any organization. Hence, it is not surprising that organizations work hard to define effective information architectures describing product structures, product related descriptive documents, releases, deliveries to customers, etc. As an example, at least twelve major different information management initiatives were carried out at Ericsson in 2000 and 2001. The foci of these initiatives were to:

- Identify Ericsson core information areas.
- Make information models for each area.
- Coordinate core information models with other models.
- Define the overall information model, for all product views.
- Define a complete product structure.

- Move commercial view into the central product management system at Ericsson.
- Define marketing view and connect to Product Data Management (PDM) systems.
- Define the PDM information system architecture of today and tomorrow.
- Improve interfaces and user interface and accessibility for each type of user.
- Implement the Ericsson core information model by defining the information flow between marketing, supply and R&D.
- Establish one common development process.

These initiatives were carried out as uncoordinated projects in spite of having considerable overlaps. It turned out that the meaning of terms used differed extensively between projects, and it was hard to see how the projects were related to each other. For example, one investigation revealed that there were about 20 different definitions of the central concept "product" in use.

It seemed to me that terms, which differed on the "surface", in fact denoted the same, underlying phenomena. Late 2002, I wrote an internal report entitled "Information Management at Ericsson, Fundamental Categories" (LTX-2002-08-15). The abstract was as follows:

Information management is a key area for Ericsson. In spite of many attempts to bring some order into this area, the situation remains in a state of disarray. We have several hundreds of uncoordinated information systems, and numerous isolated development initiatives are going on at this moment [...]. In this document, fundamental categories regarding information management at Ericsson are defined [...]. Thus, one purpose of the document is to provide common understanding about concepts regarding information management. Without such an understanding, all coordination efforts will be in vain.

Many of the ideas in this document were subsequently included in the elaborated ADT.

IS/IT Architectures

The term "IS/IT architectures" refers to the constellation of ISs and Information Technology (IT) based tools in the organization. This architecture is usually an intertwined and impenetrable maze of systems. It is virtually impossible to get a clear picture of how systems work together, if there are duplicate systems, what systems can be removed without causing problems, how to incorporate new systems, and so on. Every now and then initiatives are launched to replace or dispose systems in order to keep the number of systems down. Most often, such attempts come to nothing because it turns out that someone, somewhere, in the organization still uses the system.

Ericsson is no exception to this picture. In 2001 it was estimated that only in the sub-process Design & Verify Product there where around 250 systems in use. One employee described the architecture as "maximum entropy", illustrated by Figure 6.

Two observations captured my interest. First, it appeared that most organizations showed the same, "spaghetti" like IS/IT architecture. Although obviously not desirable, it seemed that the organizations had virtually no control over this development. It was almost as if the IS/IT architecture lived its own life within the organization.

Second, it seemed that the same pattern was repeated over and over again in the evolution of the IS/IT architecture: a movement from the varied and distributed towards the monolithic and centralized. The starting point is in general pressing practical needs somewhere in the organization. A new system is

Figure 6. A "maximum entropy" IS/IT architecture

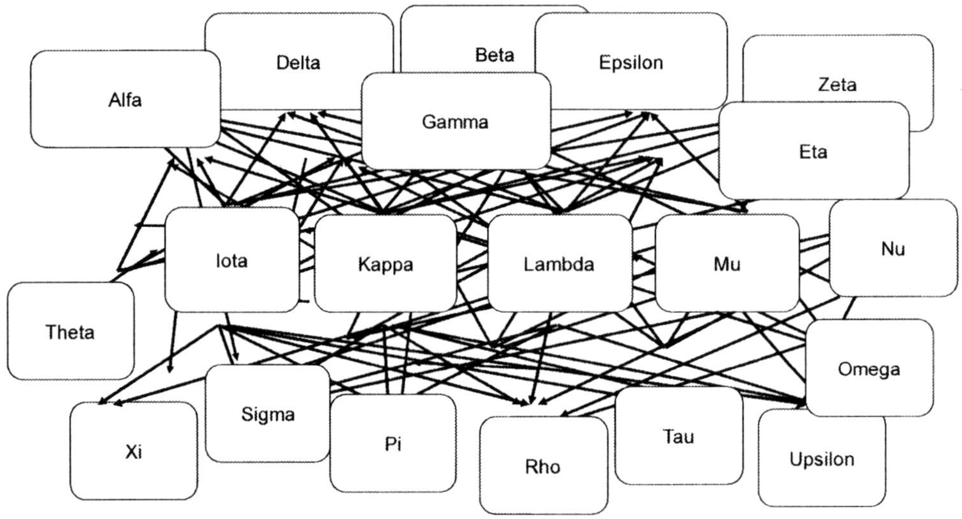

developed in-house or acquired from some vendor. After a period of adjustments, adaptations and practical experiments the system gradually becomes operational, and as time goes by, it becomes a natural and helpful means in the particular practice where it is used.

Meanwhile, other parts of the organization may hear about the system and start initiatives to develop similar support on the same system platform. After some time, several applications exist side by side, something which is conceived by management as a waste of resources since the "same" system is used at both places. The different applications built on the same platform are considered as oversights that should be corrected. Sooner or later, central initiatives are launched to consolidate the many applications into a single one. This pattern was repeated over and over again at Ericsson[3].

The interesting question is why the trajectory always seemed to go from many applications to one instead of the other way around? If the ultimate goal is to have one application only, it makes much more sense to develop that application up front rather than doing so afterwards. This is even more so since the effort of consolidation seemed to be quite significant and, in addition, often put to pieces well working applications and many years of training and learning.

These observations pointed to the need to acknowledge the role of local initiatives, while at the same time paying attention to the need for a central control. In other words, it appeared that some kind of balance between central/local control was necessary to address the problem of the messy IS/IT architectures.

Fragmented Responsibility

In any organization it is of course necessary to somehow divide the work to be done into manageable pieces. Whatever division is chosen, it is inevitable that tension arises on the border between the inside and outside of an area of responsibility.

A conspicuous insight from Ericsson was that the evolution and maintenance of business processes, information models, IS/IT and corporate business standards were organized as separate areas of responsibilities. The business process unit was responsible for Ericsson business processes such as in Figure 2. Core Information models, that is, companywide models of products and product related information were managed in a unit of its own. Corporate IS/IT had an overarching responsibility for strategies, platforms and introduction of business information systems like the product archive and ERP systems. Corporate business standards maintained business rules and principles for how products and other information entities should be identified, revised, released, etc.

The separation of responsibilities for processes, information management, IS/IT and standards caused problems since it was quite evident that these areas were interdependent. For example, a change in the definition of an information entity must be reflected in the ISs managing this entity. Another example is that the business process must be changed if the information entities the process operates on are changed. This was quite evident in the 3G domains, where processes, information, rules, and IS/IT elements had to be tightly aligned. However, as demonstrated by the 3G experience, there is a risk that such a tight integration results in the well-known phenomenon of "silos": the organization is conceived as a number of more or less self-contained units that rarely interact with each other.

Thus, it appeared to me that the distribution of responsibilities was misaligned. There is a need to conceptualize an integrated responsibility for processes, information management, IS/IT, and standards in limited organizational areas. However, there was no obvious organizational construct by which such a responsibility could be associated. The attempts related earlier to make the business process into this construct did not seem appropriate. The conclusion is evident – there is some very basic element missing in prevalent organizational discourse.

INSIGHTS AND NEEDS

This chapter summarizes and consolidates the most conspicuous practical experiences and observations that have had a major influence of the construction of the ADT.

Common Understanding

My experience from both the 3G development and my other activities at Ericsson, made it painstakingly clear that achieving common understanding is a difficult task. In order to illustrate this I will recapitulate a typical discussion, the purpose of which was to agree on a common information model for requirement management.

The first information models had evolved for several years in the work with the incremental development package and in connection with the first projects using Matrix at the S-, A- and L domains. During 2001 a project was started to unite the different information models into a common model for the C-domain. This was mainly discussed by actors in the core network area of Ericsson. During 2002 the radio area, which had been working quite separately from the core network area, began to take part in these discussions.

The information model for requirement management was a small but important part of the overall information model. In Figure 7 an example of such a model is shown.

This model should be detailed enough to specify how requirement management would be implemented in Matrix. This meant that at least the meaning of the following items had to be agreed upon:

- The types of requirements, for example: "Input Req", "Detailed Req" in Figure 7.
- The states or life cycle of requirements: "{Requirements}".

Figure 7. An information model for requirement management

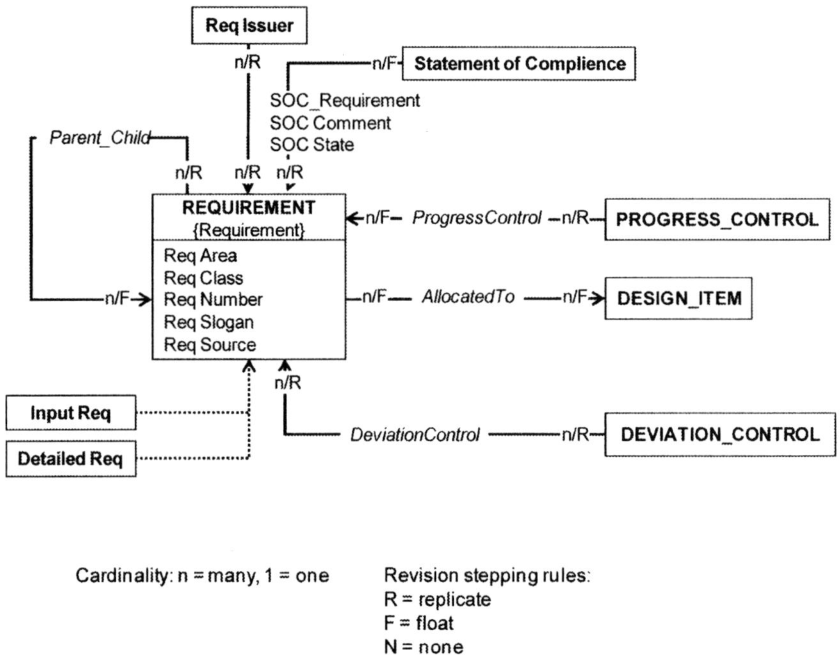

- Attributes on requirements: "Req Area, Req Class".
- Relations to surrounding items: "ReqIssuer, DESIGN_ITEM".
- Attributes on relations: "AllocatedTo, DeviationControl".
- Cardinalities on relations.
- Revision stepping rules which state how relations are treated when new revisions are instantiated of a requirement.
- Actor roles in relation to requirements, for example "requirement coordinator" or "configuration manager".
- Access rights for roles, i.e. who is allowed to do what with requirements, for example create or delete a requirement.

Two different types of requirements, "Input Req" and "Detailed Req", had been used in the S-domain for some time. One type originated from the customer ("Input Req"). The other type ("Detailed Req") was the customer requirements translated to into suitable requirements for product development. The names of the requirements were well-known and established in the S-domain but hardly outside that domain.

On April 11th, 2002, a meeting took place to discuss a common information model for requirement management. In the discussion, the model in Figure 7 was used as a starting point. The RMF (Requirement Management Framework) mentioned in the discussion was an attempt to define a common, tool independent, requirement management policy for all of Ericsson. Telelogic™ is a vendor who, among other things, sells the tool DOORS™. In the meeting, representatives from the radio area (persons A and B), the core area (persons E, F and LT) and C-domain developers (persons C, D and G) were present. The person LT is me. UAB is the organizational unit where I worked at that time. The discussion was recorded with the permission of the participants. Here is an excerpt from the discussion:

- LT: All the time at UAB, it was very useful to separate between these two types of requirements.

- A: Between what?

- LT: Between 'input' requirements and 'detailed' requirements, because they had been working like that for a long time and we couldn't do without them.

- B: But isn't it also that those input requirements are compiled from, I should say, from requirement from different issuers? Because that is also one thing to remember, that you compile several requirements into one requirement if they are similar.

- LT: Yeah, that's the task for the product management.

- B: We need to understand that too, that's the thing that can happen at each product management level to do that kind of work.

- C: OK, yes.

- E: But one thing we need here is flexibility in use of the system, because we have different needs, different history, and different practice in different organizations.

- LT: It would be nice to agree on something. If we can only agree that there is something called requirements then I think we are lost.

- B: But we can agree that we have customer requirements...

- LT: and design requirements...

- D: decomposed requirements and whatever.

- E: For me at the moment... on the level we are discussing right now we don't care about this 'detailed' requirements, decomposed or system requirements. I know what it is, for me.

- LT: Isn't it important to mirror that what you know in the model here?

- C: Yes, you have to have that in the prescription.

- LT: Or do you mean that you can take away these two types and still... or are they important to you?

- E: Yeah, they are important to me, yes. I mean, you have two different kinds of requirements.

- LT: Then I think we agree that we need two different types of requirements.

- G: We have two types of requirements.

- LT: Then it is just to change the names of those requirements.

- E: Do that if it is necessary, for me it is not that important.

- G: What it should be then is that this is a customer requirement.

- LT: Why not call it 'stakeholder' requirement? That's what at least Telelogic calls it.

- C: (laughter): Then we have another word for it!

- LT: So we can use DOORS (more laughter)

- (everybody is talking at the same time)

- LT: But change then 'detailed' requirement to 'design' requirement or system requirement or whatever so it has more of a connotation of a testable requirement.

- G: What I would like to do first of all is to get rid of this one, the parent-child relation should only be on the 'detailed' side.

- LT: But if you change one of these relations you must change the other one as well.

- C: No, but the stakeholder requirement must be connected to the 'detailed' requirement.

- G: It has to be a connection between them.

- C: How do you draw that then?

- G: I do it like this (shows on the board).

- C: Can you do that in the model? Is that possible?

- G: Yes.

- LT: Change the name of that to 'detailed' requirement before we go further we will be lost here, you can't have 'detailed' and 'stakeholder'.

- B: It's not the same.

- LT: I mean if you should change....you should change it on both levels.

- A: Actually I think I have problems with 'systems', 'input' and 'detailed', I don't even see the relation there, the logic.

- LT: What kind of names would you prefer then?

- A: (long pause) What does RMF say? (laughter)

- C: That is a good question, what does RMF say?

- G: They don't say that much.

- LT: They don't say anything about this.

- G: But stakeholder... anybody can be a stakeholder... that's wrong. A developer could be a stakeholder.

- B: But not when he is inside the project.

- E: I think it is really good to have 'stakeholder', because a stakeholder can be a customer.

- (and so on ...)

- LT: What's the reluctance of changing the name of the 'detailed' requirement?

- G: I'm not reluctant, only that no one could say the name of it!

- LT: Say 'design' requirement if that's better.

- G: I like that better.

- B: But then what do we call the ones that we name 'main requirements' then? Because that is the same level as system level?

- LT: We have requirements that can be tested and we have requirements that can't be tested.

- A: 'Product' requirements?

- G: But it could be requirements on whatever.

- E: Maybe 'product' is better.

- C: But aren't 'stakeholder' requirements put on products?

- LT: But the point about 'stakeholder' requirements is that you can't test them. We are going back again. You must have the freedom to formulate requirements that can't be tested.

- C: Give an example of a requirement that can't be tested!

- LT: "I want a car that runs fast"

- C: You can test that!

- LT: How?

- B: That's not a good requirement, you should say how fast.

- LT: Exactly, you should say how fast and that is then the design requirement.

- D: But it's just a hierarchy I think.

- LT: No, it's not, it's different.

- A: But, maybe we need both (laughter).

- G: Then we have that in the project preferences that we have a couple of requirements and that you can state whatever name you would like to be shown.

- LT: But, let's put it another way. These two types of requirements have been found good to have in the Ericsson practice and it has been found good to separate them. Can't we just leave it like that?

- F: We have to clarify that the first one is testable and the second one not testable.

- A: I mean, the advantage to call it product requirement is that it could be any type of requirement. It could be market, supply, it could be functional, it could be realization.

- LT: Well then you have quite a different thing... but the only real difference I see is really that some requirements are expressed in customer terms; other requirements are expressed in design terms or project terms. And there is a need to separate these two, believe me.

- G: Yes, I believe you and the thing is that we are discussing one thing in a model that has like 200 objects and it takes two hours. This is not the way to do it, it's impossible.

As G indicates above, after two hours of discussion, there was no consensus among eight persons about the meaning of one element (the requirement type) in an information model that may contain several hundred elements. In addition to that, these elements change due to external circumstances, new insights, new coordination situation, etc. It is quite easy to imagine the difficulties of a larger group of actors (say hundreds) to agree on the meaning of all the elements in the information model.

Furthermore, there is no objective criterion by which a "correct" model can be evaluated. There is no single answer as to which type(s) a requirement should have since this is completely arbitrary. Any proposal suggested above is equally valid in the same sense as a signifier of some signified phenomena in a social context is arbitrary. For example, the word "dog" in the English language, which signifies a quadruped with hairy fur making occasional sounds, is completely arbitrary.

Like Wittgenstein I believe that the meaning of a word is determined in the social activity among a group of actors (Wittgenstein, 1953). As long as the usage of the word in some action results in the intended consequences, the word is "correct". When misunderstandings or disagreements occur, the meaning of the word must be considered and reflected upon. This means that the only criteria by which the discussion above can be resolved is the usability of the word. If the "Input Req" type of requirement is useful in an activity domain it is "correct". This position can be seen in some of my comments above.

Consequently, disputes cannot be resolved around the table in a conference room. The information model must gain its credibility on the combat field where projects are using it to develop products. Since the model is implemented in an IS, its usability is ultimately tested by the usage of that system. If the model is detached from practice, it will become a reified phenomenon without relevance.

Thus, achieving common understanding should be up front on the agenda and treated systematically. One indication of how this was to be achieved was the iterative way of working at all sites, where information models and their implementations were tried out in practice.

Integrating Construct

Many observations pointed to the need of some organizational construct that could integrate elements like IS/IT, business processes, information structures, and business rules. Taking one of these elements, for example, business processes, as the basis for organizational analysis and change initiatives seemed to be inapt since all elements obviously were interdependent and could not be treated one at a time. Other indications of the need for an integrating construct were the glitches in the transition between state sets used in different process areas. This pointed towards some construct, which could provide both "horizontal" integration inside a process area and "vertical" integration between process areas, where the transition between these areas would be emphasized. Such a construct, I imagined, would be conceived as some kind of "kernel" structure that could be found on any level within and between organizations – much like the cell structure of living organisms.

Over time, the insight gradually grew that some kind of practice construct (Schatzki, Knorr Cetina, & von Savigny, 2001) was a viable candidate for this purpose. Eventually, the practice idea was elaborated into the activity domain, which subsequently became the integrating construct in ADT.

Contextualization

Closely related to integrating aspects is the issue of contextualization. Many observations indicated that situational and contextual aspects must be given more attention in the organizational discourse. This was most conspicuous during the 3G development, where the emergence of several coordination domains made possible an unmatched efficiency in the development of IS coordination support.

However, emphasizing situational aspects brought about consequences. The everyday fine-tuning and adaptations of systems are situational; this is where the needs are. But how could this be done in such a way that the situated did not turn out as isolated islands? After all, it is necessary to maintain some commonality between contexts. Thus, it seemed necessary to find a balanced solution satisfying both situational needs and commonality.

Contextualization is also apparent in the consolidation projects where local IT-systems were to be replaced by central ones. The trajectory was always from local to central, never the other way around. This indicated that the situational and local had precedence over the central, since it is only in the local setting that problems, needs and solutions can be worked out. This observation is supported by the depressing outcomes of efforts to develop and maintain central systems that would suit all needs. Most often, such systems resulted in various workarounds to overcome the deficits of centrally imposed systems.

Finally, contextualization seemed to play a distinctive role in characterizing things that appeared in many different organizational areas. It was quite clear, for example, that a product is differently conceived in market and development contexts. Different characteristics are emphasized depending on what is to be achieved with the product in different situations. Thus, context seemed to be necessary to include when modeling information to be managed in ISs.

Recurrent Patterns

Out of the multitude of things that I encountered in my daily work, I gradually became aware of certain patterns that appeared over and over again. At first, these patterns were mere premonitions; an annoying feeling that the immediately observed in fact expressed something more basic.

One such pattern concerned the *temporalordering* of things. Often, I came in contact with things that could be associated with time, such as processes, use cases, interaction diagrams, sequence diagrams, state sets, etc.

Likewise, another pattern had to do with how *things were related to each other*, for example, product structures, information models, anatomies, organizational charts, and the like. It appeared to me that this type of pattern expressed some kind of spatial and static dimension. Clearly these patterns were of a different kind than the temporal ones since there was no notion of time in spatial patterns.

Still another pattern was related to *how to do things*; what counted as valid and constructive actions in a particular organizational unit. Examples of such patterns, which appeared to be normative in kind, were method descriptions, standards for how to identify products, rules for how to revise a product, norms of proper conduct in the organization, organizing principles, and so on. Again, this pattern seemed to be of a different kind than spatial and temporal ones.

Finally, one type of pattern concerned the *transitions* between organizational areas. It appeared that this was a specific dimension that often was neglected as demonstrated by the experiences related in the Section *Glitches between processes*, p. 35.

Although different in kind, these patterns stood out as deeply interrelated and mutually constituting each other. For example, a process operates on some kind of information structure that signifies what information elements are affected. If this information structure is changed for some reason, the process needs to be changed as well. A change of rules for how a product is identified needs to be reflected in both the process and information structure. The same is valid for contextualization; a context change will affect the other patterns.

In short, it looked as if the immediately present in fact were signs of a deeper structure. In order to manage the apparent chaotic chain of events and things, it seemed necessary to get a better understanding of the essence of these patterns. This became the starting point for the subsequent elaboration of the patterns into what I call activity modalities in ADT – contextualization, temporalization, spatialization, stabilization and transition. These cornerstones in ADT are extensively treated in Section 3.

Enactment

The way Matrix was implemented through constant iterations was quite a break with prevailing methods for the development of large PDM systems. Such development usually progressed through a linear process of requirement collection, prioritization, analysis, implementation, testing and deployment. The most conspicuous observation with the iterative way of working was the knowledge acquisition that took place. It would have been virtually impossible to specify in advance the final capabilities of the system; most of it was developed "on the fly" so to say. This was mainly due to discoveries of new and innovative ways of working with Matrix, additional requirements included during the development, and corrections of errors.

These experiences made me realize the importance of what Orlikowski (2000) calls *enactment*; the emergence of new structures by constant engagements of people with the technology at hand. In other words, organizational artifacts must leave some imprints in the minds and bodies of actors before these become resources in the organization. A conspicuous example of this phenomenon was the attempt in 2001 to define a common information model for all sites. This work was led by a hired consultant, who from the outset tried to sell in his own model as the common one. Thus, the many years of enactment of the existing models were completely disregarded. The problem is, however, that if you put two models next to each other, there is no way to tell the enactment effort behind them. There are no historical marks on the models, they are just images. For their usability in the organization, however, enactment makes all the difference. If the consultant had succeeded in marketing his model (which he didn't), the enactment process had been forced to start from zero, something that would have had detrimental consequences.

THE PRACTICAL TRAIL: SUMMARY

In Section 1, an account of my Ericsson experience is presented. This experience was over the years molded into a certain world-view, where problems and needs were matched with an embryonic idea of how these issues could be addressed. The most urgent issues concerned:

- The achievement of common understanding about coordination.
- Finding an integrating construct that could bring processes, information architectures, business rules, IS/IT, and cooperation between organizational unit into a coherent whole.
- Paying increased attention to contextual issues.
- Being aware of the historical dimension of organizational artifacts.
- Uncover recurrent patterns of spatial, temporal, stabilizing, and transitional nature.
- Investigate the nature of enactment of coordinative means.

REFERENCES

Hammer, M., & Champy, J. (1993). *Reengineering the Corporation: A Manifesto for Business Revolution.* New York: Harper Business.

Humphrey, W. (1988). The software engineering process: definition and scope. ACM SIGSOFT Software Engineering Notes. In *Proceedings of the 4th International Software Process Workshop on Representing and Enacting the Software Process, 14*(4), 82–83.

Humphrey, W., & Kellner, M. (1989). Software Process Modelling: Principles of Entity Process Models. *CMU/SEI-89-TR-2*. Pittsburgh, PA: Carnegie Mellon University, Software Engineering Institute.

Latour, B. (1992). Technology is society made durable. In J. Law (Ed.), *A Sociology of Monsters. Essays on Power, Technology and Domination* (pp. 103–130). London: Routledge.

Orlikowski, W. (2002). Knowing in Practice: Enacting a Collective Capability in Distributed Organizing. *Organization Science, 13*(3), 249–273. doi:10.1287/orsc.13.3.249.2776

Schatzki, T. R., Knorr Cetina, K., & von Savigny, E. (Eds., 2001). *The practice turn in contemporary theory*. London: Routledge.

Taxén, L. (2005). A Socio-technical Approach Towards Alignment. *Software Process Improvement and Practice, 10*(4), 427–439. doi:10.1002/spip.245

Walsham, G. (1997). Actor-network theory: Current status and future prospects. In A. Lee, J. Liebenau, J. I. DeGross (Eds.), *Information systems and qualitative research* (pp. 466–480). London: Chapman & Hall.

Wittgenstein, L. (1953). *Philosophical investigations.* Oxford: Blackwell.

ENDNOTES

1. For a discussion about ANT and information systems research, see Walsham (1997).
2. A kind of contract between customer and supplier.
3. As described in Section 1, the introduction of the Matrix platform at Ericsson followed the same pattern.

Section 2
The Theoretical Trail

The Marxian philosophical cornerstones that have influenced the Activity Domain Theory: the concept of praxis, the dialectical relation, the epistemology of praxis, and the dialectical method. Additional cornerstones from the Russian theory of Activity: the concept of activity, the concept of mediation, and the Activity Theory view on meaning.

INTRODUCTION

How shall we understand the experience related in the first part? Are there some general traits hidden behind the seemingly disparate observations, or are these experiences only limited in time and space to one particular organization during specific circumstances? How can we bring the need for an integrating construct, contextualization, common understanding, recurrent patterns, and historical awareness into a coherent framework, which can be used for explaining observations and possibly guide us in a complex and turbulent world?

I will approach this task in two steps (see Figure 1).

The first step is to outline the basic perspective from which the ADT is derived. This perspective can, for the sake of convenience, be labeled "The Marxian conception of praxis". From the rich heritage of the Marxist praxis philosophy, I will only recapitulate such aspects that bear on the main topic of this book, the coordination of complex projects. There are four cornerstones that I will make use of: the concept of *praxis* itself, the *dialectical relation*, the *epistemology of praxis*, and the *dialectical method*. In the concept of praxis, the Marxian tradition tries to capture the fundamental nature of human activity. The dialectical relation is a way to conceive the nature of dependencies as mutually constituting

Figure 1. The path towards the ADT

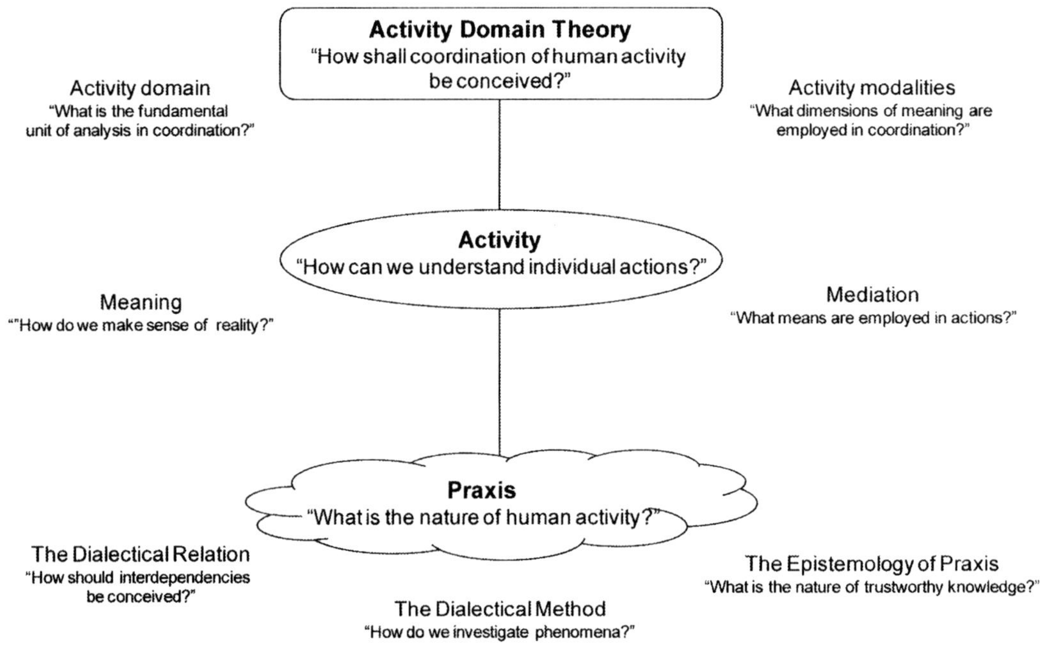

the dependent phenomena. The epistemology of praxis concerns the nature of trustworthy and actionable knowledge. The dialectical method is the Marxian approach towards investigating the underlying structures of phenomena that we experience as immediately given in everyday life. Together, these cornerstones provide an alternative conceptualization of organizational phenomena.

The second step is the elaboration of the praxis philosophy into the Russian theory of Activity (AT) done by Vygotsky, Leont'ev, Luria, and others in the 1920s (e.g. Kaptelinin & Nardi, 2006). The cornerstones I will make use of from AT are the concept of *activity*, which frames the context in which individual actions are meaningful, the concept of *mediation* (humans always put something between themselves and their object of action), and the AT view on *meaning* (how we make sense of reality).

With the praxis philosophy and the Russian theory of Activity as foundation, the ADT can be conceptualized as an elaboration of AT where coordination is in focus. The main constituents of ADT are the activity domain and the activity modalities, which are described in detail in Section 3.

Chapter 3
The Philosophical Roots

God does not work, though he creates, but man both creates and works (Saying from the Renaissance)

The notion of *praxis* was elaborated by Marx and Engels during the early years of their life-long co-operation. Praxis in the way put forward by Marx has, quite naturally, been further elaborated in many ways, and a number of works have been written on this topic. I will mainly make use of the account of praxis given by Bernstein in his seminal book "Praxis and Action" (Bernstein, 1999). Another source of inspiration have been the ideas of the Soviet philosopher Ilyenkov as explicated by Bakhurst in the equally outstanding book "Consciousness and Revolution in Soviet Philosophy: From the Bolsheviks to Evald Ilyenkov" (Bakhurst, 1991).

The heritage of Marx has hardly left any footprints in the annals of product development, and it might seem farfetched to use the ideas of a controversial thinker like Marx as a point of departure for an investigation into the nature of coordination. After all, the ideas of Marx and his forerunner Hegel have been relentlessly criticized and scorned by, for example, Popper (1945). It is but all too easy to dismiss the ideas of Marx in the light of his historicism and the way these ideas materialized in the socialist states. However, if we are able to see behind the political veil of Marxism we may be amply rewarded. It is my conviction that the ideas of the young Marx are highly relevant for coming to grips with the problems organizations face today.

In any case, we should not dismiss the potential that might be hidden in this heritage simply because its political connotations. So, let's put our blinders aside and embark on the route towards ADT!

DOI: 10.4018/978-1-60566-192-6.ch003

PRAXIS

In the Marxian tradition praxis is the nexus of human activity:

Praxis [...] is the essence of human existence in terms of producing, forming, and transforming the world. At the same time, praxis as collective productive and transforming activity, makes it possible to comprehend the social world as produced and being transformed, in contrast to viewing it as given. (Israel, 1979, p. 119)

The idea of praxis has a long philosophical history[1]. The word *praxis* originates from the Greek verb πρασσω [*prasso*]. Marx's ideas about praxis, which is the central concept in his philosophy, were originally formulated in 1845 in eleven "theses on Feuerbach" when Marx was only twenty-seven. As Bernstein points out, this document "is one of the most remarkable and fascinating documents of modern thought" (Bernstein, 1999, p. 13). The last one of these theses is the well-known statement: "The philosophers have only *interpreted* the world, in various ways; the point, however, is to *change* it" (Marx, 1852b, p. 15).

At the core of the discourse is the nature of the relation between a producer and his object of work, the product[2]. In contemporary analytical philosophy it is customary to think of someone working on something as two different ontological categories. The producer and product are separate, distinct kinds of beings. This position is precisely what is challenged by Marx. The product and its producer are *dialectically* related to each other, which means that they cannot change independently. The reason for this is that the product is seen as *the producer's activity in congealed or objectified form*. "Everything that is of fundamental importance in Marx's outlook depends on grasping this manner of viewing the relation of the objects man produces and his activity: it is essential for understanding what praxis means" (Bernstein, 1999, p. 44). A quotation from Marx illustrates this further:

A spider conducts operations that resemble those of a weaver, and a bee in the construction of her cells puts to shame many an architect. But what distinguishes the worst architects from the best of bees is this, that the architect raises his structure in imagination before he erects it in reality. At the end of every labour-process, we get a result that already existed in the imagination of the labourer at its commencement. He not only effects a change of form in the material on which he works, but also realizes a purpose of his own that gives the law to his modus operandi, and to which he must subordinate his will. (Marx, 1867)

Marx claims that the activity of human beings working together for some purpose not only brings about artifacts but also *creates the social existence of man himself.* To clarify this idea, let's consider an orchestra giving a concert. In what sense can we say that the natures of the musicians in the orchestra are "created" by this very activity? Obviously they cannot be created in a biological sense. But consider for a moment what is needed to perform. First, the musicians must have common understanding of the purpose or goal of what they are expected to achieve. Next, they must learn to use their instruments, how to read a score, etc. They must be able to follow the intentions of the conductor. Maybe there is a need for new ways of playing and invent new musical symbols to express this. Above all, the musicians must coordinate their actions in order to reach the goal. They must all interpret "the language of music" in a consorted way. Some musicians cannot play fortissimo when the score says pianissimo. All these things contribute to how the musicians understand and apprehend what they are doing. When the task is completed, the individuals in the orchestra have changed in the sense that they have acquired new skills and experiences, which in turn modify the way they see their world.

At first glance there is nothing strange about this scenario. We encounter learning, tools, cooperation, goals, methods, and a multitude of other things in everyday work, and we do not reflect on the deeper meaning of taken for granted and familiar things. We experience ourselves and the artifacts we encounter as separate things. However, the skills of the musicians, their instruments, and the coordination of their activities enable the performance. This activity is reflected back into the consciousnesses of the musicians and the artifacts as new insights, modified ways of playing, new musical concepts, improved instruments and so on. These two aspects cannot be detached from each other and treated independently:

[The] relationship between man, the producer, and the social world, his product, is and remains a dialectical one. That is, man [...] and his social world interact with each other. The product acts back upon the producer. (Berger & Luckmann, 1966, p. 78)

In organizational settings, the view of the producer and his product as dialectically related can be exemplified by the introduction of an Enterprise Resource Planning (ERP) system. In order to become useful, the ERP system has to be adapted to the workings in the organization. However, as soon as the adaptation work begins, for example by making up project plans for the adaptation, the mere plan will change the way the actors apprehend their organization. As the work proceeds, a new reality is gradually formed in the minds of the actors, a reality that did not exist before the ERP implementation. This new conception will in turn be reflected back into the adaptation work. From a praxis point of view, the conceptualization of the ERP system and the system itself cannot be detached from each other; they are dialectically related.

In summary, the essence of the praxis category is that "the very nature of man is determined by what he does or his praxis, and his products are concrete embodiments of this activity" (Bernstein, 1999, p. 44). As we shall see later on, this view has profound impacts on how we approach organizational issues in general and coordination in particular.

THE DIALECTICAL RELATION

The praxis perspective brings with it a specific way of apprehending relations between phenomena – the dialectical relation. Praxis is usually translated into English as "practice" with its cognate "practical", indicating some fluent and goal-directed performance of everyday tasks. In this doing, the practitioner is focused on carrying out the task at hand, and he is primarily not concerned with theory or reflection.

In Western philosophy, this understanding of practice is reflected in the distinction between practice and theory as two different realms, which can be demarcated and investigated as self-contained, autonomous areas, albeit related to each other. Such a *dualistic* view of the world can be dated back to Descartes who asserted that body and soul were two quite different substances. In this view, material and psychological categories are seen as related but clearly separated. Their relation to each other is *external*, by which we mean that the categories have nothing in common other than that they happen to be juxtaposed in a certain context, for example, a car being parked at a certain parking place. When two things are related externally, they are not changed by the relation. A contemporary variant of the dualistic view can be found in, for example, Habermas "world relations" between the speech act and the "subjective", "objective" and "social" worlds respectively (Habermas, 1984).

For Marx, the dualistic way of approaching reality is utterly false. The alternative, which permeates all of his works and dates back to Hegel, is the *totality* (Totalität) category. As in dualism there are pairs of opposites. However, the relationship between these pairs is *internal*, which means that opposites are different but mutually depending on and impacting each other within a totality. According to Israel (1979, p. 112) a dialectical relation is characterized by the following:

- The elements in the relation form a unity or totality.
- The elements are different, i.e., each element can be identified as something specific.
- The elements depend on each other in a contradictory way.
- The mutual dependency between the elements is not random or contingent. One element cannot be conceived without the other.
- The elements have something in common.

An example of this type of relation given by Hegel is the "master"-"slave" relation (Bernstein, 1999). The meaning of "master" is defined by referring to "slave" and vice versa. Although they may appear as distinct categories, they in fact mutually constitute each other. Without "master", there is no "slave". Moreover, the nature of their relationship evolves over time. In the beginning the slave is completely subdued to the master and the result of his toil, the products, are in complete command by the master. However, as soon as this relationship is established, the master inevitably becomes dependent on the slave for his subsistence. At a certain point in time the slave realizes this, and from that moment on the relationship changes. The slave can envisage another life as a free man and take measures to achieve this, for example by revolting. Thus, the internal relationship bears within itself the seed to its own destruction and possibly transformation to a new and qualitatively different form.

The dialectical way of understanding a relation has profound implications for how we apprehend the parts-whole relationship. Consider the following example (from Levins & Lewontin, 1985). A person cannot fly by flapping her arms, no matter how much she tries, nor can a group of people fly by all flapping their arms simultaneously. But people do in fact fly. This is a consequence of a long and tedious historical process where socially organized human activity over time has produced airplanes, pilots, landing strips, fuel, and all the other things necessary to fly. Although the biological constitution of humans has not changed, we have in fact acquired a qualitatively new individual property as a social being – we can fly. Today, man can look down on clouds from thousands of feet above the ground, while before the twentieth century, man could only look up at the clouds.

However, this is not the end of the story. It is the dialectical relation between humans, for example pilots, and products such as airplanes that make flying possible. They constitute each other. Without capable pilots – no flying; without capable airplanes – no flying. In fact, the mere categories of pilots and airplanes do not make sense in isolation. Moreover, it is not only humans that have changed according to this relation. The airplane and its parts have acquired new properties: they can fly by being parts of the airplane and the totality of flying. A jet engine would never get off the ground if it was not part of this totality. Thus, a dialectical world view carries with it an intrinsic way of apprehending the relationship between parts and the whole made up by these parts:

But the ancient debate on emergence, whether indeed wholes may have properties not intrinsic to the parts, is beside the point. The fact is that the parts have properties that are characteristic of them only as they are parts of wholes; the properties come into existence in the interaction that makes the whole. (Levins & Lewontin, 1985, p. 273)

Given praxis as the nexus of human activity and the dialectical relation as the fundamental way of conceiving the relation between actors and their products, how can we arrive at trustworthy knowledge about how to act in praxis? This is the subject of the next section.

THE EPISTEMOLOGY OF PRAXIS

The proof of the pudding is in the eating. (From Don Quixote, Cervantes, 1615)

The praxis perspective implies a certain view on epistemology, that is, the philosophical branch that investigates what reliable knowledge is and how can we acquire such knowledge – "how do we know what we know?"

The close interrelation between various processes of production makes "praxis" a central notion for epistemology. (Israel, 1979, p. 70)

In contemporary organizational discourse, knowledge is often seen as a commodity; a thing that can be packaged into artifacts and transferred between individual minds. Sometimes this view takes on rather bizarre forms:

Download knowledge directly to the brain! Today the actual learning process takes too long. In the future we will download knowledge directly to the brain. Connect in to something which contains specific know-how and transfer it over. (Framed statement hanging on the wall at Corporate IT, Ericsson, July 2000)

This commodity view of knowledge goes back to a separation between "is" and "ought", or the descriptive versus the prescriptive. "Is" should be concerned with "facts" and "ought" with values. The praxis perspective challenges the dichotomy between "is" and "ought". We cannot sharply distinguish that which is immediately given from what is constructed, inferred, or interpreted by us (Bernstein, 1999, p. 72). All observation is "theory-laden" in the sense that we must direct our attention to some part of reality, implying that other parts remain unnoticed. What we see is a consequence of our conception of reality.

This view of reality is shared by many philosophical branches, for example, phenomenology. What is specific for Marx is that he relates human cognition to the evolution of man's practical needs as manifested in praxis (ibid, p. 73). The individual is born into a social world at a certain place and time as a biological creature among others. However,

From the moment man in his onto- and phylogenesis begins to dominate the world of things intellectually - from the moment he invents instruments that can organize it and then expresses this organization in words – he finds that world already constructed and differentiated, not according to some alleged natural classification but according to a classification imposed by practical need for orientation in one's environment. (Kolakowski, 1969, p. 46)

This does not mean a retreat to absolute idealism, that is, the position that there is no objective world existing independently of mankind. On the contrary, there is always a nonhuman, naturally given and material "substrate" that man makes use of in praxis. Man does not have a unique position among other living organisms in this sense; he is a natural being among other natural beings. However, we can only "see" the external world through the window of human praxis. It is meaningless to ask what the absolute and "true" nature of this world or "substrate" is.

[Each] separate sensual impression arising in individual consciousness is always a product of the refraction of external stimuli through the extremely complex prism of the forms of social consciousness the individual has appropriated. This 'prism' is a product of social human development. Alone, face to face with nature, the individual has no such prism, and it [the prism] cannot be understood from an analysis of the relation of an isolated individual to nature. (Ilyenkov, 1960, p. 40-41, in Bakhurst, 1991, p. 151)

When we are confronted with issues such as "truth", "reality", "facts" we must realize that such issues cannot be treated as something external to praxis. On the contrary, they lose their meaning outside praxis. For example, the general law of gravity as Newton formulated it in the seventeenth century is usually taken as a "fact". However, we know that it was formulated in a particular society and time, it was written in a particular language (Latin), the writing was carried out using specific artifacts, its findings are based on the activity of previous scholars, and these findings were subsequently elaborated and modified by Einstein. Thus, what are counted as "facts" is a consequence of praxis and its development in time and space.

Consider again the ERP implementation example. Suppose the motive for implementing the ERP system is to get a common way to manage products, and to replace overlapping and fragmented legacy ISs. The starting point for such a task is usually an investigation of the "as-is" situation, by which the "facts" are established. This is followed by a phase where the "to-be" situation is envisaged, that is, the "ought" is established. Based on this information, actions are performed to transform the "as-is" state into the desired "to-be" state.

From a praxis perspective, the "facts", that is, the current constellation of ISs and the information managed in these systems are the congealed form of human activity. This form and the way it comes forth for the actors are specific for that particular organization and moment in time. In the activity of establishing the "as-is" situation by using, for example, a particular method and type of documentation, the "humanized form" of the situation will inevitably change. New artifacts will be created (the documentation), and the actors working with producing the "as-is" "fact" will see the situation in a new light. Thus, the mere act of establishing the "as-is" situation will change that situation.

The same considerations are of course valid for the establishment of the "to-be" situation and the task of implementing the ERP system according to the "to-be" vision. Carrying out these tasks is a dialectical process where both the "as-is" and the "to-be" situations will inevitably change. From a rationalistic point of view, these changes are regarded as undesirable disturbances that should be minimized by improved methods and planning techniques. However, from a praxis perspective, such changes are inevitable; they are inherent in the activity itself.

The dichotomy between the "is" and "ought" is misconceived. If we are inclined to think in these categories, we will be unprepared to manage the dynamics in human activity and unable to capture new possibilities that emerge in praxis:

[Man's] actuality and his potentiality change in the course of man's historical development. Genuinely new potentialities arise as a result of human praxis. (Bernstein, 1999, p. 70)

Issues of "truth", "reality", and similar expression for praxis-external expressions of alleged objective and fixed "facts" can only be resolved in praxis:

The question whether objective [gegenständliche] truth can be attributed to human thinking is not a question of theory but is a practical question. In practice man must prove the truth, that is, the reality and power, the this-sideness [Diesseitigkeit] of his thinking. The dispute over reality or non-reality of thinking which is isolated from practice is a purely scholastic question. (Marx, 1852b, p. 13, italics in original)

THE DIALECTICAL METHOD

Marx's dialectical method represents cognition (poznanie: the process of coming-to-know) as a movement from "abstract" to "concrete". (Bakhurst, 1991, p. 138)

Given the dialectical relation and the praxis perspective on epistemology, how can we perform an inquiry into the nature of social reality? How do we acquire well-grounded and reliable knowledge about this reality? Although our daily life may seem chaotic and exposed to a manifold of events over which we have no command, we also know that there are regularities that make life to some extent predictable. The bus arrives at the bus stop most of the times it is supposed to do so; we drive on one side of the road, names of persons seldom change, and so on. We simply have to take some things for granted, otherwise society would not be possible to uphold. The dialectical method of inquiry assumes that there is some kind of intelligible structure of social reality that it is possible to inquire into.

In everyday life man is actively producing the necessary means for survival. He manipulates and attends well-known and familiar tools, systems, institutions, etc., to reach whatever goals are at hand. The circumstances in which this is carried out are called "care" (Sorge) by Heidegger (1962). Care is the fundamental of humans "Being-in-the-world", not matter or spirit:

Dasein's facticity is such that its Being-in-the-world has always dispersed itself or even split itself up into definite ways of Being-in. The multiplicity of these is indicated by the following examples: having to do with something, producing something, attending to something and looking after it, making use of something, giving something up and letting it go, undertaking, accomplishing, evincing, interrogating, considering, discussing, determining .... (Heidegger, 1962:H56)

Usually, the circumstances in which care is carried out are taken for granted and fixed. It is not until something irregular happens that we become aware of the existence of something else that is of fundamental importance to our lives. We experience the *phenomena* but not the *essence*: the totality category in dialectical thinking. For example, in our daily life we use money, a piece of paper or metal, without ever reflecting upon what the fundamental character – the essence – of money is. The essence reveals itself, for example at wartime when hyperinflation may occur, or in moments of stock-market breakdowns such as during the depression in the 1930s.

The aspects of things that are most important to us are hidden because of their simplicity and familiarity. (One is unable to notice something - because it is always before one's eyes.) The real foundations of his enquiry do not strike man at all. Unless that fact has at some time struck him. - And this means: we fail to be struck by what, once seen, is most striking and most powerful. (Wittgenstein, 1953:§129)

The essence appears to us as something alien to us as humans, as detached from our own activity or *alienated* as Marx put it. However, even if we are not aware of the essence until everyday life breaks down, both care and the essence are consequences of human action:

Since care is the entanglement of the individual in social relations seen from the perspective of the involved subject, it also amounts to trans-subjective world seen by that subject. Care is the world in the subject. The individual is not only that which he considers himself or the world to be: he is also a part of the situation in which he plays an objective trans-individual role of which he may be quite unaware. (Kosík, 1976, p. 38)

The familiarity of everyday life means that we cannot lay bare the totality category directly: "familiarity is an obstacle to knowledge" (Kosík, 1976, p. 46). In order to do so, we must start from what is immediately given in care and work our way through until we grasp the intelligible structure or essence in its entirety. The central thought figure in this endeavor is "*the ascent from the abstract to the concrete*". In order to understand what Marx meant by this I will recapitulate the meaning of these terms in non-dialectical thinking.

The Concrete and Abstract in Non-Dialectical Thinking

In everyday language we think of the concrete as something immediately perceivable and sensuously accessible. I can feel the touch of the keys as I am writing this sentence and I can sense the pleasant smell of newly brewed coffee in the cup next to me. Concrete is "particular, sensually perceived things or their perceptual images" (Ilyenkov 1960, pp. 14-15, in Bakhurst 1991, p. 144). Abstract, on the other hand, is something that is derived from the concrete. Abstract is "the general forms of things, their identically repeated qualities and lawlike relations, expressed in terms, names, and numbers" (Ilyenkov 1960, p. 15, in Bakhurst 1991, p. 144). The abstract does not have a concrete individual existence of its own that can be named. Its purpose is to classify and typify the manifold of sensory impressions into meaningful commonalities. Knowledge has a hierarchical character where at its base "the boundless sea" of sense particulars is and at its summit stands our most general concepts and beliefs. An example from contemporary software development is the Object-Oriented modeling and design (e.g. Rumbaugh,

Blaha, Premerlani, Eddy, & Lorensen, 1991), where class hierarchies are erected with the most general types at the top, and subtypes inheriting the properties of the supertypes lower down in the hierarchy.

The non-dialectical philosophy, in particular the empiricist one, conceives of cognition, or coming-to-know, as a movement from the concrete to the abstract. The starting point of cognition is the formation of concepts from data of perception in the individual. By observing the world, we successively acquire knowledge by constructing even more elaborate and abstract conceptions of it. Out of raw, sensuous and unimpaired data – a clean sheet without preconceptions - we arrive at an enlightened view of the world. An example of this process is described by Mackie:

I see a white piece of paper at a particular time and place, and notice that it resembles in colour other pieces of paper, cups of milk, fields covered with snow, and so on; I pay attention to the feature in which it resembles these other things and pay no attention to the shape or size of the piece of paper or its surroundings or even to the time at which I see it; I remember this feature and associate the word "whiteness" with it ... and I am thus ready to use the same word 'whiteness' with respect to that same feature in any other things at any other places and times ... (Mackie, 1976, p. 110, in Bakhurst, 1991, p. 145)

The Concrete and Abstract in Dialectical Thinking

In dialectical thinking, the concrete and abstract mean something wholly different than in non-dialectical thinking. "Concreteness" is a characteristic of the intelligible structure which makes up the essence:

[A concrete object is a] "system," "totality," or "whole" composed of individual phenomena integrally related to each other; that is, the parts are pictured as essentially related to one another, their natures constituted by their position and role within the whole. Concreteness may be seen as a function of the strength of the relations between components: The tighter the bonds of mutual determination between its parts, the more concrete the whole. (Bakhurst, 1991, p. 139)

Thus, the object of inquiry is a "whole" where the parts are related internally to each other and mutually constitute the whole and the parts – a *concrete totality*. The structure of the whole is based on the diversity of its parts, and tensions between parts give rise to the dynamic evolution of both the whole and its parts.

In every concrete totality there is a particular component that has a unique position in the sense that it determines the other. This component is called the *concrete universal* of the totality, and can be thought of as an elementary form of the whole (Bakhurst, 1991, p. 140). The concrete universal is the germ, the central category which determines the evolution and structure of its own progress and of the whole. The task of the dialectical method is to reproduce the concrete totality and its "seed", the concrete universal. The inquiry starts from the abstract by which is understood:

[Anything] picked out, isolated, existing "on its own," in relative independence from everything else, - any "side," aspect or part of a real whole, any determinate fragment of reality or of its reflection in consciousness. (Ilyenkov 1967, p. 192, in Bakhurst, 1991, p. 141)

An everyday example of a concrete universal is given by Marx: the "commodity". In capitalist economies, the commodity is the universal around which everything revolves and the concrete totality emerges. The abstract expressions of this universal are the detached views of the economy that is presented in our daily life. For example, the collapse of the financial market in 2008 and 2009 is explained in terms of psychology, greedy brokers, deregulated markets, lack of control, etc. Every time similar collapses occur, assurances are issued that the "we have learnt from events and this will not happen again". The concrete totality of the capitalist economy and its underlying structure is seldom disclosed, which implies that abstract phenomena remain unexplained.

On the surface, the two positions of the non-dialectical and dialectical methods seem to be compatible. Both methods start with an observation of the given, sensuously given data. The non-dialectical method generalizes theories from these data, while the dialectical method finds the concrete totality from the very same data. There is however a major difference between the two. The non-dialectical method assumes that a picture of the world is given to each individual by perception of raw, sense experience. However, the point of departure for an inquiry can never be a clean sheet, unimpaired by a preconception of what the world is. According to Marx we must realize that cognition does not start with "unprocessed sense experience but a conception of the world *inherited* 'ready-made' from the community of which he or she is a member" (Bakhurst, 1991, p. 150). As a child is born into a certain community in time and space, this conception of the world is appropriated by the child by reproducing the practices of that community. "This is not a conscious process, but the process of becoming conscious" (ibid, p. 150).

Every formation of a concept is possible only if the individual already possesses a conceptual schema. The mere observation of an object in the immediate environment presupposes some schema in order to identify it. After all, we cannot attend everything that happens around us, big or small. Some basis must exist from which we judge resemblances between objects, decide that they have some features in common, and evaluate them as suitable stepping stones for abstraction. Thus, the dialectical method represents a different way of how we can achieve knowledge about the reality that confronts us in everyday life. This does not mean a dismissal of the non-dialectical method. On the contrary, the capability of man to form abstractions is a prerequisite for recognizing the individual:

The world as it is conveyed to man's knowledge and as it is communicable in language is a world composed of abstractions; one arrives at the individual only through the intermediary of abstract and general concepts. (Kolakowski, 1969, p. 45)

For the purpose of this book, the importance of the dialectical method lies in the fact that it provides a constructive way of approaching the issues that confronted me in the Ericsson practice. Together with the other cornerstones – the praxis point of departure, the dialectical relation and the epistemology of praxis – the dialectical method provides a solid ground for the ADT. However, before we can dwell on the intricacies of ADT, we need to add another building block: the concept of "activity" in Russian Theory of Activity. This is done in the next chapter.

REFERENCES

Bakhurst, D. (1991). *Consciousness and Revolution in Soviet Philosophy: From the Bolsheviks to Evald Ilyenkov*. Cambridge, UK: Cambridge University Press.

Berger, P., & Luckmann, T. (1991). *The Social Construction of Reality*. London: Penguin Books.

Bernstein, R. (1999). *Praxis and Action*. Philadelphia: University of Pennsylvania Press.

Habermas, J. (1984). *The Theory of Communicative Action,* (Vol. 1, Reason and the Rationalization of Society). Boston: Beacon Press.

Heidegger, M. (1962). *Being and Time.* J. Macquarrie & E. Robinson (Trans.). New York: Harper & Row.

Ilyenkov, E. V. (1960). *The Dialectics of the Abstract and the Concrete in Marx's "Capital".* S. Syrovatkin (Trans.). Moscow: Progress, 1982. Retrieved Oct 7, 2008, from http://www.marxists.org/archive/ilyenkov/works/abstract/index.htm

Ilyenkov, E. V. (1967). Problema abstraktnogo i konkretnogo v svete "Kapitale" Marksa [The Problem of the Abstract and Concrete in the light of Marx's *Capital*]. *"Kapital" K. Marksa. Filosofiya i sovremennost'* [*Marx's "Capital": Philosophy and the Present*]. Moscow. Israel, J. (1979). *The language of dialectics and the dialectics of language.* New York: Humanities Press.

Kaptelinin, V., & Nardi, B. (2006). *Acting with Technology - Activity Theory and Interaction Design.* Cambridge, MA: The MIT Press.

Kolakowski, L. (1969). *Toward a Marxist humanism: essays on the left today.* New York: Grove Press.

Kosík, K. (1976). *Dialectics of the concrete*. Dordrecht: Reidel.

Levins, R., & Lewontin, R. (1985). *The Dialectical Biologist.* Cambridge, MA: Harvard University Press.

Liedman, S.-E. (1998). *I skuggan av framtiden.* [In the shadow of the future]. Finland: Albert Bonniers Förlag (in Swedish).

Mackie, J. (1976). *Problems from Locke.* Oxford, UK: Oxford University Press.

Marx, K. (1852b). Theses on Feuerbach. In *Karl Marx and Frederick Engels, Selected Works, Vol. 1* (pp. 13–15). Moscow: Progress Publishers, 1966.

Marx, K. (1867). *Capital, I, chapter 7.* New York: International Publishers. Retrieved Sept 22, 2008, from http://www.ecn.bris.ac.uk/het/marx/cap1/index.htm

Popper, K. (1945). *The open society and its enemies.* London.

Rumbaugh, J., Blaha, M., Premerlani, W., Eddy, F., & Lorensen, W. (1991). *Object-Oriented Modeling and Design.* Upper Saddle River, NJ: Prentice-Hall International, Inc.

Wittgenstein, L. (1953). *Philosophical investigations.* Oxford, UK: Blackwell.

ENDNOTES

[1] Praxis was used by Aristotle to denote actions which are goals as such. The ability to realize such goals he called *phro'nesis* or wisdom. *Poi'esis* denotes actions which are realizing goals outside the actions. The ability to realize these goals is called *techne*. Abilities which are a prerequisite for *theoria*, theoretical knowledge, is called *episte'me* (e.g. Liedman, 1998:266).

[2] Words like "man", "himself", "he" and other similar expressions are used in the original works to express general features of humankind. I have chosen to adhere to this rather than finding more appropriate, gender indifferent expressions in the book.

Chapter 4
Activity Theory

In this chapter I will give an account of some ideas in the Russian Theory of Activity or Activity Theory (AT) that have influenced the Activity Domain Theory (ADT): *activity*, *mediation*, and *meaning*. The activity domain in ADT is a direct descendant from "activity" in AT, while "mediation" and "meaning" are necessary prerequisites for the activity modalities construct in ADT.

The AT was an attempt to apply the ideas of Marx and Engels to psychology in the early decades of the new socialist state, the Soviet Union. The front figure in this pioneering movement was the Russian psychologist and semiotician L. S. Vygotsky (1896-1934) together with his collaborators A. N. Leont'ev (1903-1979) and A. R. Luria (1902-1977). Other prominent researchers in this spirit were V. N. Vološinov (1895-1936) and M. M. Bakhtin (1895-1975). With the advent of the Stalinist era the momentum of the AT was more or less crushed. However, small but marginalized groups kept the ideas of AT alive. One of the most prominent philosophers was the previously mentioned E. Ilyenkov (1924-1979). During the last couple of decades, AT has gained a renewed momentum among Western researchers and been further developed by the works of M. Cole, J. Wertsch and Y. Engeström and others.

THE CONCEPT OF ACTIVITY

The central concept in AT, the "activity" (German: Tätigkeit; Russian: deyatel'nost'), has a specific meaning that differs from how activity is usually understood in English. Activity was first introduced by Leont'ev as a fundamental unit in his investigations of the early manifestations of the mind in the human evolutionary history:

DOI: 10.4018/978-1-60566-192-6.ch004

I will call the process of activity the specific processes through which a live, that is, active relation of the subject to reality is realized, as opposed to other types of processes. (Leont'ev, 1981, in Kaptelinin & Nardi, 2006, p. 55)

Activities cannot exist without *objects*: "Any activity of an organism is directed at a certain object; an 'objectless' activity is impossible" (Leont'ev, 1981, in Kaptelinin & Nardi, 2006, p. 55). Activity in this sense is equally applicable to all organisms that engage in its immediate "life-sphere", whether they merely respond to signals like ticks, modify their environment like spiders or use tools like apes using sticks to catch termites. With the evolution of neural networks in organisms came the possibility of activity mediated by *representations* of phenomena, for example, heeding calls warning for predators. However, the significance of these representations rarely stretches beyond the immediate situation in space and time that the organism encounters.

With the human mind, a qualitatively new level of the psyche is reached. Representations can signify situations beyond here and now: past times, future events, places far away, and so on. The survival of a human being is not solely determined by physical and biological things of such as the availability of food, shelter, etc., but also of the social reality the individual is born into. Ontogenesis, i.e. the development of the individual, implies the appropriation of meaningful, referential concepts and signs that have evolved historically over time in a specific *cultural* setting. In particular, the individual must learn to master the language of its social milieu.

With the cultural dimension, activity becomes a social phenomenon in which humans join forces to fulfill social *needs*, such as supplying food to all. This brings forward another dimension of activity – the division of labor. Rather than having everyone doing all and the same tasks, the collective effort is distributed to different individuals, each proficient in performing a specific task. This implies that the actions of each individual need to be *coordinated* with the others. Thus, the division of labor brings about a more efficient way of fulfilling the social need at the expense of an extra effort to coordinate individual actions.

With the division of labor, the object of the activity becomes more sophisticated. The reason for the activity as such is fulfilled by the coordination of individual actions. However, individual actions may, taken one by one in isolation, be apprehended as contradicting this reason. This was illustrated by Leont'ev by the famous example of the activity of hunting. Consider the roles of the beaters in this activity. Although the obvious reason for the activity is to get food, the actions of the beaters drive the quarry away. Taken out of the context of the activity, these actions appear to be meaningless if not downright misdirected.

In AT these different aspects of activity are distinguished as follows. The activity itself is said to be oriented towards a *motive*, which is "the object, which stimulates, excites the subject. It is the object that the subject ultimately needs to attain" (Kaptelinin & Nardi, 2006, p. 62). The actions needed to achieve the motive of the activity are directed towards *goals*. "Goals are conscious; we are typically aware of the goals we want to attain" (ibid). Thus, the goals of the beaters in the hunting example are to drive the quarry away, and the activity is motivated by the need to get food for all individuals belonging to the group. The motive of the activity determines the structure of the actions, that is, what kind of actions are meaningful and how these are to be coordinated.

The significance of the concept of activity for ADT is that activity provides a common basis for organizational units at different levels such as organizations, business units, groups, teams, etc. In ADT, these are conceptualized as activity domains, which should be seen as activities where coordinative aspects are emphasized.

MEDIATION

A key concept in AT is *mediation*, which refers to "the idea that humans always put something else between themselves and their Object of work" (Bødker & Bøgh Andersen, 2005, p. 362). The motivation for paying attention to mediation in connection with ADT is that mediation provides a way to approach the issue of coordinative means; by which means do humans coordinate their actions? In everyday life it is clear that we use two seemingly different types of means. If I am going to meet someone I can, for example, talk to this person and decide on time and place, or we could simply exchange e-mails. In order to meet at the agreed time and place, we may use clocks and maps. Thus, both language and more materially oriented means may be used for coordination. At first glance, these types appear to be of a very different nature. However, in order to approach the construct of activity modalities, I want to examine what properties they have in common rather than emphasizing their differences.

The most obvious examples of mediation are various sorts of tangible tools such as hammers, drills, axes, etc., aiming at changing the material world for some purpose. Such tools were called "technical" by Vygotsky. However, from the point of making a change in the world, there are other mediational means available for humans. Vygotsky extended the concept of meditation to "psychological tools", by which he referred to "various systems for counting; mnemonic techniques; algebraic symbol systems; works of art; writing; schemes, diagrams, maps, and technical drawings; all sorts of conventional signs, and so on" (Vygotsky, 1981, p. 137, cited in Wertsch, 1991, p. 93). Psychological, or "semiotic", tools are unique to humans and mediate what Vygotsky called "higher mental functions". These functions are determined by the ontogenetic development of an individual, that is, they are learnt in a particular social environment in time and space.

Languages in general and so called "speech acts" in particular, are obvious examples of semiotic tools. The expression "speech acts" was coined by Bühler (1933), and further developed by Austin (1962) and others to express the fact that language is used, not only to describe something, but also to achieve something; to act. Speech acts can be of different kinds such as promising, declaring, and requesting something. Examples of speech acts are "I promise to see you tomorrow", "The United States is now at war with Germany", and "Close that window, please!" Speech acts complement more obvious material means in order to achieve something. In contrast to technical tools, semiotic tools are inherently social; language cannot be conceived as instrumental for a single individual in isolation, whereas a technical tool can. For Robinson Crusoe, his ability to construct technical tools was necessary for survival. Language, however, was not relevant until Friday came along.

Characteristics of Semiotic Tools

In order to investigate similarities between technical and semiotic tools, I will start by examining some features of language and signs in general. A written or spoken human language is considered to be the sign system *par excellence*. One of the main controversies in language philosophy concerns language as an individual speech act (*parole*) versus language as a system (*langue*).

In daily life, individuals are engaged in situated, fluent, more or less random verbal (and nonverbal) communication. A particular utterance is initiated by a speaker and taken up by one or several hearers, which in turn become speakers and hearers again. The words and sentences are generated by and directed to individuals. The meanings of the words may be crystal clear or diffused and so on. Thus, language has an inherent individual/subjective character. On the other hand, language has a trans-individual character.

It can be abstracted into a system of language parts like subject, verb, pronoun, etc., and structured in formal grammars. As a system, language seems to be very stable during the lifetime of an individual, yet it changes over time. Language has a truly social/objective character.

In linguistic thought, two trends can be identified where the essence of language is placed at either one of these poles. The trends are labeled *individual subjectivism* and *abstract objectivism* respectively by Vološinov (1986). Individual subjectivism is based on the following principles (ibid, p. 48):

- Language is activity, an unceasing process of creation (energia) realized in individual speech acts.
- The laws of language creativity are the laws of individual psychology.
- Creativity of language is meaningful creativity, analogous to creative art.
- Language as a ready-made product (ergon), as a stable system (lexicon, grammar, phonetics), is, so to speak, the inert crust, the hardened lava of language creativity, of which linguistics makes an abstract construct in the interest of the practical teaching of language as a ready-made instrument.

The most prominent representative of this trend is von Humboldt (1999), who puts the emphasis on language, not as an object or product (Greek ergon) but as an activity (energeia) constantly renewed in interchanges among speakers.

De Saussure (1966) is a major contributor to the second trend: abstract objectivism, which is based on the following principles (Vološinov, 1986, p. 57):

- Language is a stable, immutable system of normatively identical linguistic forms which the individual consciousness finds ready-made and which is incontestable for that consciousness.
- The laws of language are the specifically linguistic laws of connection between linguistic signs within a given, closed linguistic system. These laws are objective with respect to any subjective consciousness.
- Specifically linguistic connections have nothing in common with ideological values (artistic, cognitive, or other). Language phenomena are not grounded in ideological motives. No connection of a kind natural and comprehensible to the consciousness or of an artistic kind is obtained between the word and its meaning.
- Individual acts of speaking are, from the viewpoint of language, merely fortuitous refractions and variations or plain and simple distortions of normatively identical forms; but precisely these acts of individual discourse explain the historical changeability of linguistic forms, a changeably that in itself, from the standpoint of the language system, is irrational and senseless. There is no connection, no sharing of motives, between the system of language and its history. They are alien to each other.

Vološinov rejects abstract objectivism as the actual mode of being of linguistic phenomena: "This system cannot serve as a basis for understanding and explaining linguistic facts as they really come into being" (Vološinov, 1986, p. 82). The problem of abstract objectivism is that it rejects the speech act as something individual. On the other hand, the problem of individual subjectivism is that the utterance becomes something confined to the individual psychic life of the speaker. Thus individual subjectivism is unable to explain the inert characteristics of language.

The truth according to Vološinov can be found only in a *dialectical synthesis* of these two trends. The speech act and its product – the utterance – are *social phenomena*. The mode of being of linguistic phenomena and signs in general is intersubjective and relational:

A word is a bridge thrown between myself and another. If one end of the bridge depends on me, then the other depends on my addressee. A word is territory shared by both addresser and addressee, by the speaker and his interlocutor. (Vološinov, 1986, p. 86)

Vološinov goes even further and suggests that the very constitution of our minds occurs in the interplay between the external physical and social reality, and the internal psychological reality:

By its very existential nature, the subjective psyche is to be localized somewhere between the organism and the outside world, on the borderline separating these two spheres of reality. [...] the organism and the outside world meet here in the sign. (Vološinov, 1986, p. 26)

Similarities Between Semiotic and Technical Tools

Given the view of semiotic tools as bridging the inner and external realms, the natural question to ask is whether technical tools can be apprehended in a similar way. The ideas of Vološinov are closely related to a long philosophical debate about the nature of "ideal" phenomena (Bakhurst, 1991). The ideal refers to the state of non-material properties in the material world, for example, thoughts, beliefs, sensations, values, meaning, etc. (ibid, p. 175). In order to understand these phenomena Ilyenkov starts with the *artifact*, that is, something created by human activity for a particular purpose. A hammer, for example, is obviously a material thing made out of wood and metal. However, it is also something that has an intended use purpose (or possibly several purposes); it is a meaningful thing in a certain context.

For Ilyenkov, this reasoning is valid also for semiotic tools. A word or a diagram are "brute physical entities" (ibid, p. 185) in the sense that a spoken word is materialized as sound waves, a written word as dots on paper, a diagram as connected pixels on a computer screen, and so on. Words, whether spoken or in written text are in fact artifacts in the very same sense as a hammer: they appear in and mediate human activity, and they are material phenomena.

Both technical and semiotic tools are dependent on the particular context and purpose of the tool. If I am cutting wood with an axe, the properties of the axe shape the way of using it. For example, it is pointless to try to cut the wood by using the back side or the flat side of the axe. The axe is designed to be used in a particular way. Thus, a mediational means brings with it a certain way of interacting with reality in order to produce the intended effect; in this case cut wood. However, this does not exclude other possible uses of the tool. The back side of an axe can be used as a hammer to drive in nails, and the sharp side of an axe has been used to kill people.

The same characteristics are valid for semiotic tools. Language is inherently contextual and purposeful. The same term, say 'lie', has two quite different meanings depending on the context: "a statement that deviates from or perverts the truth" or "to be located or situated somewhere". Vološinov expressed this as follows:

Every stage in the development of a society has its own special and restricted circle of items which alone have access to that society's attention and which are endowed with evaluative accentuation by that attention. Only items within that circle will achieve sign formation and become objects in semiotic communication. (Vološinov, 1986, pp. 21-22)

Obviously, there are many similarities between technical and semiotic mediational means. Both are material phenomena, they are used in some context for a purpose; to achieve something. They set "foot-prints" in both the external environments and in the mind, they are intended to be used in certain ways and in most cases, their effects must be commonly interpreted by those impacted by their usage.

However, there are basic differences as well. A drawing of an airplane is not an airplane (cf. the famous painting of Magritte: Ceci n'est pas une pipe, René Magritte, 1926). A drawing cannot fly (unless someone folds it to become a paper plane!). Ideal phenomena signify something else; they stand for something outside themselves, and are not the same as that they stand for:

Signs (...) are particular, material things; and (...) any item of nature, technology or consumption can become a sign, acquiring in the process a meaning that goes beyond its given particularity. A sign does not simply exist as part of a reality - it reflects and refracts another reality. (Vološinov, 1986, p. 10)

It is obvious that the concept of mediation is a fundamental aspect of human activity. According to Wertsch (1991), action and mediational means are so deeply intertwined that it is more appropriate to speak of "individual(s)-acting-with-mediational-means" than rather "individual(s)" alone when referring to the agent of action (Wertsch, 1991, p. 12). By acting with mediational means, humans transform what is given by nature into artifacts that fulfill social needs such as food, shelter, objects of pleasure and so on. In this process, not only the external environment is transformed, but also the internal minds of the individuals using the tools. When learning how to use a tool or a language construct, the individual gradually appropriates their properties and comes to an understanding about how to act it in order to achieve the desired result. Thus, the nature of a mediational means is, so to say "stamped" on both the external environment and the internal minds of those using the means.

The constitution of our inner worlds and the external reality as we apprehend it, are not separate and detached from each other. In the language of dialectics, they are *internally* related, that is, one cannot be changed without the other being changed; they mutually constitute each other. An inevitable consequence of this is that the characteristics of our cognitive system will be reflected in the way we apprehend the world. There is some kind of congruence between our inner world and the social reality we experience.

So, how should we conceive material and mediational means? Is there a common ground for them or shall we see them as separate? One answer is given by Vygotsky himself, who regarded technical tools and psychological tools as two focal areas of the concrete universal *meaning* (Bakhurst, 1991). As explained above, the term "concrete universal" refers to the basic "germ" of a concrete totality in which the elements are internal, or dialectically related. Thus, semiotic and technical tools must be analyzed as an indivisible unit. Although the obvious differences between technical and semiotic tools, there is a deeper affinity between them: they are mediational means mediating human action, and as such they are intrinsically bound to the activity in which they are meaningful. By acting with mediational means, the world emerges as meaningful to humans. So, let's take a closer look at the elusive concept of meaning.

MEANING

Meaning has been suggested as fundamental for understanding the human mind. For example, Bruner claims that "The central concept of a human psychology is meaning and the processes and transactions involved in the construction of meanings" (Bruner, 1990, p. 33). Furthermore, meaning is intrinsically related to culture and human action:

[C]ulture and the quest for meaning within culture are the proper causes of human action. The biological substrate, the so-called universals of human nature, is not a cause of action but, at most, a constraint upon it or a condition for it. (Bruner, 1990, p. 20)

Meaning seems at first glance to have a strange double nature. On the one hand, meaning seems to be a property of the things that surround us. When we hear a spoken word, read a text, see a chair, and so on, we know what these things mean. Meaning appears to be an inherent quality of the things that surround us. Thus, there is an "objectivist" account of meaning; it seems to be "out-there", existing on its own, regardless of each particular individual.

On the other hand, it is clear that meaning is quite different in nature than the properties of the physical, natural world. We cannot characterize meaning by properties that can somehow be measured by some instrument; we cannot "see" meaning, "smell" it, "touch" it, and so on. If meaning cannot be considered part of the physical world, we are forced to locate the origin of meaning in the mind of the individual, that is, to give a "subjectivist" account of meaning. In this view, meaning is an intrinsic property of our brains that somehow is "projected" onto the physical world. Although meaning is a distinct subjective phenomenon, our minds befool us to apprehend meaning to be anchored "out-there". The appeal of the subjectivist account is that it we can conceive a meaningless, pure objective and physical world without humans. When the human race has vanished from the earth, meaning ceases to exist.

Thus, in accounting for the nature of meaning we seem to be forced to take stand with either one of two incompatible positions – either the objectivist or the subjectivist one. Ilyenkov saw this dilemma as an impetus to find a different way of conceiving meaning that incorporates the insights of both positions. The key issue is, according to Ilyenkov, 'to lift the ban on anthropocentricity' that says that a property is "objective if it is *there anyway* independently of us" (Bakhurst, 1991, p. 179; italics in original). Objectivity is something that is intrinsically related to humans. It is meaningless to ask if something is "objective" without humans.

If we loosen this criterion of objectivity, i.e., that something is part of objective reality only if it can be understood without reference to humans, it is possible to reconcile the two positions. The key maneuver here is to ground the genesis of meaning neither in the heads of the individual, nor in the physical elements of our world, but in the *activity* of humans. According to Ilyenkov, it is human activity that idealizes nature (Bakhurst, 1991, p. 180). Without activity, meaning can neither be constructed, nor maintained.

Meaning is the generalization of reality that is crystallized and fixed in its sensuous vehicle, i.e. normally in a word or a word combination. This is the ideal, mental form of the crystallization of mankind's social experience and social practice. The range of a given society's ideas, science, and language exists as a system of corresponding meanings. Meaning thus belongs primarily to the world of objective, historical phenomena. And that must be our starting point. (Leont'ev, 1981, p. 226, referred by Tolman, 2004)

Ilyenkov summarizes the nature of meaning and other ideal phenomena in two key tenets (Bakhurst, 1991, p. 180):

- Ideal phenomena can have objective existence in the world.
- They owe this existence to human activity.

The Commonality of Meaning

Besides being grounded in activity, meaning has an individual – collective dimension. Meaning becomes real because of human activity, that is, meaning is subjective in character. Yet, meaning acquires a truly objective status, again due to the activity of humans. Thus, meaning is "independent of any particular individual and is thus transindividual, but [it] exist only through the activity and reason of individuals" (Kosík 1976, p. 146).

Through interaction with its environment, an individual gradually constructs a meaningful world ranging from the meaning of near-sensory impressions to, in due time, abstract symbols in a particular cultural-historical setting. In order to construct meanings, an individual needs to interact with other individuals. This is formulated by Vygotsky in his genetic law of development:

Every function in the child's cultural development appears twice: first, on the social level, and later, on the individual level; first, between people (interpsychological), and then inside the child (intrapsychological). (Vygotsky, 1978, p. 57)

Thus, the construction of meaning is inextricably related to social interaction between individuals in a particular culture. This interaction brings about institutionalized aspects of meaning that stabilizes the social system:

All social interaction is situated interaction - situated in time and space. It can be understood as the fitful yet routinised occurrence of encounters, fading away in time and space, yet constantly reconstituted within different areas of time-space. The regular or routine features of encounters, in time as well as space, represent institutionalized features of social systems. (Giddens, 1984, p. 86)

The institutionalization of meaning was expressed by Leont'ev (1978) as a differentiation between personal "sense" and objective "meaning". Objective meaning refers to the meaning of a word given in a lexicon. Sense refers to the translation of objective meaning into an internal, idiosyncratic impression for each individual. In the literature, it is customary to describe objective meaning as "shared" indicating that meaning can somehow be split and distributed among individuals. However, from the discussion above it is clear that there is no such thing as shared meaning in a strict sense. This has also been pointed out by Boland (1996) and Walsham (2005). In this book I will refer to objective meaning as "communal" meaning to emphasize that objective meaning has a contextual dimension. "By virtue of this actualization in culture, meaning achieves a form that is public and communal rather than private and autistic" (Bruner, 1990, p. 33).

An important consequence of the routinization of meaning construction is that different objective meanings evolve in different cultures separated in time and space. What people in one culture understand as their common language differs between cultures. Thus, *communal meaning is relative to the culture in which it is constructed,* an observation that will be of outmost importance for the structure of the activity modalities to be discussed in Section 3.

In summary, the genesis of meaning is human activity. Meaning has an individual, idiosyncratic as well as a social facet, and is at the same time both a subjective, individual phenomena manifested in the minds of human beings and an objective, social phenomena. The construction of meaning takes place in cultures or social systems that are situated in time and space:

Meaning, however, also exists as a fact of the individual consciousness. Man perceives the world and thinks about it as a social, historical entity; he is armed and at the same time limited by the ideas and knowledge of his time and his society [...]. Man does not know the world like a Robinson Crusoe making independent discoveries on an uninhabited island. He assimilates the experience of preceding generations of people in the course of his life; that happens precisely in the form of his mastering of meanings and to the extent that he assimilates them. Meaning is thus the form in which the individual man [appropriates] generalized and reflected human experience. (Leont'ev, 1981, p. 226, in Tolman, 2004)

THE THEORETICAL TRAIL: SUMMARY

In Section 2, the theoretical roots of ADT are described. The motivation for this part of the book is that I find it important to be specific about the grounds that ADT rests on. Only so can the reader evaluate the validity of arguments and implications put forward in the book. It is a question of being sincere about the transparency of assumptions that might otherwise remain hidden.

The fundamental conceptualization of ADT is found in the Marxian notion of praxis. Praxis frames the nexus of human activity; a framing that is subsequently refined and articulated by the notion of "activity" in the Russian Theory of Activity, and the construct of the activity domain in the ADT.

From the praxis philosophy, four cornerstones are pertinent to the ADT: the praxis notion itself; the dialectical relation; the epistemology of praxis; and the dialectical method. These cornerstones provide a coherent thinking that permeates the perspective taken throughout the book. From the Russian Theory of Activity further cornerstones are collected: the construct of activity; the concept of mediation; and the view of meaning. Mediation and meaning are prerequisites for the elaboration of the activity modalities in ADT. These modalities signify dimensions by which meaningful means, both technical and semiotic, mediate the coordination of human activity.

REFERENCES

Austin, J. L. (1962) *How to do things with words*. Oxford, UK: University Press.

Bakhurst, D. (1991). *Consciousness and Revolution in Soviet Philosophy: From the Bolsheviks to Evald Ilyenkov*. Cambridge, UK: Cambridge University Press.

Bødker, S., & Bøgh Andersen, P. (2005). Complex Mediation. *Human-Computer Interaction, 20*, 353–402. doi:10.1207/s15327051hci2004_1

Boland, R. J. (1996). Why shared meanings have no place in structuration theory: A reply to Scapens and Macintosh. *Accounting, Organizations and Society*, *21*(7/8), 691–697. doi:10.1016/0361-3682(96)00014-1

Bruner, J. (1990). *Acts of Meaning*. Cambridge, MA: Harvard University Press.

Bühler, K. (1933). *Die Axiomatik der Sprachwissenschaften*. Einleitung und Kommentar von Elisabeth Ströker. Frankfurt am Main (1969).

De Saussure, F. (1966). Cours de Linguistique Générale. de Mauro, T., (Ed.). Paris: Payot.

Giddens, A. (1984). *The Constitution of Society: Outline of the Theory of Structuration*. Los Angeles: University of California Press.

Kaptelinin, V., & Nardi, B. (2006). *Acting with Technology - Activity Theory and Interaction Design*. Cambridge, MA: The MIT Press.

Kosík, K. (1976). *Dialectics of the concrete*. Dordrecht: Reidel.

Leont'ev, A. N. (1978). *Activity, consciousness, and personality* (M.J. Hall, Trans.). Englewood Cliffs, NJ: Prentice-Hall. Original work published 1975.

Leont'ev, A. N. (1981). *Problems in the development of the mind*. Moscow: Progress Publ.

Tolman, C. W. (1988). The basic vocabulary of Activity Theory. *Activity Theory, 1*, 14–20. Retrieved Feb 10, 2008, from http://www.comnet.ca/~pballan/AT2.htm

Vološinov, V. N. (1986). *Marxism and the Language of Philosophy*. London: Harvard University Press.

Von Humboldt, W. (1999). *On language: on the diversity of human language construction and its influence on the mental development of the human species*. M. Losonsky (Ed.), P. Heath (Trans.). Cambridge: Cambridge University Press.

Vygotsky, L. S. (1978). *Mind in Society – The development of higher Psychological Processes*. M. Cole, V. John-Steiner, S. Scribner, & E. Souberman (Eds.). Cambridge MA: Harvard University Press.

Vygotsky, L. S. (1981). The genesis of higher mental functions. In J. W. Wertsch (Ed.), *The concept of activity in Soviet psychology*. Armonk, NY: M. E. Sharpe.

Walsham, G. (2005). Knowledge Management Systems: Representation and Communication in Context. *Systems, Signs & Actions, 1*(1), 6–18. Retrieved August 4, 2005, from http://www.sysiac.org/

Wertsch, J. V. (1991). *Voices of the mind: a sociocultural approach to mediated action*. Cambridge, MA: Harvard University Press.

Section 3
The Activity Domain Theory

The constitution of the activity domain. The activity domain perspective of enactment, coordination, transformative and coordinative capabilities, and dependencies between resources. The activity modalities. The cognitive grounding of the Activity Domain Theory. The operationalization of the Activity Domain Theory. Positioning the Activity Domain Theory against complex systems theory and other action-oriented theories. The practical and theoretical trails in hindsight. The practical and theoretical trails described in Section 1 and Section 2 of the book form the general background against which the Activity Domain Theory (ADT) emerged. I will for the time being put these trails aside, and describe the ADT on its own merits in Section 3. In the last chapter, the threads back to the practical and theoretical roots are reconstructed.

INTRODUCTION

The scope of the ADT is the coordination of human activity in social settings. In order to capture the general ideas in ADT an example from past times may be used: that of mammoth hunting (see Figure 1):

When looking at this scene some things immediately come to mind. First, the hunt can be seen as a context, a situation, with the mammoth in focus. The mammoth is the point of attention around which everything else revolves: hunters, bows and arrows, actions, possibly shouting and screaming, and so on.

From the ADT perspective, the whole context framing relevant things and actions is the *activity domain*. The mammoth is the *work object* of the activity domain towards which the hunters' actions are

Figure 1. An illustration of an activity domain (Bryant & Gay, 1883. Original wood engraving by E Bayard)

directed. The people who carry out actions are called *actors*. There is also a clearly identifiable *motive* in the scene: to kill the quarry, presumably in order to get food but maybe also for clothing, making arrowheads, and the like. By killing the quarry, social *needs* motivating the activity is fulfilled.

It can also be perceived in the background of the illustration that some actors, the beaters, have started a fire and make noises to scare the quarry away. The mammoth escapes in a direction where other hunters wait to circumvent the quarry and kill it. It is only in the light of the activity, the whole social framing of actions, that the beaters' actions stand out as meaningful. Just scaring the quarry away does not make sense if other hunters are not waiting to ambush it somewhere else.

It is obvious that the activity is a social one. Hunters have to coordinate their actions by shouting, pointing, making other gestures, and so forth. If every hunter attacked the mammoth one by one without coordination, the result would be disastrous. In order to coordinate actions, certain commonality must exist about how the hunters conceive the activity as a whole. Thus, there is a *contextual* dimension of the activity, which is determined by the work object in focus and the motive of the activity. The contextual dimension frames the activity in terms of what is relevant or not, and how relevant things are characterized. For example, if the motive of the activity would be just watching the movements of the mammoths for understanding how they wander in the surroundings, then other things, different from the hunting activity, would be relevant, for example, noticing markers in the geography for later use.

Besides contextualization, the hunters must have a common *spatial* understanding about how relevant phenomena are related to each other. In order to succeed with the hunt it is vital to know the position of the mammoth in the landscape, possible geographical formations that may influence the outcome (for example, the river in the background that prevents the mammoth from escaping), the weather, the visibility, etc. Without a spatial orientation coordinated actions cannot be achieved.

Next, actions have to be carried out in a certain order. First, the quarry is located, then it is scared away, then it is attacked and killed. Thus, there is a *temporal* dimension of the activity; a dimension that is intrinsically related to the spatial dimension. For example, the actions of the hunters must be coordinated with how the quarry moves in the terrain.

Another aspect of the activity is how to carry out actions. Shooting an arrow without aiming is not good practice. Moreover, there might be better or worse parts of the mammoth to aim for. The arrows must not be shot in directions that might wound other hunters, and so on. In every activity, there are rules, norms, traditions, narratives, etc., which routinize actions and distinguish meaningful actions from meaningless ones. These elements provide a *stabilizing* dimension of the activity, without which social cooperation would be impossible.

Finally, the hunting activity is not isolated. The brought-down quarry will be cut into pieces and prepared to eat. This is done in a cooking activity, which in turn has a motive and a work object (which happens to be the same as for the hunting activity: the mammoth). Other related activities might be manufacturing weapons and weapon parts from the bones and the tusks of the mammoth. When several activities cooperate, certain issues must be resolved in the transition between the activities, such as how to share the quarry between hunters and cooks, or decide how many ready-made arrow heads will be returned for a certain amount of food. Thus, there is a *transitional* dimension of the activity (or more precisely; between the activities).

Chapter 5
The Constitution of the Activity Domain

The purpose of the activity domain construct is to capture the essence of the coordination of human activity. As indicated by the mammoth hunting example, this essence is assumed to be invariant to the various forms activities take in time and place. The same basic constitution of activity applies to the collection of edible roots at the dawn of humankind as to the production of sophisticated products and services by global enterprises of today. The invariance of the fundamental features of the activity domain implies that its constitution can ultimately be traced back to the phylogenetic evolution of the human species. We have developed the faculty of cognizing constituting elements of activities, and this faculty is inherent in all healthy human beings.

In everyday life, we constantly encounter things that can be characterized as activity domains. When I go to the barber shop to get my hair done, when I buy a new car from the car dealer, when I contact the electric company to complain about my latest bill; in all these instances I interact with an activity domain. Usually we speak about these domains as organizations, companies, offices, and the like, which are the perceivable manifestations of the underlying activity domain constitution. This constitution is equally applicable to any forms of organized work, such as teams, groups, business units, companies, enterprises, or any other constellation of organized work settings. Thus, in ADT the activity domain is regarded as the "kernel" construct in organizational contexts; as the DNA of the firm, so to speak.

DOI: 10.4018/978-1-60566-192-6.ch005

The constitution of the activity domain is illustrated in the conceptual model in Figure 1:
The basic features of the activity domain are as follows:

- The *activity domain* is a work setting (team, group, business unit, etc.) whose existence is *motivated by* its capability to produce some *outcome*. The outcome may be a product or a service that is capable of fulfilling some need in other domains. For example, the motivation for the existence of a hospital is to cure ill people, or, in the case of private hospitals to make a profit for shareholders.
- The outcome is achieved through the actions of socially organized *actors*, who transform a *work object* into the outcome. In the hospital example, the work object is ill people, and the outcome is, hopefully, cured people. This is valid for both types of hospitals. However, although the work object is the same, the different motives of public and private hospitals will have a direct influence on the constitution of the domain. For example, in private hospitals, the first priority is to make sure that the patient can pay the bill.
- In order to transform the work object, actors *enact resources* consisting of means and skills. Enactment refers to the fact that the potential capabilities of means become resources only when the actors have drawn these into the social fabric of the activity domain and collectively learnt how

Figure 1. A conceptual model of the activity domain

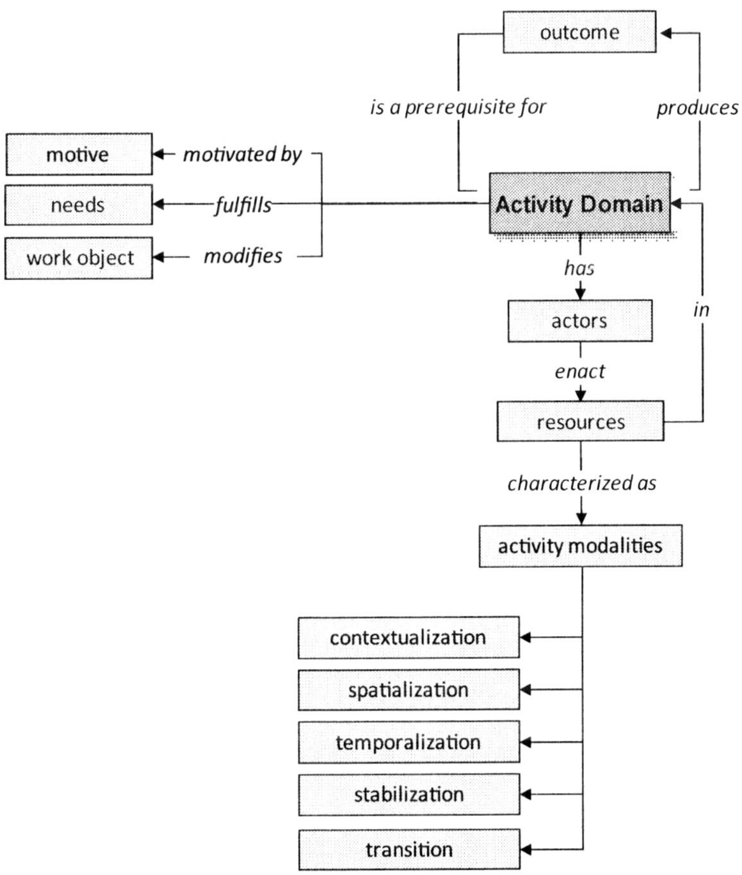

to use them. Thus, capabilities of means and actors must always be related to the work object and motive in order to become useful resources in the domain. Knowing how to cut down a tree with an axe is of little use when sailing a boat in an archipelago. Resources in hospitals are routines, surgical instruments, medical supplies, etc., as well as the skills of actors using these means.

- Resources are characterized by the *activity modalities*. These modalities denote fundamental, inherent dimensions of capabilities that humans draw on in making sense of the world. The activity modalities are as follows:
  - Contextualization: cognizing the framing of the activity
  - Spatialization: cognizing spatial structures
  - Temporalization: cognizing temporal orderings
  - Stabilization: cognizing valid and meaningful actions
  - Transition: cognizing cooperation across activities
- The outcome of one domain may be the *prerequisite* for another domain, which means that the activity domain construct is *recursive* and *scalable*. The same structure is found in every domain. In the hospital example, one domain may be the operating room, which provides a newly operated patient to another domain, the recovery room. Both these domains have the same basic structure as shown in Figure 1.

The activity domain and its constituting elements are considered to be *internally* interrelated, meaning that they are distinct but mutually interdependent. Thus, a change in one of them impacts all the others. For example, changing the spatial structure of the domain means that new actions or a different ordering of actions must be enacted. Thus, the temporal structure of the domain will also change.

The notions of activity, work object, motive and need are all appropriated unchanged from Activity Theory (AT) into the ADT (see Section 2). By introducing the term "domain" between "activity" and "theory", I want to indicate both a kinship with and a demarcation form the AT. This means that many aspects included in AT are not considered in ADT; above all the issue of contradictions, which is an inherent part of AT. On the other hand, the focus on coordination in ADT is less evident in AT; in particular the coordination between activity systems.[1]

ASPECTS OF THE ACTIVITY DOMAIN

One motivation for introducing the activity domain into the organizational discourse is the need for an intermediate construct between the individual and the organization. In social science, an ongoing debate concerns the relation between "macro" and "micro" perspectives (e.g. Wiley, 1988). Macro perspectives utilize systemic and trans-individual concepts, while micro perspectives focus on individually related constructs in analyzing the firm. These perspectives are often referred to as different levels of inquiry (ibid), and can be traced in pairs of opposites like "individual knowledge" – "organizational knowledge", "individual capability" – "organizational capability", "personal memory" – "organizational memory", and the like. In line with this, the firm is conceptualized as an institution for integrating knowledge possessed by *individual* specialists:

If the strategically most important resource of the firm is knowledge, and if knowledge resides in specialized form among individual organizational members, then the essence of organizational capability is the integration of individuals' specialized knowledge. (Grant, 1996b, p. 375)

The conception of knowledge integration exposed in these and similar statements foregrounds the individual and decontextualizes the intra-organizational view on knowledge. Knowledge construction occurs at the individual level, and there is a direct, transparent link between the individual/micro level and the organizational/macro level. Nothing shields the individual in her everyday doings from the events taking place at the organizational boundary.

Such a view of knowledge integration unfortunately pushes two important aspects into oblivion: 1) it leaves the question of what the knowledge *is about* unattended, and 2), from AT we know that individual actions can only be understood in relation to the activity in which the individual actions are carried out. The organizational level is in general too distant to be the immediate determinant of individual knowledge. These two aspects are central in the activity domain based view of coordination I wish to pursue here.

Enactment of Resources

As is well known, the debate on the nature of knowledge is an ever-lasting one. The literature on knowledge in organizations still "presents sharply contrasting and at times even contradictory views of knowledge" (Brown & Duguid, 2001, p. 198). Traditional assumptions about knowledge "offer a compartmentalized and static approach to the subject" (Blackler, 1995, p. 1021). It is common to see knowledge as "*embodied, embedded, encultured* and *encoded*" (ibid, italics in original).

As an alternative "knowledge (or, more appropriately, *knowing*) [may be] analyzed as an active process that is *mediated, situated, provisional, pragmatic* and *contested*" (ibid, italics in original). A similar position has been advocated by Orlikowski (2002) who proposes to use "organizational knowing" instead of "organizational knowledge" to emphasize that knowing is *enacted* in practice:

Knowledgeability or knowing-in-practice is continually enacted through people's everyday activity; it does not exist "out there" (incorporated in external objects, routines, or systems) or "in here" (inscribed in human brains, bodies, or communities). Rather, knowing is an ongoing social accomplishment, constituted and reconstituted in everyday practice. (Orlikowski, 2002, p. 252)

The enactment view of knowledge brings with it a connotation of "enactment towards what?" Enactment must be directed to something. Organizations are intentionally created to fulfill social needs. Consequently, knowledge in organizations is used for productive purposes:

Organizations are not basically knowledge systems, but systems that produce something of value to the society. [...] Only when the knowledge-creation process is set into the context of an organization's activities, does the understanding of the knowledge processes help us understand organizational learning. (Virkkunen & Kuutti, 2000, p. 297)

The primary role of the firm is in the "application of existing knowledge to the production of goods and services" (Grant, 1996, p. 112). This position is also emphasized by Burstein & Linger (2003), who maintain that knowledge must be seen in relation to the task at hand. Instead of focusing on a philosophical discussion of the nature of knowledge, where the target of inquiry is "knowledge" in general, the principal question to ask should be: "What kind of knowledge is needed in order to produce whatever the organization produces? What is the knowledge about?"

This view is foregrounded in AT by emphasizing the work object of the activity. Work objects can be material or intangible things as long as they can be shared for manipulation and transformation by the actors. The foregrounding of the work object implies that individual knowledge is utilized in doing something with something; knowledge and means are enacted to transform the work object into something that fulfils social needs.

In order to capture the enactment view of knowledge, the practice-related notion of *capability* is employed in ADT. A decisive argument for concentrating on capability is that *capability can be equally applied to both humans and non-human things*. According to The American Heritage® Dictionary of the English Language, capability is defined as follows:

- A talent or ability that has potential for development or use. Often used in the plural: *a student of great capabilities*.
- The capacity to be used, treated, or developed for a specific purpose: *nuclear capability*.

Through enactment, a recurrent interaction with available means, human capabilities and potential capabilities of means are mobilized in transforming the work object towards the outcome. A capability, then, becomes a resource in the very moment it can be associated with a purpose: that of fulfilling the social needs motivating the existence of the domain.

Coordination Revisited

Coordination is at the same time both self evident and elusive. In everyday life we constantly coordinate our actions. When, for example, someone misses an agreed appointment, the lack of coordination is obvious. In organizations, the need for coordination arises from the collective nature of the work:

Ultimately, all differences between companies in cost or price derive from the hundreds of activities required to create, produce, sell, and deliver their products or services, such as calling on customers, assembling final products, and training employees. (Porter, 1996, p. 62)[2]

The coordination of these activities is at the heart of the organization:

The assumptions that there are gains from specialization in knowledge acquisition and storage, and that production requires the input of a wide range of specialized knowledge, restates a premise which, either explicitly or implicitly, is fundamental to all theories of the firm. [...]. Given the efficiency gains of specialization, the fundamental task of the organization is to coordinate the efforts of many specialists. (Grant, 1996, p. 113)

As a consequence, "the primary task of management is establishing the coordination necessary for this knowledge integration" (Grant, 1996, p. 120). Thus, coordination is a core organizational issue regardless of the particular purpose of the firm. In the literature, several attempts to define coordination can be found. For example, Larsson (1990) lists nineteen definitions, and Malone & Crowston (1994) identify eleven definitions. Malone & Crowston also emphasize the multidisciplinary nature of coordination; the study of coordination must draw on disciplines such as organisation theory, management science, computer science, economics, linguistics and psychology (ibid, p. 88).

The difficulties of conceptualizing coordination bring about problems for operationalizing coordination, that is, what means or mechanisms may be employed to manage coordination. For example, Mintzberg suggests five such mechanisms: mutual adjustment by informal communication, direct supervision, standardization of work processes, standardization of the output of the organisation, and standardization of skills needed to perform a certain task (Mintzberg, 1983). Grant (1996) points to the following mechanisms for integrating specialized knowledge: rules and directives, sequencing, routines, and group problem solving and decision making. Furthermore, in line with other scholars (e.g. Bryan-Kinns, Healey, Papworth, & Vaduuva, 2007; Hemingway, 1998; Kraut & Streeter, 1995; Smircich, 1983) Grant states that all coordinating mechanisms depend on "the existence of common knowledge for their operation" (Grant, 1996, p. 115). Coordination from this point of view would focus on means to manage meaning and shaping interpretations.

However, in spite of all efforts to come to grips with the phenomenon of coordination, it appears that the basic concern remains:

Although widely addressed, organization theory lacks a rigorous integrated, well developed and widely agreed theory of coordination. (Grant, 1996, p. 113)

The ADT is an attempt to develop a theory for coordination from an alternative perspective, that of the activity domain. It is my ambition that this reconceptualization of coordination shall advance the understanding of coordination.

Transformative and Coordinative Capabilities

Given that a firm employs a multitude of human and non-human capabilities to achieve its goals, the question arises what kinds of capabilities are needed. Complex products such as telecom systems consist of many parts implemented in various technologies, such as application specific integrated circuits, software programs, optical fibers, radio technology, and so on. This indicates that a firm should be considered as a constellations of activity domains; each one with its particular work object.

In each domain, actors perform various actions directed to the common work object. It follows that two fundamentally different kinds of capabilities are needed: capabilities to transform the work object into an outcome, and capabilities to coordinate transformative actions. In order to analytically distinguish these two kinds of resources, the activity domain may be comprehended as having two, dialectically interrelated activity modes: the *transformative* one and the *coordinative* one. These two modes can be seen as two activities, however, with the transformative as the primary one since this mode is directly related to the work object and the motive of the domain. Consequently, coordination is seen as an activity in itself:

Thus, by entering into cooperative work relations, the participants must engage in activities that are, in a sense, extraneous to the activities that contribute directly to developing the product or service. That is, compared with individual work, cooperative work implies an overhead cost in terms of labor, resources, time, etc. The obvious justification for incurring this overhead cost, and thus the reason for

Figure 2. The two modes of the activity domain

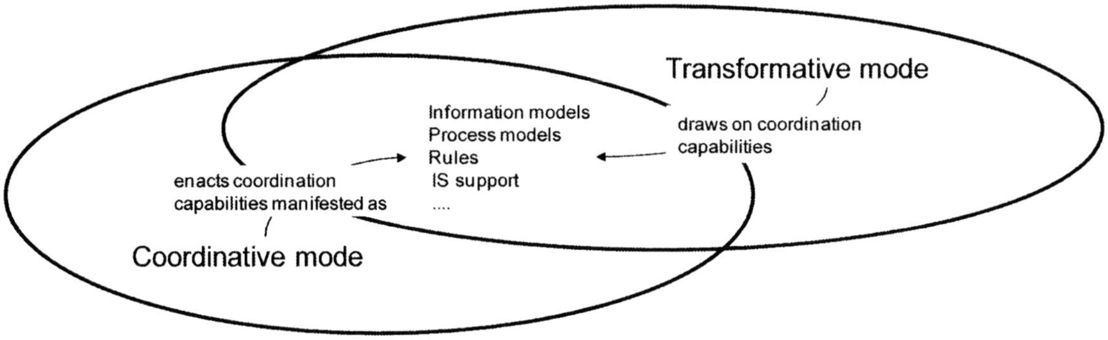

the emergence of cooperative work formations is, of course, that workers could not accomplish the task if they were to do it individually. (Schmidt, 1990; in Schmidt & Bannon, 1992, p. 8)

An example may clarify the relation between the transformative and coordinative modes. Let's consider a requirement on a car. The content of such a requirement may be: "The car shall consume less than 0.5 liters per 10 km at a cruising speed of 100 km per hour". The work object – the car – must fulfill this requirement. In addition, there might be a multitude of other requirements on form, safety, exhaust limits, and the like. In order to coordinate the actions in the transformative mode, there is a need to keep track of all the requirements, preferably in a requirement management tool. To achieve this, certain coordinative capabilities must be enacted, such as uniquely identifying each requirement; deciding on a set of states indicating what state a requirement is in (for example, whether the requirement is fulfilled or not); defining attributes characterizing the requirement; agreeing on relationships to other items such as what specific customer has issued the requirement, and so on. In the coordinative mode, only such coordinative aspects of requirements are relevant, not the actual content of the requirement. The content, on the other hand is highly relevant in the transformative mode but not in the coordinative mode.

The outcomes in the coordinative mode are information models, process models, rules, information system (IS) support, etc.; items that are relevant for coordinating actions. Several examples of such outcomes are given in Section 1 in terms of information models and IS implementations of these models. In the transformative mode, these coordinative capabilities are utilized in the transformation of the work object. Thus, the activity domain unfolds through an ongoing focal shift between the coordinative and transformative modes, in which transformative and coordinative capabilities are enacted. This is illustrated in Figure 2.

Coordination Across Activity Domains

Since activity domains differ with respect to their work objects and motives, meaningful transformative and coordinative actions in one domain may be meaningless in other domains. Each domain enacts its own particular world-view of what actions are relevant and useful. Even if two domains have the same kind of work objects and motives, like two automobile workshops servicing the same make of car, they would still enact domain specific capabilities. This is the main reason why it is necessary to introduce the activity domain as an intermediate construct between the individual and the organization. The de-

Figure 3. The organization as a constellation of activity domains

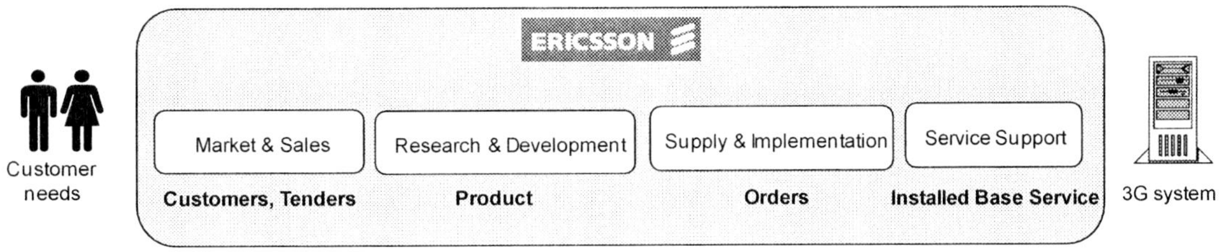

terminant of individual capabilities and meaningful actions is the activity domain, not the organization. In fact, the only case when an intermediate realm is not needed is when there is no division of labor, that is, when there is but one actor in the organization. This would be the case for a one-person activity such as a painter, a sculptor, etc.

Therefore, it is necessary to take into account coordination *between* activity domains in addition to coordination *within* domains. Consider the example with the car: one activity domain may be providing the motor of the car, while another provides the chassis. It is evident that the chassis has to be in place before the motor can be attached to the chassis. This dependency between activity domains needs to be managed as well as dependencies between actions within each domain.

These considerations indicate a key tenet in ADT: The coordination of actions within an activity domain *is in principle not different from the coordination between activity domains*. The coordination between domains takes place in an overarching domain with its own work object and motive. This overarching domain may be the firm itself. Thus, the transformative and coordinative modes are found in every activity domain, regardless of the particular position of the domain in a constellation of domains.

In Figure 3 this idea is exemplified for a telecom system provider such as Ericsson. The Ericsson Company is considered as an activity domain that provides 3G systems to some customer. In order to do so, a number of activity domains are mobilized: Market & Sales working with customers and tenders, Research & Development working with the product, Supply & Implementation working with orders, and Service Support working with servicing the installed base of products. Each of these domains enacts different transformative and coordinative capabilities, and each domain can in turn employ the capabilities of other domains.

The recursive nature of the activity domain means that domains may employ other domains as resources. This is illustrated in Figure 4. An activity domain (A) is employing four other activity domains (a1, a2, a3, and a4). Besides utilizing these domains for transformational purposes, there may be transformational capabilities that are specific for domain A. All transformational capabilities need to be coordinated by employing coordinative capabilities, again specific for domain A. This pattern is then repeated in each of the other domains, which in turn may employ other domains, and so on.

Dependencies Between Resources

The view of activity domains as capabilities that may become resources for other domains accentuates the importance of controlling the *dependencies* between resources. A missing or erroneous resource needed in a domain means that the domain cannot fulfill its goal. The dependencies between resources can be illustrated in a dependency map such as the one in Figure 5.

Figure 4. The recursive nature of the activity domain

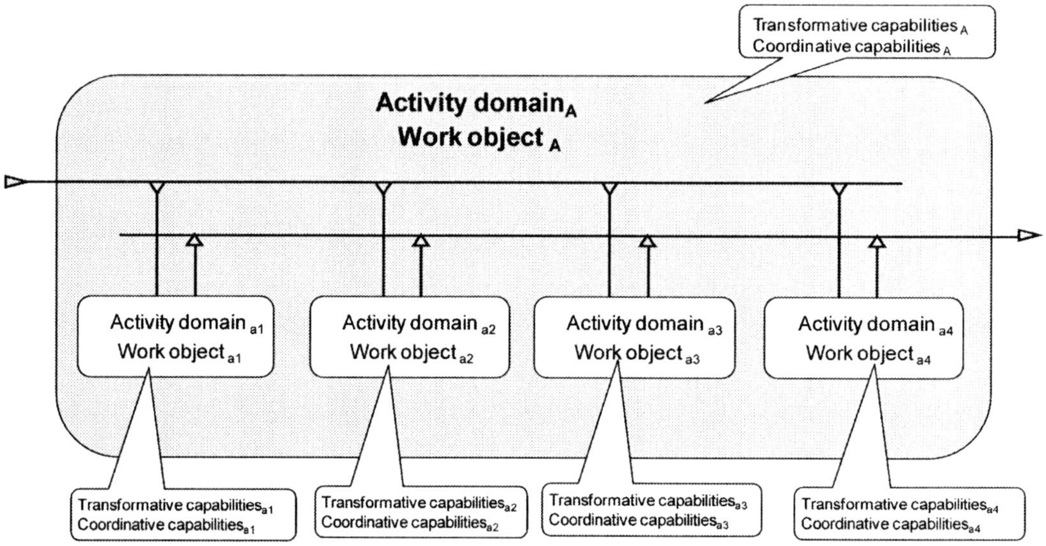

Figure 5. The dependencies between resources

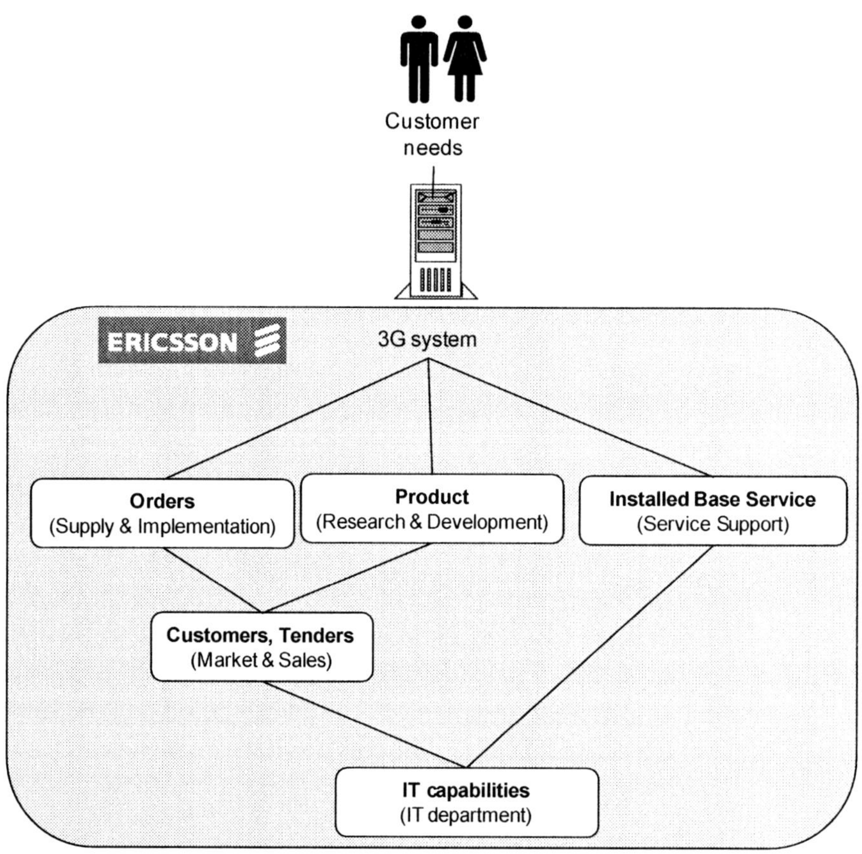

The map shows that the 3G system fulfilling the customer's needs is realized by the Research & Development domain developing the product, the Supply & Implementation domain providing order management resources, and the Service Support domain providing services on the installed base of products. The Research & Development and Supply & Implementation domains both need the resources of managing customers and tenders provided by the Market & Sales domain. The resources provided by the IT department domain are needed by all other domains in the organization, indicating that if this domain fails for some reason, the entire organization will come to a halt.[3]

THE CONGRUENCE OF MIND AND ACTIVITY

If we for a moment consider the world of artifacts that surrounds us, is it easy to see that artifacts are constructed according to the measures of mankind. There is an obvious physical congruence between our bodies and the size, shape and structure of our artifacts (see Figure 6). Chairs aimed for sitting on are not 2 meter high; pencils are not 1 meter long and weigh 5 kilos; doors are not half a meter high, and so on. This is so evident that we do not think about it in everyday life. Cars, trains, buildings, books or whatever artifacts constructed, are all providing affordances that we can utilize for our benefit. This is valid also for symbols like the alphabet, traffic signals and the like.

But the external world is not infinitely malleable. Only physical stimuli accessible to our sensory organs can become meaningful sensations. Stimuli not directly accessible to human senses, for example, ultraviolet light and high-pitch sounds, can be made meaningful only through some translation and

Figure 6. Mini-horses or gigantic tables?

processing. Consequently, the products of human activity reflect the physical constitution of man as it has evolved during the phylogenetic evolution of the human species, as well as the constitution of the physical world as it exists without human interference.

If the physical expressions of human activity are constructed to fit humans, it is near at hand to expect that the same is valid for our construction of symbolic or ideal phenomena. Our social realities are organized the way they are because our minds have a certain, biologically inherited constitution. As an example, consider the concept of time. Time is obviously fundamental to human activity. Without a notion of time and artifacts signifying time, it would be impossible to coordinate human actions. Humans suffering from a deprived sense of time, such as ADHD (Attention-Deficit/Hyperactivity Disorder) patients, have difficulties in organizing their lives. We can experience time precisely because we have this faculty due to the phylogenetically achieved constitution of our cognitive system.

The idea that the human mind does not have its own structure and logic of development, distinct from the structure of object-related activity is a recurrent theme in AT (Stetsenko, 1999). A prominent advocate of this line of thought was Rubinstein, who proposed a theory of *constitutive relationism* (Meacham, 1999). This theory claims that "mind is determined by its relationships to all material phenomena and is thus at the same time both an active reflection of the material world and a higher nervous activity" (ibid, p. 138). Mind "must be understood in relation to other aspects of the material world, including both the inner matter of the brain and the outer material world of society, culture, and history" (ibid, p. 137). Zinchenko, another scholar in AT, emphasizes the dialectical nature of mankind and its world: "What is considered mental, or subjective, is objective at the same time" (Zinchenko, 2001, p. 138). The mind, just like culture, does not have its own enclosed territory, but is "situated instead at the borders between own and not-own" (ibid, p. 139).

The consequence of this position is that the structure of social reality will develop in *congruence* with the structure of object-related activity. It is not only our physical artifacts that are constructed according to the measures of mankind; *this is valid also for our mental constructions* such as meaning and language. This is a premise I will pursue throughout the book. As it turns out, this premise is fundamental for the construct of activity modalities that will be discussed in the next chapter. The very idea of activity modalities is based on the assumption that there are certain dimensions along which human activity is coordinated, and that these dimensions – the activity modalities – do not emerge at random. On the contrary, they emerge because they are deeply rooted in the neural and cognitive structure of our brains.

The idea of congruence implies that mind and activity are internally related to each other. Neither exists without the other; they mutually constitute each other, and yet, each can be identified as something specific. For analytical purposes it is convenient to apprehend this unity as two intertwined processes of enactment – *objectification* and *objectivation* – that are constituted in two realms – in the human mind and in societal praxis. The first process of enactment, objectification ("Vergegenständlichung"; Kosík, 1976), is directed outwards, and concerns the transformation of the world into meaningful artifacts such as tools, institutions, organizations, etc. The second process, objectivation ("Objektivierung"; ibid, p. 131), is directed inwards, and concerns the construction of meaning. Together, these processes result in what AT calls "functional organs", which are defined as "functionally integrated, goal oriented configurations of external and internal resources" (Kaptelinin, 1996, p. 50).

In order to achieve communal meaning, i.e., meaning that enables coordinated actions, the individual needs to be integrated in a trans-individual whole as one of its elements. This incorporation transforms the individual: "The subject abstracts from his subjectivity and becomes an object and an element of the system" (Kosík, 1976, p. 50). Thus, the essence of objectivation is the construction of communal meaning necessary to perform coordinated actions.

The interconnection of objectified and objectivated praxis of mankind, labeled as substance, objective spirit, culture or civilization, [...] forms the historically attained 'reason' of society, which is independent of any particular individual and is thus trans-individual, but which really exists only through the activity and reason of individuals. The objective social substance, in the form of materialized production forces, language and forms of thinking, is independent of the will and consciousness of individuals, but exists only through their activity, thinking and language. (Kosík 1976, pp. 145-146)

Enactment apprehended in this way can be illustrated by a concrete example – the activity of playing a string quartet. First, there are obvious objectified elements involved like the instruments and the musical score. Each individual actor/player has to appropriate her instrument by a long and intense interaction with it. Technical and musical abilities must be learnt. However, in order to bring forth music the musicians cannot act one by one. They have to appropriate communal meaning of scores, notes, tuning procedures, performance manners, etc. In short, they have to be integrated in a trans-individual whole – the activity of playing – where they start playing at the same time, use the same phrasing and dynamics, etc. Without going through this objectivation process, the musicians cannot coordinate their actions.

THE ACTIVITY MODALITIES

So far, I have argued that there is some kind of congruence between mind and activity. The next step is to inquire into the nature of this congruence with the specific focus on the coordination of human activity. This inquiry needs to be grounded in insights related to both mind and activity. As the point of departure I will take the recurrent patterns that I experienced at Ericsson that somehow seemed to appear over and over again in various disguises (see Section 1): contextual dependency, spatial relationships between things, temporal ordering of actions; proper ways of acting; and transition between contexts.

Since these recurrent patterns are closely related to the coordination of activities it is near at hand to regard them as signs of basic modes of behavior humans draw on when coordinating their actions. I will call these modes or dimension *activity modalities*, where "modality" is apprehended as "a modal relation or quality; a mode or point of view under which an object presents itself to the mind" (Webster's 1913 Dictionary, 2008). Through enactment, the activity modalities, which I will call *contextualization, spatialization, temporalization, stabilization* and *transition*, are manifested as objectified elements in activity, and congruent, objectified elements in the minds of actors in the activity domain.

Contextualization

It is clear that human action is framed by the situation where the action takes place. Our visual system simplifies a visual scene into foreground and background. Certain things are attended in the foreground while other things remain unattended in the background (Jackendoff, 1983, p. 42). Thus, a capability to contextualize appears to be innate. We act according to what comes into the focus of our attention. Memories of earlier experiences are compared with the situation at hand, and possible alternative actions are evaluated. Sometimes the situation is completely new, but in most cases the current situation is similar to situations in the past to which we can relate. In addition, contextualization appears to be strongly related to the capability of humans to classify things into categories or types. In order for something to appear as something of the kind so and so, the ability to contextualize is necessary:

In general, representing object concepts involves a process of conceptual partitioning, in which the mind extends a boundary around a portion of what would otherwise be a continuum of space, and ascribes to the contents within the boundary the property of being a single-unit entity. In such a partitioning process, contents that are perceptually salient, such as those having a clear boundary or those identifiable by shape, would be identified and ascribed quickly and frequently. (Chen, 2003, p. 965)

Although the notion of context is intuitively understandable, an unambiguous definition of context is less straightforward. In fact, there seems to be no common understanding of what context is and how it is used (Kofod-Petersen & Cassens, 2006). For our purposes we may regard "context" as the set of circumstances and conditions which surround and determine an idea, theory, proposition, or concept (Gershenson, 2002). An important observation regarding context is put forward by Jensen:

[To] create knowledge is to use information for a productive purpose in a certain context. All productive knowledge is contextual [...]. This means that all (productive) knowledge is linked to and dependent upon an organizational context. In order to understand learning and accumulation of knowledge in an organization, it is necessary to define the content of the organizational context. (Jensen, 2005, p. 54)

Thus, in a discourse about knowledge and capabilities in organizations, it is vital to identify the different contexts inherent in the organization. For example, if we consider the development of, say a mobile telecommunication system, we know that this task needs to be divided into contexts such as development of software, hardware, mechanics, and radio. Each of these contexts requires specific competences determined by the constitution of the work object.

An aspect of system development where contextualization is particularly conspicuous concerns the notion of "functionality". Systems are developed and tested according to functional specifications. Usually, functionality is taken to be an inherent property of the system. However, a closer examination reveals that the function of a system is inextricably associated with its context. For example, the intended function of a chair is undoubtedly its ability to provide something to sit on[4]. In the context of sitting, some properties of the chair will be salient such as softness and sitting area. However, in other contexts such as using the chair to step on to reach something, the intended function of the chair is irrelevant. Here other characteristics become important, for example, the height and stability of the chair.

Enactment in the contextualization modality entails to the construction of activity domains in which actions are meaningful. This construction implies at least two things: framing and context dependency of meaning. Framing is mainly determined by the motive of the domain and the work object. Only those things that are relevant for the purpose of the activity acquire meaning. A direct consequence of this is that the same object is enacted differently in different contexts (Parsons, 1996). Consider, for example, the life cycle of a product. From its inception to its disposal, the product passes through a number of different work contexts such as marketing, design, manufacturing, distribution, maintenance, and finally, scrapping. Although the product is recognized as a particular individual throughout its life cycle, it will be characterized differently in each of the contexts. When marketed, properties like appearance, price, availability, etc., are relevant. When manufactured, the manufacturability of the product is in focus. When disposed, recycling and environmental concerns are emphasized, and so on.

From a coordination point of view, contextualization plays a fundamental role since several activity domains need to be aligned in all organizations in order to produce the outcome, except maybe for the trivial case of only one work context. I fully endorse the view of Augier, Shariq, & Vendelø concerning the importance of contextualization:

We claim that knowing how context emerges and transforms is of paramount significance if we want to understand how people create, use and share knowledge. (Augier, Shariq, & Vendelø, 2001, p. 125)

Spatialization and Temporalization

As with contextualization, consensual definitions of spatialization and temporalization do not exist[5]. Spatial schemata seem to be at the core of cognitive structures and form a basis for many less concrete domains:

Spatialization serves the user's need to organize abstract concepts coherently [...] and to ground them in familiar experiences [...]. Consequently, because spatial experience is so fundamental to humans, spatial metaphors act as fundamental sense makers for abstract domains. (Fabrikant, 2000, p. 68)

The point here is that there seems to be no difference between physical and abstract spatialization. A capability to spatialize is needed to make sense of both the map and its "natural" correspondence. For the purpose of this book I will use spatialization as the enactment of spatial constructs, whether abstract or concrete, where spatial should be understood in the sense of "relating to, occupying, or having the character of space; of or relating to facility in perceiving relations (as of objects) in space" (Merriam Webster, 2008).

I will use a similar definition of temporalization as the enactment of temporal constructs, whether abstract or concrete, where "temporal" should be understood in the sense "of or relating to time as distinguished from space; of or relating to the sequence of time or to a particular time" (Merriam Webster, 2008). Examples of spatial structures in the organizational context are information models, object-oriented models, data models, product structures, conceptual models, etc. Likewise, examples of temporal structures in organization are business process models, interaction diagrams, event diagrams, use cases, etc.

In spite of Einstein's discovery of the unity of space and time as space-time, we have a strong sense of distinct spatial and temporal dimensions in our daily life. Time is experienced as something qualitatively different from space:

The world of everyday life is structured both spatially and temporally. [...]. Temporality is an intrinsic property of consciousness. The stream of consciousness is always ordered temporally. (Berger & Luckmann, 1991, p. 40)

Yet, even if we experience time and space as separate entities, we still find them closely intertwined. For example, in the sport of orienteering, the goal is to run as quickly from one control to another with the aid of a map and a compass. First, you have to orient yourself on the map. Where am I and where am

I supposed to go? This act presumes a spatial capability; to literally orient oneself in the environment. The map can be seen as a spatially oriented means to achieve the orientation task.

Next, I need to figure out a sequence of routes taking me from the start to the goal. This presumes a temporal capability; an ordering of events. First I will run towards the big stone along this path, then I cross the field aiming for the hill, then I run along the west side of the small marsh, and finally I head straight north to the goal. When I start on the first route I count my steps – a temporal dimension – and follow the direction of my compass – a spatial dimension. Now and then I check the map for a correspondence with the nature surrounding me to make sure I am on the right track.

In everyday life we can find an abundance of situations similar to the orienteering one. We constantly orient ourselves spatially and perform actions in an ordered manner. In doing so, we make use of various means devised to reflect one dominant modality, either spatiality or temporality such as maps, signposts, arrows, route descriptions, etc. (spatialization), and clocks, calendars, notification devices, etc. (temporalization).

A striking example of the interdependence between spatialization and temporalization can be found in a musical score. The staff provides a common reference for the notes where the placement of notes vertically indicate the pitch; a spatial dimension, and the successive ordering of notes with their time duration (whole-notes, half-notes, quarter-notes, etc.) indicate a temporal dimension. As we shall see later on, most of the mediational means used to signify the temporal or spatial dimensions manifest both modalities, albeit usually with one modality as the dominant.

The enactment of spatial and temporal structures enables the coordination of activity. For temporal structures, Orlikowski & Yates expressed this in the following way:

People in organizations experience time through the shared temporal structures they enact recurrently in their everyday practices [...]. Whether implicitly or explicitly, people make sense of, regulate, coordinate, and account for their activities through the temporal structures they recurrently enact. (Orlikowski & Yates, 2002, p. 686)

This is equally valid for spatial structures, and spatial and temporal structures complement each other:

Time and space provide contrasting perspectives on events. A temporal perspective highlights the sequence of transitions, the dynamic changes from segment to segment, things in motion. A spatial perspective highlights the sequence of states, the static spatial configuration, things caught still. Capturing the temporal and the spatial at once seems elusive; like waves and particles, the dynamic and the static appear to complement each other. (Zacks & Tversky, 2001, p. 19)

Stabilization

In social situations a balance between order and disorder must be maintained:

Every social situation lives in order and disorder; every social situation holds a moment of order and a moment of disorder. But at total disorder, there is no social life and at total order there is just a social petrification. Man lives his life between the impossible possibilities chaos and petrification, but in an endless number of ways. The question is then what composition, combination or mixtum compositum is the happy one? (Nilson, 1976, p. 10, my translation)

The need to maintain a balance implies that there is some kind of stabilizing elements in every persistent situation or context. Such elements may be principles rooted in nature, which result in death or serious diseases if violated, for example, consuming lethal food or falling off a cliff. For humans, there may be social conventions codified in religious or juridical norms. Some stabilizing elements are learnt in social interaction and internalized without reflection. Violating the rules in a "language game" in the Wittgensteinian sense means that an actor either does not understand the rule or deliberately chooses not to participate in the game. Traditions and culture are necessary to keep a society together. Together, the stabilizing elements constitute an *ideology*, that is, a wide-ranging system of belief or thought that prescribes what phenomena are taken for real and which actions are considered valid. This ideology is, for example, expressed as activity domain specific, local languages (Bechky, 2003; Goodwin & Goodwin, 1996).

Rules may be more or less comprehensive in a particular situation or context. A completely rule-free situation approaches a state of chaos or disorder, while a completely rule-controlled situation is approaching a state of petrification. At either end points coordination of actions is impossible. Somewhere in between there is optimal rule coverage:

An illustrative example of stabilization comes from Lakatos's (1974) anticipation of science as research programs (see Chalmers, 1976 for a comprehensive description). In order for scientists to fully explore a certain scientific line of research, they cannot constantly be engaged in debates regarding the basic assumptions and methods of a research program. Lakatos calls these basic assumptions the "hard core" of the program. These are guarded from falsification by a "protective belt" of auxiliary hypotheses. As long as the hard core can be protected, the research program can be explored to its full potential. The moment that the hard core cannot be sustained, the research program has to be abandoned. The hard core can be apprehended as a "stabilizing core" consisting of fundamental elements and assumptions in the particular scientific community.

In the organizational realm we can see manifestations of stabilization in many areas. For example, in organizations there is a contradiction between continuity and change. If the organization is too laden with respect for old traditions it may be slow to implement necessary changes. On the other hand, if the organization changes too quickly, the gradual building of human and material resources may be impeded. Sawhney & Prandelli have proposed a governance mechanism for managing distributed innovation called a "community of creation" with the purpose of blending the benefits of hierarchies and markets by offering a compromise between too much structure and complete chaos (Sawhney & Prandelli, 2001).

In a large and distributed organization like Ericsson, there will always be a need to adapt enterprise strategies to specific circumstances at the local design centers around the world. At the same time, there must be enterprise wide common rules for how to identify products and documents, rules of conduct for how to approach customers, compulsory legislative norms, and so on. The balance between strict control and decentralization is often difficult to maintain, and the organization tends to oscillate between these extreme points from time to time. In either case, the ability to adapt to changes will be low, as illustrated in Figure 7. Thus, there is a need for overall stabilizing elements such as standards in the organization. It is important to note, however, that such elements do not impede coordinated action. On the contrary, they are prerequisites for coordination provided they yield a proper balance between order and disorder.

Figure 7. Balancing between chaos and petrification

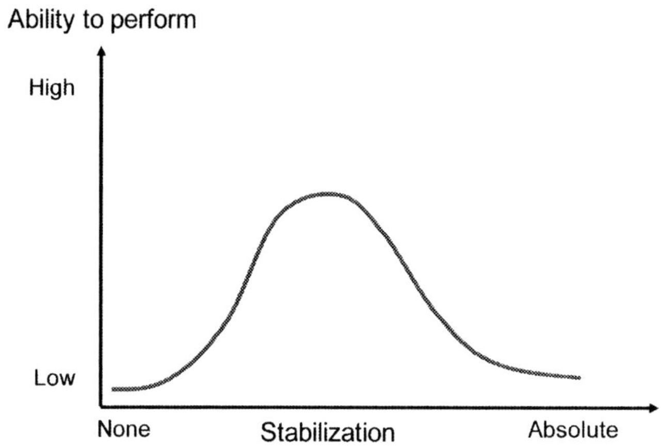

Enactment along the stabilization modality results in the emergence of stabilizing elements, a stabilizing core or an ideology, in a domain. These elements have the function of "... reducing the infinite number of things in the world, potential or actual — to a moderate number of well-defined varieties" (March & Simon, 1958, p. 181). The ideology is manifested as communal meaning of conventions, norms, values, habits, routines, rules, methods, standards, domain specific languages, etc., which enable a habitual performance of actions. The elements in the ideology should be regarded as "programs of action - not only a form of 'rule-guided behavior,' but more precisely as context specific, experience-based, action rules" (Grandori & Kogut, 2002, p. 225).

The ideology is in principle idiosyncratic to the domain, which means that, taken to its limits, there are no two identical ideologies. In the organizational realm, this means that there are no two identical organizational units. Every organization enacts its own ideology, even if it shares its motive and work object with other organizations. This is a consequence of the enactment process where actors need to acquire communal meaning of how to coordinate their actions.

The importance of the division of labor is that it also organizes our occupational identities, how we see the word, how we learn. (Grandori & Kogut, 2002, p. 227)

Even if exactly the same objectified elements (rules, process models, information structures, etc.) where used in two organizations, the objectivated elements in the minds of the actors, would still differ[6].

Every stage in the development of a society has its own special and restricted circle of items which alone have access to that society's attention and which are endowed with evaluative accentuation by that attention. Only items within that circle will achieve sign formation and become objects in semiotic communication. (Vološinov, 1986, pp. 21-22)

In summary, rules and stabilizing structures are necessary in any social situation. Stabilization means that each domain develops its own stabilizing core. However, in general, domains have to interact or cooperate. This presumes some kind of transitional element between ideologies.

Transition

Enactments in the contextualization and stabilization modalities bring about a context, the activity domain, and a domain specific ideology. However, the idiosyncratic nature of the ideology does not mean that different domains cannot coexist or cooperate. Some elements of the ideology may be common to several domains, but in general, these vary according to their motives and character of their work objects. A necessary condition for interaction is that the ideologies are in some sense commensurable; there must be some way of actors to arrive at communal meaning about how to coordinate the outcomes of each domain. This is where the transition modality plays a dominant role.

"Transition" in the sense I use it here is derived from the lexical definition "passage from one state, stage, subject, or place to another" (Merriam Webster, 2008). The enactment along the transitional modality manifests a capability of actors in one domain to cooperate with actors in other domains. Transition plays a fundamental role in ADT, since it enables the definition of the activity domain as a *recursive* construct. Through transitional elements, the outcome of a domain can be interpreted in the internal ideology of another domain and utilized for its own purposes. This is a definite advantage when analyzing different constellations of cooperating activity domains. There is no need to phrase such an analysis in terms like "levels", "horizontal and vertical coordination", "internal and external coordination" and the like, which is commonly the case (e.g. Christensen & Lægreid, 2007). Instead, the same recurrent analysis grid, as expressed by the activity domain constituents, can be used for any domain, regardless of its place in a constellation of domains.

As with the other modalities we are surrounded by manifestations of transition in our daily life. Dictionaries, outlet adaptors, currency exchange tables, passports, passport controls, gatekeepers, door keys, airlocks, etc., are all examples of transitional elements. Likewise, in the organizational realm, we can find an abundance of transitional elements such as contracts, agreements, interface specifications, mapping between different article codes, rules for conversion between analog and digital signals, compilers for software languages, etc.

In general, when our attention changes from one focus to another, we are familiar with the foregrounds and backgrounds. However, when we encounter something completely unknown, alien concepts properties and relations present themselves to our minds. In such a situation, we need to construct or enact some transitional elements in order to make sense of the new situation. Imagine, for example, that you have hired a car somewhere abroad; a car that you have not driven before. After a stop you need to engage the reverse gear, but no matter how you try, you cannot find out how to do it. You have to look in the instruction book to find it out. The instruction book is a transitional element that enables you to act in a new context[7].

The inevitable differentiation of the work immediately raises the issue of how these units – activity domains – shall work together, or be integrated (Lawrence & Lorsch, 1967). Differentiation and integration are opposites that somehow need to be reconciled. An illustrative example of how this can achieved is given by Curtis, Krasner, & Iscoe (1988) in their field study of the software design process for large systems. In many projects one or two persons, usually a senior engineer, took on a prime responsibility for the project. These persons possessed expertise knowledge, both in the application domain of the project and the implementation domain. For example, in the application domain of military avionics software, knowledge about flight control, navigation, electronic countermeasures, and target acquisition is needed. In the implementation domain, knowledge about the programming language, its control structures, the processor architecture, algorithms and data structures is necessary.

The experts, the "exceptional designers", had the capacity to bridge the application and the implementation domains in the sense that they could see the impact in the implementation domain of a particular requirement from the application domain. Likewise, they could convey possibilities and constraints from the implementation domain to the application domain, for example, how the capacity of the processor impacted the application. This means that exceptional designers can be seen as transitional elements, which in turn implies that such elements are not restricted to artifacts. In fact, since transitional elements have to be interpreted in the domains affected and mapped or translated between them, cognitive aspects are always associated with transition (as, for that matter, with the other modalities).

Border crossing between social contexts has been extensively researched in the subject area of "boundary objects" introduced by Star (1989). Boundary objects refer to objects that serve as an interface between different "communities of practice" (Wenger, 1998). Such objects are shared by several communities but viewed or used differently in each of them. Boundary objects work because they contain sufficient detail to be understandable by both parties; however, neither party is required to understand the full context of use by the other. In this way, boundary objects serve as point of mediation and negotiation.

However, in spite of the obvious focus on situatedness in organizational discourse, boundary objects have received less attention in industrial settings (Bannon & Bødker, 1997; Lutters & Ackermann, 2007). In particular, the issue of translation and interpretation of different communal meanings in different communities seems to be underresearched. From a transitional point of view it seems that boundary objects do not have the property of relating the internal and external facets of an activity domain. Boundary objects are one and the same for all contexts, just differently interpreted.

The Activity Modalities: Unity of Opposites

We experience and act in the world from sensory impressions provided through our senses. Although the modes in which these impressions are presented to us, they must be integrated into a coherent whole:

Our perception of the world is not only based on vision, but on complex sensory syntheses drawing on all sensory modalities (vision, audition, taste, smell, the haptic sense, and proprioception [the ability to sense the position and location and orientation and movement of the body and its parts – my remark]. (Bærentsen, 2000, p. 43)

The essence of the activity modality construct is that *the integrated, coherent perception we create out of sensory impressions can be characterized along the activity modalities*. Our cognitive system is set up in such a way that impressions mediated through sensory modalities are "cross-coupled" to activity modalities. Through enactment, we create different social realities in activity domains. The enactment brings about two, dialectically related forms of manifestations: objectivated elements in the brains and bodies of actors, and objectified elements in the domain. These manifestations are characterized by the activity modalities in the ADT. This is illustrated in Figure 8.

The activity modalities are meant to signify distinct modes of coordinating actions in the same way as sensory modalities indicate distinct modes of sensory impressions; modes that must be integrated in order to provide a complete conception. Coordination restricted to a particular subset of modalities will necessarily result in a deprived coordination capability, in much the same way as a loss of a sensory modality will deprive an individual of the full capability of experiencing the world. Using a business process model only to coordinate actions is a limitation in the same sense as using only one sense for acting in

Figure 8. Integration of sensory modalities and activity modalities

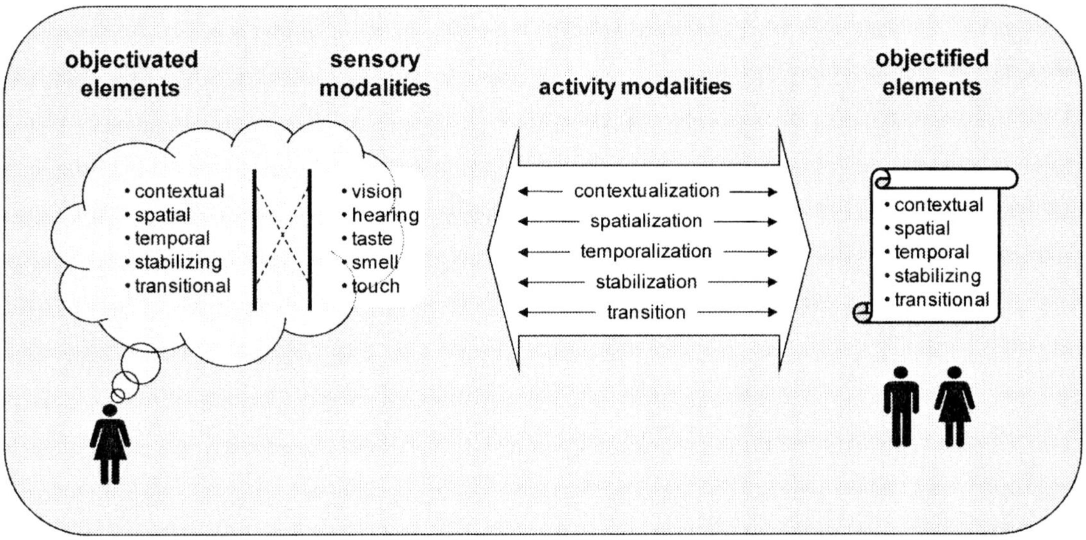

the world. The activity modalities are inextricably intertwined. Contextualization implies transition; the need to cross domains. Temporalization, the ordering of actions, presumes a spatial orientation in the domain. Stabilization impacts all the other modalities in the sense that spatial, temporal and transitional manifestations must acquire communal meaning in the domain.

An example of the interdependencies between modalities is given by Yakura who has investigated visual artifacts that represent time (Yakura, 2002). In particular, she has researched timeline artifacts like Gantt charts that compare what was done with what was planned over time in a project (see Figure 9).

Besides being an obvious manifestation of temporalization, the Gantt chart is also a manifestation of transition since it plays the role of a boundary object between various groups/units carrying out the activities shown in the chart.

The degree to which different sensory modalities are engaged varies with the context of action. It is likely that more modalities are engaged in immediate action in the nearby physical environment, for example, when the action of taking cover from a thunderstorm is triggered by smelling the ozone, see-

Figure 9. An example of a Gantt chart (Taxén & Lilliesköld, 2007. ©Elsevier. Used with permission)

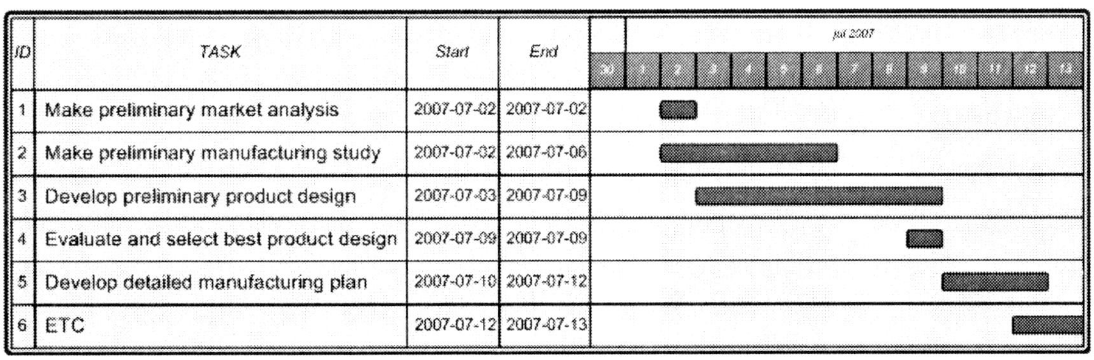

ing the lightning, hearing the thunder, and feeling the rain on the skin. When coordinating actions in the organizational context, it appears that mainly vision and hearing are engaged. This is, quite naturally, a consequence of the properties of mediational means used. For example, images signifying process models and information structures are visual in character. However, even so the activity modalities must be seen as an integrated whole:

Eventually, I would like to explore how the narrative and the information processing modes of cognition form a genuine union, but in this paper, I would like to focus on how these different modes are being informed at a deep level by our basic concepts of space and time. I will propose that the information processing mode takes an inherently spatial approach to cognizing situations, and that the narrative mode takes an inherently temporal approach. As in the genuine union model, both are needed, but one dimension, the spatial dimension, has dominated and suppressed the temporal dimension. (Boland, 2001, p. 14)

MEDIATIONAL MEANS AND ACTIVITY MODALITIES

According to ADT, the activity domain and the activity modalities are fundamental for the coordination of human activity. As discussed in Section 2, activity is *mediated*, i.e., "humans always put something else between themselves and their Object of work" (Bødker & Bøgh Andersen, 2005, p. 362). The purpose of this section is to convey how the activity modalities are related to various kinds of mediational means. As the point of departure I take the concept of the "tool".

Tools are an intrinsic part of humankind. When the hominids started to walk upright, the hands were set free for other purposes than support for moving around. This was a decisive step in the evolution from ape to man. It opened up for an intricate interplay between the human organism and her environment which is still going on. Through hundreds of thousands of years of labor, the human hand evolved into a tool which is infinitely more sophisticated than that of any simian:

When the little child performs the perfect grip of the tweezers she starts her path as a tool maker; she becomes Homo faber, man as craftsman, the tool making animal. (Liedman, 2001, p. 180, my translation)[8]

The first tools were created and controlled by the hand, but the opposite is also valid: "... the hand is not only the organ of labor, it is also the product of labor" (Engels, 1954). Moreover, the ability to produce tools played an important role for the increase of the brain and the low positioned larynx, which in turn were prerequisites for the development of language. The development of new tools goes hand in hand with the construction of a new language in order to use these tools, which in turn can give rise to new tools, etc. There is an intricate relationship between tools and language:

A tool is also a mode of language. For it says something, to those who understand it, about the operations of use and their consequences. . . . In the present cultural setting, these objects are so intimately bound up with intentions, occupations and purposes that they have an eloquent voice. (Dewey, 1991, p. 52, referred in Miettinen & Virkkunen, 2005, p. 443)

Conversely, language "can be likened to a tool by which reality becomes tangible. 'Tools' thus become equivalent to every mean that is used for some purpose (Liedman, 2001, p. 176, my translation).

Besides being used for transformative purposes, tools are also used for coordinative purposes. It is obvious that language is imperative for coordination. Besides being engaged in gossip and small talk, we constantly use what Austin calls "speech acts" to coordinate our actions: "by saying something, we do something" (Austin, 1962, p. 94). By speech acts Austin means acts such as describing things, requesting someone to do something, asking someone for advice, declaring state of affairs ("I pronounce you man and wife!"), making promises to carry out tasks, and so on.

A complementary view of the coordinative aspects of language is given by Halliday, who distinguishes between *pragmatic* and *mathetic* functions of language. He suggests six basic functions of language: instrumental ("I want"), regulatory ("do as I tell you"), interactional ("me and you"), personal ("here I come"), heuristic ("tell me why"), and imaginative ("let's pretend") (Halliday, 1974, p. 100). The pragmatic function comprises the instrumental, regulatory, and to some extent, the interactional functions. These are involved in coordination of actions when communal meaning is established:

The instrumental is language as a demand for goods and services, in the satisfaction of material needs; the regulatory is language used to control the behaviour of those around, and adapt it to one's wishes. (Halliday, 1974, p. 100-101)

The mathetic functions have to do with the construction of communal meanings:

[The mathetic function is] language enabling the child to learn about his social and material environment, serving him in the construction of reality. This function is realized, in the first instance, through the child's observing, recalling, and predicting the objects and events which impinge on his perceptions. (Halliday, 1974, p. 112)

The mathetic function of language can be seen as the enactment of communal meaning in an activity domain; a meaning that enables the coordination of actions in the domain. As the enactment proceeds, the pragmatic function of language becomes more attenuated. The importance of these functions varies depending on what phase of evolution the domain is in.

Technical and Semiotic Tools from the Activity Modality Perspective

As been discussed in Section 2, the relationship between tangible, "material" means like hammers, and more intangible "immaterial" means like speech acts has been the subject of much contemplation. Such means that directly change the world – hammers, drills, machines, chemicals, etc., – were called "technical tools" by Vygotsky, while those means that change the world indirectly – through communication, signs, symbols, etc., – were called "semiotic tools". Both kinds of tools follow similar trajectories: through enactment they become capabilities in purposeful human activity.

In ADT, the categorization of means as technical and semiotic is downplayed. Both types of means are characterized along the dimensions given activity modalities – contextualization, spatialization, temporalization, stabilization, and transition. Thus, mediational means are seen from an alternative perspective as follows.

Contextualization

The capability of a mediational means is inherently contextual. What can be achieved depends on the dispositional properties of the mean in a certain context:

To see an action as available for choice is to notice that it is afforded by the current situation. An affordance of a situation, crudely stated, is a way that a situation lends itself to being used. [...] We can change the affordances of an object merely by altering its context. [...]. An affordance, as we shall use the term then, is a dispositional property of a situation defined by a set of objects organized in a set arrangement, relativized to the action repertoire of a given agent. (Kirsh, 1995, p. 43)[9]

Consider the contextual dimension of the action of nailing. A hammer can be used to drive in nails, kill somebody or as a plummet by attaching a string to it. On the other hand, a hammer is utterly useless to cut down a tree. In spite of being the same object, the capability of the hammer to turn into a resource is dependent on the situation at hand.

Concerning semiotic tools, language is of course intrinsically contextual. The same term, say "lie", has two quite different meanings depending on the context: "a statement that deviates from or perverts the truth" or "to be located or situated somewhere". The central concepts in the language view of Wittgenstein (1953) are language games, utterances, meaning, action, language rules and context. The concrete situation in which the utterance is uttered is where the meaning of the utterance must be sought. For Wittgenstein, the context is not just the linguistic context, i.e., the relationships between the linguistic elements in a sentence, but *everything* which is relevant in the situation such as, for example, who utters the utterance and the body language of the speaker.

An extreme proponent of contextuality is Vološinov:

The meaning of a word is determined entirely by its context. In fact, there are as many meanings of a word as there are contexts of its usage. (Vološinov, 1986, p. 79)

For Vološinov, verbal communication can never be understood and explained outside a connection with a concrete situation. None of our descriptions of the world is total, and new aspects can always be discovered. The way we describe something is not only determined by inherent qualities of the things but also of how we relate to them and in what context. Blue means quite different things to a physicist and an artist.

Spatialization and Temporalization

What about spatialization and temporalization? Continuing with the hammer example, spatialization is at play when a carpenter frames a work situation in terms of placing the nail in the exact position on the plank, positioning himself in relation to the plank, judging the influence of other things in the situation that might intervene with his actions (such as the safety of the scaffold he might be standing on), and so forth. Temporalization is at work in laying out the order of actions: grasping the hammer, placing the nail, hitting the nail, replacing the hammer in the toolbox, etc.

In language, spatialization and temporalization are at play in noun phrases ("[$_{NP}$ The boat that I saw coming into the harbor at nine o'clock] has just left") and verb phrases ("Mary [$_{VP}$ sent me a nice birthday present]"). Something is talked about and something happens – object and action. Spatialization and temporalization can also be recognized in speech act theory where the terms "locutionary" and "illocutionary" acts are used (e.g. Austin, 1962; Love, 1997; Searle, 1969). The locutionary or propositional act tells you something that you can understand or interpret within a context; what is talked about (spatialization). The illocutionary act is the act performed in saying something (temporalization).

Stabilization

Stabilization is present in the skill the carpenter has acquired in nailing. There is a preferred way of nailing that will be learnt and refined through repeated actions. It is all too well known what might happen with the thumb when an inexperienced person tries to hit the nail! Other manifestations of stabilization are, for example, rules that stipulate the wearing of safety glasses in order to avoid splinter, the wearing of safety lines when working on high altitudes, and so on. Stabilization is also present in standardization of nail sizes, quality classes of nails and hammers, and the like.

Stabilization is evident in the grammar of a language: there are certain rules of composition that need to be enacted if utterances are to be intelligible; the syntax of the language has to be followed. Also, the semantics of a language can be seen as a manifestation of stabilization. In every activity domain, the communal meaning of terms is one aspect of stabilization.

Transition

Transition is perceptible when the carpenter goes to the hardware store to buy more nails or a new hammer. He needs to explain his needs to the merchandiser, and he is dependent on other activity domains to have produced the nails and transported them to the store. Moreover, the payment of the merchandise is dependent on common understanding about the exchange value of the money used in the transaction.

For language, transition is most evident in the translation from one language to another, for example, between Swedish and English, but also between domain specific languages.

Summary

Technical tools can, at least in principle, be used by individuals acting on their own as in the example with the carpenter above. Semiotic tools, however, apply only when two or more actors are working together towards a common goal. In such situations, communication becomes imperative. Thus, there is a qualitative difference between technical and semiotic tools concerning one or several actors.

The activity modalities, on the other hand, are indifferent in this respect. It is straightforward to generalize the one actor situation to situations where several actors have to coordinate their actions. Take, for example, the activity of drilling of a hole in a rock in order to blast it. Before the advent of drilling machines this was done by two people. One actor was hitting a hand drill with a sledge hammer, and another actor held the drill and turned it between each hit by the hammer. Both actors needed to have common understanding about the position and slant of the drill, its position on the rock, etc. (spatialization). The actor hitting the steel had to do so in between the turning of the steel by the other actor. This coordination was often amplified with the singing of a work song, indicating the pulse of the hits (temporalization). Both actors had to agree on how the combined hitting – turning should be carried out; at what frequency the steel was to be hit, how much the steel should be turned in between the hits, etc. (stabilization). Their activity (drilling a hole) had to be coordinated with the activity of loading the hole with dynamite and the blasting. This required that the hole had a particular width and depth, that is, the hole was a transitional element between the activities (transition).

The characterization of mediational means along the activity modalities sidesteps the thorny question of the fundamental differences between technical and semiotic tools, and concentrates on their roles in mediating coordinative actions. Although these two groups of mediational means are different (an image of a drill is not the same as the drill), they also have much in common, and are often used interchangeably in achieving a certain goal. For example, the first atomic bomb that was dropped on Hiroshima August 6, 1945 was certainly a technical tool in the sense that it completely destroyed the city and its inhabitants. The dropping of the bomb, however, would not have come about if it wasn't for President Truman ordering the attack – a semiotic tool in the form of a speech act: "Go ahead!"

In summary, I suggest that the activity modality perspective provides an alternative way of conceiving mediational means, which overcomes some inherent problems with the separation of means in technical and semiotic tools. The activity modalities are equally applicable to individual actions and consorted actions by several actors. This strengthens the arguments that the activity modalities situations are indeed derived from innate, cognitive abilities.

REFERENCES

Augier, M., Shariq, S. Z., & Vendelø, M. T. (2001). Understanding context: its emergence, transformation and role in tacit knowledge sharing. *Journal of Knowledge Management*, *5*(2), 125–136. doi:10.1108/13673270110393176

Austin, J. L. (1962) *How to do things with words*. Oxford, UK: Oxford University Press.

Bærentsen, K. (2000). Intuitive User Interfaces. *Scandinavian Journal of Information Systems*, *12*, 23–60.

Bannon, L., & Bødker, S. (1997). Constructing common information spaces. In J. A. Hughes, W. Prinz, T. Rodden, & K. Schmidt (Eds.), *Proceedings of the Fifth Conference on European Conference on Computer-Supported Cooperative Work (Lancaster, UK, September 07 - 11, 1997) ECSCW* (pp. 81–96). Norwell, MA: Kluwer Academic Publishers.

Bechky, B. A. (2003). Sharing meaning across occupational communities: The transformation of understanding on a production floor. *Organization Science*, *14*(3), 312–330. doi:10.1287/orsc.14.3.312.15162

Berger, P., & Luckmann, T. (1991). *The Social Construction of Reality*. London: Penguin Books.

Blackler, F. (1995). Knowledge, Knowledge Work and Organization: An Overview and Interpretation. *Organization Studies*, *16*(6), 1021–1046. doi:10.1177/017084069501600605

Bødker, S., & Bøgh Andersen, P. (2005). Complex Mediation. *Human-Computer Interaction*, *20*, 353–402. doi:10.1207/s15327051hci2004_1

Boland, R. J. (2001). The Tyranny of Space in Organizational Analysis. *Information and Organization, 11*, 3–23. doi:10.1016/S0959-8022(00)00007-2

Brown, J. S., & Duguid, P. (2001). Knowledge and Organization: A Social-Practice Perspective. *Organization Science, 12*(2), 198–213. doi:10.1287/orsc.12.2.198.10116

Bryan-Kinns, N., Healey, P. G. T., Papworth, D., & Vaduuva, A. (2007). Cues to Common Knowledge. In L. Bannon, I. Wagner, C. Gutwin, R. Harper, & K. Schmidt (Eds.), *ECSCW'07: Proceedings of the Tenth European Conference on Computer Supported Cooperative Work* (pp. 43–447), September 24–28, Limerick, Ireland.

Bryant, W. C., & Gay, S. H. (1883). *A Popular History of the United States* (Vol. I). New York: Charles Scribner's Sons.

Burstein, F., & Linger, H. (2003). Supporting post-Fordist work practices. A knowledge management framework for supporting knowledge work. *Information Technology & People, 16*(3), 289–305. doi:10.1108/09593840310489395

Chalmers, A. F. (1976). *What is this thing called science?* St Lucia, Australia: University of Queensland Press.

Chen, X. (2003, December). Object and Event Concepts: A Cognitive Mechanism of Incommensurability. *Philosophy of Science, 70*, 962–974. doi:10.1086/377381

Christensen, T., & Lægreid, P. (2007). *The Challenge of Coordination in Central Government Organizations*. Bergen, Norway: Stein Rokkan Centre for Social Studies. Retrieved May 2, 2008, from http://hdl.handle.net/1956/2434

Curtis, B., Krasner, H., & Iscoe, N. (1988). A Field Study of the Software Design Process for Large Systems. *Communications of the ACM, 31*(11), 1268–1287. doi:10.1145/50087.50089

Dewey, J. (1991). 'Logic'. In J. A. Boydston (Ed.), *The Theory of Enquiry. The Later works of John Dewey*, (Vol. 12). Carbondale and Edwardsville, IL: Southern Illinois University Press.

Engels. F. (1954). On the part played by labour in the transition from ape to man. *Dialectics of Nature*. Moscow: Foreign Languages Publishing House. Retrieved June 4, 2008, from http://www.marxists.org/archive/marx/works/1876/part-played-labour/index.htm

Engeström, Y. (2001). Expansive Learning at Work: toward an activity theoretical reconceptualization. *Journal of Education and Work, 14*(1), 133–156. doi:10.1080/13639080123238

Fabrikant, S. I. (2000). Spatialized Browsing in Large Data Archives. *Transactions in GIS, 4*(1), 65–78. doi:10.1111/1467-9671.00038

Gershenson, C. (2002). Contextuality: A Philosophical Paradigm, with Applications to Philosophy of Cognitive Science. *Philosophy of Cognitive Science Essay*. Brighton, UK: University of Sussex, COGS.

Gibson, J. J. (1979). *The Ecological Approach lo Visual Perception*. Boston: Houghton Mifflin.

Goodwin, C., & Goodwin, M. H. (1996). Seeing as situated activity: Formulating planes. In Y. Engeström, D. Middleton (Eds.), *Cognition and Communication at Work* (pp. 61–95). Cambridge, UK: Cambridge University Press.

Grandori, A., & Kogut, B. (2002). Dialogue on Organization and Knowledge. *Organization Science, 13*(3), 224–231. doi:10.1287/orsc.13.3.224.2774

Grant, R. M. (1996). Toward a Knowledge-Based Theory of the Firm. *Strategic Management Journal, 17*, Winter Special Issue, 109–122.

Grant, R. M. (1996b). Prospering in Dynamically-competitive Environments: Organizational Capability as Knowledge Integration. *Organization Science, 7*(4), 375–387. doi:10.1287/orsc.7.4.375

Halliday, M. (1974). A Sociosemiotic Perspective on Language Development. *Bulletin of the School of Oriental and African Studies. University of London. School of Oriental and African Studies, 37*(1), 98–118.

Hemingway, C. (1998). Towards a Socio-Cognitive Theory of Information Systems: An Analysis of Key Philosophical and Conceptual Issues. In *The Workshop Proceedings of Information Systems: Current Issues and Future Changes.* Boston: Kluwer. Retrieved June 2, 2008, from http://citeseer.ist.psu.edu/261417.html

Jackendoff, R. (1983). *Semantics and Cognition*. Cambridge, MA: MIT Press.

Jensen, P. E. (2005). A Contextual Theory of Learning and the Learning Organization. *Knowledge and Process Management, 12*(1), 53–64. doi:10.1002/kpm.217

Jones, K. S. (2003). What Is an Affordance? *Ecological Psychology, 15*(2), 107–114. doi:10.1207/S15326969ECO1502_1

Kaptelinin, V. (1996). Computer-Mediated Activity: Functional Organs in Social and Development Contexts, In B. A. Nardi (Ed.) *Context and consciousness. Activity theory and human-computer interaction* (pp. 45–68). Cambridge, MA: MIT Press.

Kirsh, D. (1995). The intelligent use of space. *Artificial Intelligence, 73*(1–2), 31–68. doi:10.1016/0004-3702(94)00017-U

Kofod-Petersen, A., & Cassens, J. (2006). Using Activity Theory to Model Context Awareness. In T.R. Roth-Berghofer, S. Schulz, & D.B. Leake (Eds.), *MRC 2005 (Modeling and Retrieval of Context), LNAI 3946* (pp. 1–17). Heidelberg: Springer-Verlag.

Kosík, K. (1976). *Dialectics of the concrete*. Dordrecht: Reidel.

Kraut, R., & Streeter, L. (1995). Coordination in Software Development. *Communications of the ACM, 38*(3), 69–81. doi:10.1145/203330.203345

Lakatos, I. (1974). Falsification and the Methodology of Scientific Research Programmes. In I. Lakatos, & A. Musgrave (Eds.), *Criticism and the Growth of Knowledge* (pp. 91-196). Cambridge, UK: Cambridge University Press.

Larsson, R. (1990). *Coordination of Action in Mergers and Acquisitions - Interpretative and Systems Approaches towards Synergy*. Dissertation No. 10, Lund Studies in Economics and Management, The Institute of Economic Research. Lund, Sweden: Lund University Press.

Lawrence, P., & Lorsch, W. (1967). Differentiation and Integration in Complex Organizations. *Administrative Science Quarterly*, *12*(1), 1–47. doi:10.2307/2391211

Liedman, S.-E. (2001). *Ett oändligt äventyr - Om människans kunskaper.* [An eternal adventure – About the knowledge of mankind]. Stockholm, Sweden: Albert Bonniers Förlag (in Swedish).

Love, N. (1997). Integrating Austin. *Language Sciences*, *19*(1), 57–65. doi:10.1016/0388-0001(95)00027-5

Lutters, W., & Ackerman, M. (2007). Beyond Boundary Objects: Collaborative Reuse in Aircraft Technical Support. *Computer Supported Cooperative Work*, *16*(3), 341–372. doi:10.1007/s10606-006-9036-x

Malone, T., & Crowston, K. (1994). The Interdisciplinary Study of Coordination. *ACM Computing Surveys*, *26*(1), 87–119. doi:10.1145/174666.174668

March, J. G., & Simon, H. A. (1958). *Organizations*. Cambridge, MA: Blackwell Publishers.

Meacham, J. (1999). Riegel, Dialectics, and Multiculturalism. *Human Development*, *42*(3), 134–144. doi:10.1159/000022619

Merriam-Webster On Line. (2008). Retrieved October 5, 2008, from http://www.merriam-webster.com/

Miettinen, R., & Virkkunen, J. (2005). Epistemic Objects, Artefacts and Organizational Change. *Organization*, *12*(3), 437–456. doi:10.1177/1350508405051279

Mintzberg, H. (1983). *Structures in Fives: Designing Effective Organizations.* Upper Saddle River, NJ: Prentice-Hall.

Nilson, G. (1976). *Oordning/ordning: studier i kärlekens villkor*. [Order/disorder: studies in the conditions of love]. Göteborg: Bokförlaget Korpen (in Swedish).

Orlikowski, W. (2002). Knowing in Practice: Enacting a Collective Capability in Distributed Organizing. *Organization Science*, *13*(3), 249–273. doi:10.1287/orsc.13.3.249.2776

Orlikowski, W., & Yates, J. (2002). It's About Time: Temporal Structuring in Organizations. *Organization Science*, *13*(6), 684–700. doi:10.1287/orsc.13.6.684.501

Parsons, J. (1996). An Information Model Bases on Classification Theory. *Management Science*, *42*(10), 1437–1453. doi:10.1287/mnsc.42.10.1437

Porter, M. (1996). What Is Strategy? *Harvard Business Review*, (November-December): 61–78.

Reber, A. S. (1995). *The Penguin dictionary of psychology*. London: Penguin Books.

Sawhney, M., & Prandelli, E. (2001). Communities of Creation: Managing Distributed Innovation in Turbulent Markets. *IEEE Engineering Management Review, First Quarter*, 6–24.

Schmidt, K. (1990). *Analysis of Cooperative Work. A Conceptual Framework.* Risø National Laboratory, DK-4000 Roskilde, Denmark, [Risø-M-2890].

Schmidt, K., & Bannon, L. (1992). Taking CSCW Seriously: Supporting articulation work. *Computer Supported Cooperative Work, 1(1–2),* 7–40. Retrieved April 4, 2008, from http://citeseer.ist.psu.edu/schmidt92taking.html

Searle, J. R. (1969). S*peech acts. An essay in the philosophy of language*. London: Cambridge University Press.

Smircich, L. (1983). Concepts of Culture and Organizational Analysis. *Administrative Science Quarterly*, *28*(3), 339–358. doi:10.2307/2392246

Star, S. L. (1989). The Structure of Ill-Structured Solutions: Boundary Objects and heterogeneous Distributed Problem Solving. In L. Gasser & M. N. Huhns (Eds.), *Distributed Artificial Intelligence* (pp. 37–54). London: Pitman Publishers, Ltd.

Stetsenko, A. (1999). Social Interaction, Cultural Tools and the Zone of Proximal Development: In Search of a Synthesis. In S. Chaiklin, M. Hedegaard, & U.J. Jensen (Eds.), Activity *Theory and Social Practice: Cultural-Historical Approaches* (pp. 235–252). Aarhus, Denmark: Aarhus University Press.

Taxén, L., & Lilliesköld, J. (2008). Images as action instruments in complex projects . *International Journal of Project Management*, *26*(5), 527–536. doi:10.1016/j.ijproman.2008.05.009

Virkkunen, J., & Kuutti, K. (2000). Understanding organizational learning by focusing on "activity systems." *Accounting . Management and Information Technologies*, *10*(4), 291–319. doi:10.1016/S0959-8022(00)00005-9

Vološinov, V. N. (1986). *Marxism and the Language of Philosophy*. London: Harvard University Press.

Webster's 1913 Dictionary (2008). Retrieved Sept 4, 2008, from http://humanities.uchicago.edu/orgs/ARTFL/forms_unrest/webster.form.html

Wenger, E. (1998). *Communities of Practice. Learning, Meaning and Identity*. Cambridge, UK: Cambridge University Press.

Wikipedia (2008). Retrieved Aug 6, 2008, from http://en.wikipedia.org/wiki/Homo_faber

Wiley, N. (1988). The Micro-Macro Problem in Social Theory. *Sociological Theory*, *6*(2), 254–261. doi:10.2307/202119

Wittgenstein, L. (1953). *Philosophical investigations*. Oxford, UK: Blackwell.

Yakura, E. K. (2002). Charting Time: Timelines as Temporal Boundary Objects. *Academy of Management Journal*, *45*(5), 956–970. doi:10.2307/3069324

Zacks, J., & Tversky, B. (2001). Event Structure in Perception and Conception. *Psychological Bulletin*, *127*(1), 3–21. doi:10.1037/0033-2909.127.1.3

Zinchenko, V. (2001). External and Internal: Another Comment on the Issue. In S. Chaiklin (Ed.), *The Theory and Practice of Cultural-Historical Psychology* (pp. 135–147). Aarhus, Denmark: Aarhus University Press.

ENDNOTES

[1] This topic is gaining momentum in the study of the so called third generation of AT (e.g. Engeström, 2001).

[2] The different usage of the terms "action" and "activity" may easily be confusing. In English, these terms are used indiscriminately; as more or less equivalents. For example, Merriam-Webster defines action as "the most vigorous, productive, or exciting activity in a particular field, area, or group <they itch to go where the action is — D. J. Henahan> (Merriam-Webster On Line, 2008). Activity on the other hand is defined as "vigorous or energetic action: liveliness" (ibid). When I use "activity" I always refer to the AT sense of the term.

[3] Due to ease of reading, the direct dependencies from the IT department to Supply & Implementation and Research & Development are suppressed in the diagram. These dependencies are indirectly present through the Market & Sales domain.

[4] In Gibson's theory of perception (Gibson, 1979), human relates to the work through what he calls affordances which stands for the "invitational" quality of a percept or an event. Thus, a part of the affordance of a hammer is its graspability, of a chair its sit-on-ability. In a sense, affordance refers to the intrinsic properties of items and events (Reber, 1995).

[5] For a thorough discussion of spacing and timing in organizations, see the special issue on this topic in Organization (2004, 11(6)).

[6] This is surely one of the reasons why it is so hard to realize benefits of scale by merging two organizations, or for that matter, to outsource parts of an organization to another organization. In such cases, the enactment process has to start anew, which is a tedious and unwieldy task.

[7] This is actually an example I have experienced myself. It turned out that you needed to disengage a particular lock by pressing a button on the gear shaft. In order not to offend anyone, I withhold the make of the car.

[8] Homo faber (Latin for "Man the Smith" or "Man the Maker"; in reference to the biological name for man, "Homo sapiens" meaning "man the wise") is a concept articulated by Hannah Arendt and Max Frisch. It refers to humans controlling the environment through tools (Wikipedia, 2008).

[9] The concept of affordances is still discussed by scholars (e.g. Jones, 2003). For the reasoning here, it suffices to comprehend "affordances" as roughly equal to "mediational means".

Chapter 6 Cognitive Grounding

A crucial element in the ADT is the assumption of congruence or correspondence between observable "traces" of the activity modalities in our environment, and the cognitive constitution of humans. In this chapter, I will provide some arguments in favor of this assumption. Given the enormous amount of research in cognitive science, these arguments can only of a "rhapsodic" character; a first indication of paths to follow in further research.

THE GÄRDENFORS MODEL OF COGNITION

I will make use of a model of our cognitive system proposed by Gärdenfors (2000). This model is structured in three levels – the symbolic, conceptual and connectionist ones – and provides a gradual transition from the outward interactions with the world towards the neuronal network in the brain.

Gärdenfors's knowledge interest is to model representations of the cognitive system for explanatory and constructive purposes such as developing artificial neuron networks. According to him, there are two dominating approaches for modeling the cognitive system: the symbolic and the associative ones. A particular type of association is connectionism, where the information carrying elements are modeled on the neural network in the brain.

DOI: 10.4018/978-1-60566-192-6.ch006

This has resulted in two branches of research concentrating on either the symbolic or the connectionist approach. In the symbolic approach, the brain is regarded as a Turing machine that manipulates symbols according to certain rules. The connectionist approach models the brain as a system of connected neurons that respond to stimuli of various kinds. These two approaches should be seen as complementary rather than separate, since the symbolical representation presumes the connectionist one in the sense that the nervous system is a prerequisite for symbol formation.

However, Gärdenfors claims that important aspects of the cognitive system such as conceptacquisition, i.e., how concepts are learnt and internalized, cannot be adequately captured by either one of these approaches. It is also unclear how to model the transition between the connectionist and the symbolical levels since the representations used are very dissimilar. The step from neuron models to symbols becomes too wide. To this end, Gärdenfors proposes a third, intermediate level: the *conceptual* one. Thus, Gärdenfors winds up with three levels (see Figure 1): the symbolical, conceptual and connectionist ones.

Gärdenfors proposes that the structure of the conceptual level is geometrical; hence the subtitle of the book: "The geometry of thought". Human beings have a cognitive capacity for generating ordering relations of stimuli based on what Gärdenfors calls quality dimensions (ibid, p. 6). Examples of such quality dimensions are temperature, weight and spatial dimensions like height, width and depth. Some quality dimensions are culturally dependent, e.g. time which may be conceived as linear or circular (ibid, p. 28).

The quality dimensions can be associated with geometrical, topological or ordering structures (time being one of them), and they provide the basis for building what Gärdenfors calls conceptual spaces. The cognitive elements on the conceptual level are concepts that are spatially related to each other (ibid, p. 5). Due to the structure of quality dimensions is it possible to talk about distances along these dimensions. Such distances in the conceptual space are closely related to similarity judgments,

The conceptual level provides an appropriate mechanism for grounding the symbolic level, which would otherwise have to be grounded directly in the connectionist level. A major point in Gärdenfors's contribution is that such a direct grounding cannot be convincingly argued for:

Figure 1. The model of cognition suggested by Gärdenfors

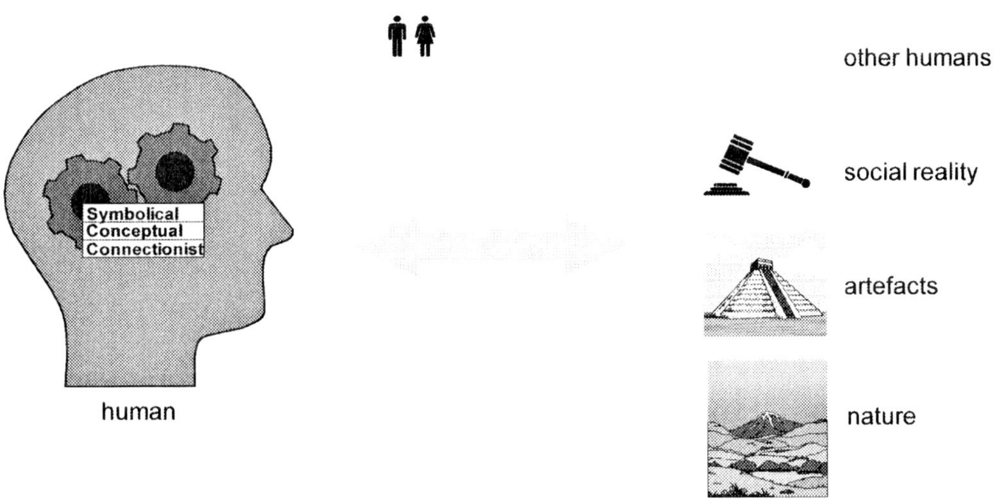

[The] symbolic and connectionist representations are not sufficient for the aims of cognitive sciences; many problems are best handled by using geometrical structures on the conceptual level. (ibid, p. 3)

The many dimensions of neural network converge into regular structures on the conceptual level through a reduction of neural dimensions to quality dimensions (ibid, p. 240). Thus, the conceptual level is a basis for the uppermost, symbolic one.

The conceptual structure makes it possible to inquiry into the nature of meaning. According to Gärdenfors, social meanings emerge from the heads of individuals:

The core idea is that meanings of linguistic expressions are mental entities – meanings are elements of the cognitive structure in the heads of the language users. (ibid, p. 154)[1]

Clearly, this position differs from the one taken in this book, where activity is regarded as the genesis of all cognitive effects. Gärdenfors's position runs into problems regarding semiosis (the formation of signs) and the question of how the mental entities came into the head to begin with. Another problem is the social nature of meaning: how is communal meaning among individuals possible? Gärdenfors realizes this difficulty:

A fundamental assumption of this analysis, however, has been that the conceptual structure belongs to some individual language user: the meanings of words reside in the heads of individuals. On the other hand, it is also obvious that language is a social phenomenon. [...]. If each person can mandate his own cognitive meaning, [...] how can it be established that we talk about the same things? [...] the communicative question is a genuine problem for cognitivist theories of semantics. (ibid, pp. 189-190)

The way Gärdenfors tackles this problem is to start from the observation that any association between a thing and a concept must take place in the brain. There is nothing in the external world that can hold this association. Therefore, in a communicative situation where a speaker and hearer are communicating about things, their cognitive systems must be aligned to some degree. This imposes certain constraints on individual cognitive representations; constraints that are, according to Gärdenfors, related to the "linguistic power structure":

I argue that the social meanings of the expressions of a language are indeed determined from their individual meanings – the meanings the expressions have for the individuals, together with the structure of linguistic power that exists in community [...]. The linguistic powers concern who is the master of meaning – who decides on what is the correct meaning of an expression in a society. (ibid, p. 197, italics in original)

One assumption put forward by Gärdenfors is that the situation will drive the representations towards the most economical way of achieving a reasonable communal meaning. Social interactions will result in a set of "communicable references", which are located in the brains of both speaker and hearer as cognitive structures. This stance leads to a chicken-and-egg problem: are cognitive structures prerequisites for, or emergent results of successful communication? The answer given by Gärdenfors is "both":

My "sociocognitive" position can be summarized as follows: meanings are not in the head of a single individual, but they emerge from the conceptual structures in the heads of the language users together with the linguistic power structure. (ibid, p. 202, italics in original)

This position is in essence not far from the praxis perspective, although ontologically and epistemologically different. On the other hand, the praxis perspective is quite compatible with the position that meanings are elements of the cognitive system of humans.

Shifting the Focus: From Heads to Interactions

Gärdenfors's model provides a well-grounded point of departure for further analysis. However, in order for the model to comply with the praxis perspective, the focus needs to be shifted from heads to social interactions:

[Consciousness] itself can arise and become a viable fact only in the material embodiment of signs [...]. Understanding is a response to a sign with signs [...]. Signs emerge, after all, only in the process of interaction between one individual consciousness and another [...]. Consciousness becomes consciousness only [...] in the process of social interaction. (Vološinov, 1986, p. 11)

The position taken by Vološinov, to which I adhere, is that social interaction is the foundation for the emergence of meaning. It is still valid that the association between a thing and a concept is confined to the brain. However, the formation of the association is achieved through the dialectical encounter with the environment, rather than through some reified, external agent like the "linguistic power structure":

The organizing centre of any utterance, of any experience, is not within but outside - in the social milieu surrounding the individual being. (Vološinov, 1986, p. 93)

In addition to a more interaction-oriented model, I will interpret "symbolic" and "connectionist" somewhat differently. The symbolical level in Gärdenfors (2000) refers to symbol manipulation according to some rules. This representation comes from Artificial Intelligence (AI). Since this kind of representation is not of interest here, I will conceive the symbolic level as a *linguistic* level[2]. By "linguistics" I refer to the study of human speech including the units, nature, structure, and evolution of language. Some sub-fields in this area are phonology and phonetics (sounds of language), semantics (meaning), syntax (grammar), etc.

Another point where I wish to depart from Gärdenfors concerns his notion of "level". This conception suggests a strict, hierarchical ordering that I want to avoid. It is true that the neural apparatus of a particular organism at a particular instant in time may appear to be more basic than concept or symbol formation. However, without interaction the cognitive, genetically inherited system of an organism would not have evolved in the first place. When placing interaction in focus, the evolution at the different levels becomes rather a matter of long-term or short-term interaction. Since I want to stress the formative aspects of the model, I will use "stratum" instead of "level" since "stratum" has connotations of history, sedimentation, diffusion, and the like. Moreover, "stratum" indicates a more fuzzy transition between levels, which is in line with the dialectical perspective.

Figure 2. The model of interaction (adapted after Gärdenfors, 2000)

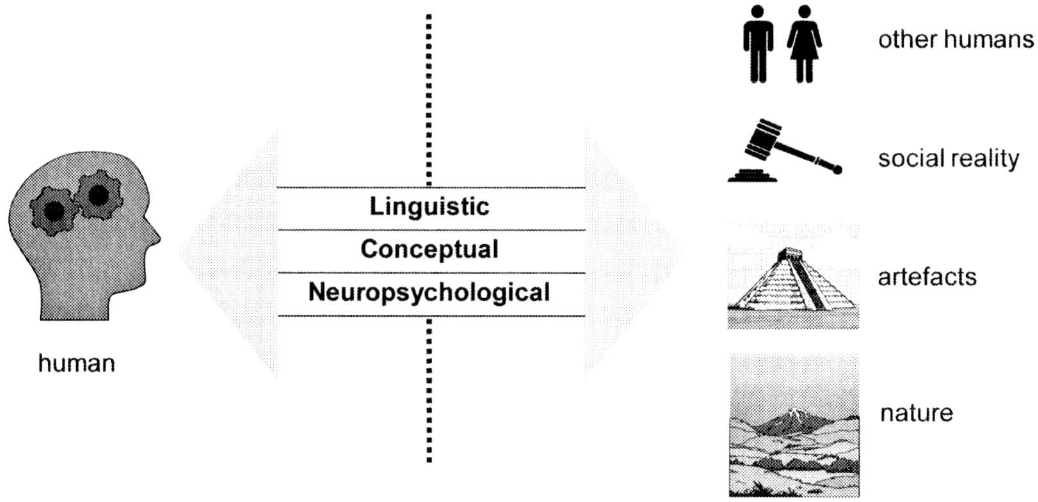

I will retain the connectionist focus on the properties of neural network in the brain. However, in order to include general neurophysiological aspects, I will call this level the "neuropsychological stratum". In this stratum I will also include broader aspects of the "biological substrate" underlying the neural network in the brain.

To summarize: The Gärdenfors model of the cognitive system of individual humans is interpreted as a model of the social interaction between a human individual and her environment. The different strata (neurophysiological, conceptual and linguistic) represent a gradual focal shift from the biological realm to the social realm, with the social as the dominant one:

In the lower strata of behavioral ideology, the biological-biographical factor does, of course, play a crucial role, but its importance constantly diminishes as the utterance penetrates more deeply into an ideological system. Consequently, while biographical explanations are of some value in the lower strata of experience and expression (utterance), their role in the upper strata is extremely modest. (Vološinov, 1986, p. 93)

The resulting interaction model is illustrated in Figure 2.

As with any model, this one is a theoretical construct with a particular purpose in mind: to inquire into the cognitive grounding of the activity modalities. The model should not be apprehended as a representation of "real" interactions in nature. The cognitive apparatus of organisms still resides in the nervous system in the brain. There is nothing in the external world which materializes interactions as such. We cannot point to something and say "This is an interaction of the kind so and so". However, by shifting the focus from the head to interaction, I want to emphasize the manifested, tangible expressions in our environment of these interactions without detaching their anchoring in the brain. Thus, from a scientific point of view, the model is basically instrumental (Chalmers, 1976).

In the following sections, the adapted model will be used as a background for discussing arguments supporting the activity modalities and other basic constructs in ADT.

THE LINGUISTIC STRATUM

The first step is to inquiry into traces of the activity modalities in the linguistic stratum. To this end, I will make use of the theory of meaning proposed by Jackendoff (1983). His knowledge interest is two-fold (ibid, p. 3):

- What is the nature of meaning in human language, such that we can talk about what we perceive and what we do?
- What does the grammatical structure of natural language reveal about the nature of perception and cognition?

According to Jackendoff, these two questions cannot be separated. He suggests that there is a single level of mental representation where linguistic, sensory and motor information are compatible: the *conceptual structure* (ibid, p. 17); very much in line with Gärdenfors. Unless there is such a level, it would be impossible to use language to report sensory input. We could not speak about what we see, hear, touch, smell, etc. Likewise, linguistic information must be compatible at some level with information conveyed to the motor system. For example, an utterance like "Please close that window!" presumes an interaction of visual, linguistic and motor information.

Like Gärdenfors, Jackendoff claims that some kind of conceptual representation is needed as a bridge between the linguistic/symbolical stratum and the neural/connectionist stratum. However, while Gärdenfors models the conceptual stratum geometrically, Jackendoff proposes that the conceptual structure is characterized by a finite set of conceptual well-formedness rules that are universal and innate (ibid, p. 17). In order to explore the properties of these rules, two issues need to be addressed: What is the nature of the information that language conveys, and what is this information about? The first is concerned with sense or intension – the second with reference or extension.

These questions are approached from a discussion about what we see when we perceive something. From ambiguous figures like the famous Wittgensteinian "duck-rabbit" in Figure 3, it is clear that the brain overlays such figures with an organization that is not present in any physical sense.

There is no reason why this mechanism should be confined only to the vision modality. Thus, the world as we perceive and experience it is "unavoidably influenced by the nature of the unconscious processes for organizing environmental input. One cannot perceive the 'real world as it is'" (ibid, p. 26). It follows that it is necessary to distinguish between the source of the environmental input and the world as experienced. Jackendoff calls the former the "real world" and the latter the "projected world". We

Figure 3. Duck-rabbit (Wittgenstein, 1953)

have conscious access only to the projected world – the world as unconsciously organized by the mind. We can talk about things only insofar they have achieved mental representation through these processes of organization. Hence, "information conveyed by language must be about the projected world" (ibid, p. 29). The information in the conceptual structure is the single level of mental representation onto which and from which all peripheral information is mapped. In this perspective, meaning is apprehended as the connection of sensory inputs to entities in the projected world.

In order to analytically distinguish the real world form the projected one, Jackendoff introduces a meta-language. References to the projected world are surrounded by # #. Entities in the conceptual structure, which give rise to the projected world, are designated in CAPITALS. Real-world entities have no particular marking. However, in this section I will surround real-world entities by * *. Moreover, the notion of different worlds may appear odd, since all phenomena discussed are parts of the same world. Therefore, in the following I will refer to the different worlds of Jackendoff as "realms".

To illustrate this, consider again Figure 3. By interacting with other people in an English speaking society, two entities in the conceptual structure are formed in the mind: RABBIT and DUCK. The real-realm electromagnetic radiation from the illustration may be labeled *duck-rabbit*. This physical input on the retina is processed by the nervous system and mapped to either RABBIT or DUCK in the conceptual structure. From this structure, either the entity #rabbit# or #duck# may be projected into awareness. However, both projections cannot take place simultaneously. Thus, the same real-realm input may give rise to different experiences in the projected realm. If, for example, #duck# is projected, this entity may be transformed via correspondence rules to the linguistic and motor systems, resulting in the utterance "That is a duck!", possibly accompanied with a pointing gesture.

Consequently, the sense of the linguistic expression consists of expressions based on the conceptual structure. The reference of the expression takes place in the projected realm, not the real-realm. This view has far-reaching consequences. The traditional view of truth and reference as relations between statements and facts in the real-realm cannot be upheld, since the direct connection between language and this realm does not exist.

The single level of mental representation provided by the conceptual structure indicates that "semantic structures could be simply a subset of conceptual structures – just those conceptual structures that happen to be verbally expressible" (ibid, p. 19). Thus, the study of semantics and grammar of natural language should give evidence for the organization of the conceptual structure and the conceptual well-formedness rules. This is discussed in the following section.

Ontological Categories

An obvious feature of the projected realm is that it includes #things# that have some kind of spatial and temporal integrity. These #things# are projected from a corresponding mental representation in the conceptual structure. Jackendoff calls a unitary piece of mental representation a *conceptual constituent* (Jackendoff, 1983, p. 42). These are surrounded in the Jackendoff formalism with square brackets [ ]. Thus, in order to account for the fact that humans can perceive a #rabbit#, a corresponding conceptual constituent [RABBIT] must have been established in the conceptual structure.

Utterances like "I have read that book (*pointing*)" relate linguistic and visual information. Such utterances are examples of what has been coined "pragmatic anaphora" (ibid, p. 48). It turns out that there is a set of grammatically different classes of pragmatic anaphora, which can be identified by different types of grammatical phrases. The utterance above is clearly about [THING]s, and the part "that book" can be identified as a noun phrase (NP). However, an utterance like "Your book is there (*pointing*)" is about a [PLACE], which is recognized by a prepositional phrase (PP). [PLACE] and [THING] are clearly of separate kinds. Such different types of basic categories are called *ontological categories*.

Through an elaborate linguistic discussion, Jackendoff identifies a set of ontological categories, which combine to perform meaning functions. The categories, which are of prime interest for my purposes, are as follows:

[THING]
[PLACE]
[PATH]
[EVENT]
[STATE]
[PROPERTY]
[DIRECTION]
[ACTION]

The semantic structure of a sentence is built from a hierarchical composition of the conceptual constituents, each belonging to an ontological category. These categories are realized by syntactically mapping them according to well-formedness rules to phrasal categories such as sentences (S), noun phrases (NP), verb phrases (VP), adjective phrases (AP), adverbial phrases (AdvP) and prepositional phrases (PP). For example, the sentence "John walked into the room" is mapped between syntactical and conceptual structures as follows:

Syntactic: $[_{S}\ [_{NP\ John}]\ [_{VP}$ walked $[_{PP}$ into $[_{NP}$ the room]]]].

Conceptual: $[_{Event}$ GO $([_{Thing}$ John], $[_{Path}$ TO $[_{Place}$ IN $[_{Thing}$ the room]]])].

In this way, Jackendoff tries to establish a connection between the lexical categories and phrases used in different, ordinary languages, and the conceptual structure common to all humans.

In later works, Jackendoff includes "thematic roles" as part of the conceptual structure (Jackendoff, 1990). A thematic role can then be seen as a relationship that ties a term with an event or a state, establishing a semantic relationship between a predicate (e.g. a verb) and an argument (e.g. the noun phrases) of a sentence. Examples of thematic roles are:

[AGENT] (The instigator of an action)
[ACTOR] (The willful instigator of an action)
[GOAL] (What the action is directed towards)

Jackendoff also discusses categorization as an essential aspect of cognition. Categorization refers to the ability to judge that a particular thing is or is not an instance of a specific category (Jackendoff, 1983, p. 77). The ability to categorize is indispensable for using previous experiences to guide the interpretation of new experiences. In order to discuss matters concerning categories and instances, Jackendoff refers to the thing being categorized as [TOKEN] and the category as a [TYPE]. The latter is the information that is created and stored when an organism learns a category.

How can the theory of Jackendoff be related to the activity modalities and the ADT? First, it can be noted that basic action elements in ADT have corresponding ontological categories: "action" - [ACTION], "goal" - [GOAL], "actor" - [ACTOR]. The role of agents is represented by a binary function CAUSE that have the structure

$[_{Event}$ CAUSE $([_{Thing}$ x], $[_{Event}$ y)]].

This means that the agent is not necessarily acting willfully as in the example "The wind blew the paper away". The case of an intentional actor is seen as a special case of [AGENT].

Mapping to Activity Modalities

With the theoretical apparatus of Jackendoff at hand, it is possible to relate this to activity modalities as follows.

Spatialization

In order to carry out actions, the actors need to know what kind of #things# are relevant in the activity domain. Spatialization manifests communal meaning of these #things#, how they are characterized, how they are related to each other, and in what state or condition they are. Spatialization can be associated with the conceptual structure as follows:

[THING]s are related to other [THING]s.
[THING]s have [PROPERTIES].
[THING]s have [STATE]s.
[DIRECTION]s provide orientation.

Temporalization

Temporalization manifests communal meaning of how actions should be coordinated. Temporalization can be associated to the conceptual structure as [EVENT]s which changes [STATE]s or [PROPERTIES] of [THING]s. It can be noted that spatialization and temporalization are interdependent, something that is often ignored in business and enterprise modeling efforts (cf. Parsons, 1996).

Stabilization

The ability to categorize is central to achieve stabilization. Stabilization affects [TYPE]s and [TOKEN] s in the conceptual structure and is constantly adjusted in interaction with the environment:

A processing model of cognition must include an active component that continually seeks to adjust and reorganize conceptual structure in an effort to maximize overall stability. (Jackendoff, 1983, p. 149)

Contextualization

Contextualization concerns the construction of a context in which actions are meaningful. Contextualization implies at least two things: framing and context dependency of meaning. Framing is mainly determined by the motive and the work object of the activity. Only those #things# that are relevant with respect to the motive and work object become meaningful in the context. This means that different PROPERTIES can be associated with the same THING depending on the context (cf. Parsons, 1996).

Contextuality is inextricably related to the classification of THINGS into categories or TYPES. Our visual system simplifies a visual scene into foreground and background. Certain #things# are attended in the foreground while other *things* remain unattended in the background (Jackendoff, 1983, p. 42). Thus, a capability to contextualize appears to be innate.

Figure 4. Image-schema for "over" and "under" (adapted after Gärdenfors, 1992, p. 99)

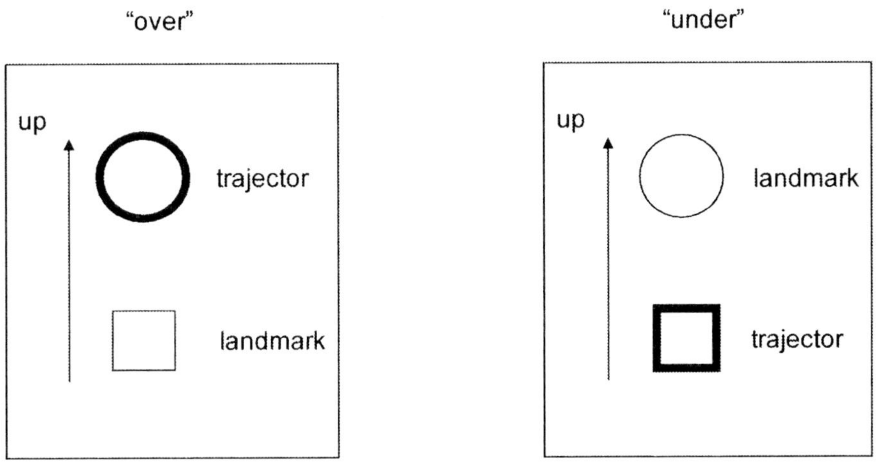

There seems to be no direct correspondence in the conceptual structure to contextualization. However, the notion of SEMANTIC FIELD is used to describe a set of lexemes that are related in some way. For example, a SEMANTIC FIELD concerning air travel would include reservation, flight, travel, buy, price, cost, fare, rates, meal, plane, etc. This can be apprehended as a list of relevant THINGS in a context such as an activity domain with the motive of providing air travel services.

Transition

It appears that there is no ontological category that directly corresponds to transition in the conceptual structure. Transition is however apparent, for example, in conceptual constituents involved in translating different languages into each other (such as when DOG in English is mapped to the HUND in Deutsch). This might indicate that transition, and its counterpart contextualization, are very basic human abilities that influence linguistic faculties without necessarily making them explicit in our language. Thus, the view of activity modalities as phylogenetically inherited capabilities is strengthened.

Summary

In summary, Jackendoff proposes that cognition emerges from a level of mental representation where language and sensory modalities are compatible. This level contains major ontological categories such as [THING], [PLACE], [DIRECTION], [EVENT], etc. These categories characterize the distinctions among major classes of #entities# that we act as though the #world# contains. The total set of ontological categories "must be universal: it constitutes one basic dimension along which humans can organize their experience, and hence it cannot be learned" (Jackendoff, 1983, p. 56).

The analysis shows that it is indeed possible to associate the ontological categories with at least a subset of the activity modalities. Thus, the theory of Jackendoff provides some arguments for the cognitive anchoring of these modalities.

THE CONCEPTUAL STRATUM

The next step is to inquiry into traces of activity modalities in the conceptual stratum.

Contextualization

To begin with, contextualization is an inherent property in this stratum: "...the salience given to various aspects of a concept may vary depending on the context. The meaning of a concept is not static, but changes with the context in which it is used" (Gärdenfors, 2000, p. 102). "The first problem when representing a concept is to decide which are the relevant domains" (ibid, p. 102). "Domains" in the terminology of Gärdenfors can be understood as characteristics of concepts. For example, an apple may be characterized by the domains color, shape, texture, and taste. The association of domains with concepts is context dependent. Consider, for example, a grand piano. For someone moving the piano, the weight domain is probably a salient domain, but for a recital performer, the pitch and timbre of the tones are more likely salient domains. Thus, what is a core characteristic in one context may become a marginal characteristic in another context[3]. Moreover, actions play a major role in the contextualization of domains: "The addition of new domains is often connected with new forms of *actions* that require attention to previously unnoticed aspects of concepts" (ibid, p. 103, italics in original).

Spatialization and Temporalization

Since the symbolic level presumes the conceptual one, the question arises how these are related to each other. This issue has been addressed by the branch of semiotic theory called cognitive semantics. The gist of this theory is to regard semantics as a relation between linguistic expressions and a cognitive structure. The semantic elements are constructed from geometrical or topological structures. Gärdenfors (1992) describes the use of so called "image-schemas" to represent perception, memory, and semantics. These image-schemas have an inherent spatial structure and are supposed to be represented in the cognitive structure of the individual. Such schemas – "paths," "up-down," "front-back," "part-whole," "centre-periphery" and the like – are considered to be "among the most fundamental carriers of meaning" (Gärdenfors, 2000, p. 162). For example, the meaning of "over" and "under" can be represented by the image-schemas given in Figure 4.

"Over" is apprehended as a relation between two objects. One object is in focus: the trajector (marked with a bold periphery). The other object is called the landmark. The only difference between "over" and "under" is that the focus changes. The image-schema shows that there is a certain relation between two objects which is valid during a certain time period. In this way, a plausible mechanism can be conceived, which relates the conceptual level to the linguistic one.

Image-schemas have an inherent spatial structure. For example, Chen argues that

There is consensus among cognitive scientists that intraconceptual relations within object concepts are in essence spatial. For example, Lakoff (1987) suggests that image-schema lie at the core of object concepts. Image-schema are schematic, spatial images that constantly recur either in our everyday bodily experience or in various orientations and relations [...]. Conceptualization of an object concept involves a process of "spatialization," in which image-schema are used to map metaphorically the spatial structures of physical space into a conceptual space. Thus, internal structures of object concepts should be understood in terms of spatial relations. (Chen, 2003, p. 965)[4]

Gärdenfors puts forward that a representation of temporalization is needed as a complement to conceptual spaces:

Conceptual spaces are static in the sense that they only describe the structure of representations. A full model of cognitive mechanisms not only includes the representational form, but also a description of the processes operating on the representations. (Gärdenfors, 2000, p. 31)

Chen claims that there is "… an ontological difference between object and event concepts: the former are spatial but the latter temporal. Experiments from cognitive sciences further demonstrate that the mind treats object and event concepts differently" (Chen, 2003, p. 962). In addition Langacker states that

It is however necessary to posit a number of "basic domains" that is, cognitively irreducible representational spaces or fields of conceptual potential. Among these basic domains are the experience of time and our capacity for dealing with two and three-dimensional spatial configurations. (Langacker, 1987, p. 5, quoted in Gärdenfors, 2000, pp. 161-162)

Congruence

Indications of congruence between individual cognition and culture can be found in the criticisms that have been raised about the alleged "universalistic" conception of the mind, disregarding cultural influences. Such conceptions have tended to obscure the socio-cultural dimensions of human cognition (Hampe, 2005). In contrast,

[Cognitive] models and schemas – including image schemas – can be seen both as expressions of universal principles at work in individual cognition and as properties of an underlying, "institutionalized" cultural "world view" […]. Consequently, a "naturalistic, biologically informed approach to human cognition" does not necessarily preclude "the recognition of the constitutive role in it of culture". (Hampe, 2005, p. 6)

To summarize, it is possible to find evidences supporting the activity modality construct also in the conceptual stratum.

THE NEUROPSYCHOLOGICAL STRATUM

The last step is to look for evidences of activity modalities in the neuropsychological stratum. The biological "substrate" for the stratified model enables an organism to interact with its environment through its cognitive system. One element of the substrate is the sensorimotorical system. Sensations are immediate sensory impressions like tastes, scents, colors, tickles, pain, etc. The biological role of sensations is to direct the organism's attention to what is happening to it at this very moment. Sensory receptors provide information to the organism's nervous system where it is processed in order to take appropriate action. Even very simple organisms have sensations. However, this might not necessarily lead to learning. Complicated, genetically programmed patterns of actions may be carried out without any learning taking place. The character of the interaction depends on the type of sensation in the interface between the sensory receptor and the environment. External cues or signals are converted or transduced into information that the organism uses to survive.

Action

Another element of the biological substrate is perceptions. The role of perceptions is to inform the organism of what is about to happen in the environment, not only what is happening right now that the sensations inform about. According to Gärdenfors (2000), perceptions of advanced organisms are associated with a type of simulators in the brain, by which different types of actions can be anticipated before they are carried out[5]. The information from the sensory receptors is fed into the simulator and amended with other signals which might emerge from stored representations rather than sensory receptors. An interpretation of the situation is carried out and the relevant action is performed. Simulators presume a learning capability, which means that more complex nervous systems are involved than pure sensations. However, the perceptual simulators are always related to sensory impressions.

Activity Modalities

It appears that the brain is structured to distinguish between temporal and spatial information. There is evidence that certain brain regions are tuned to temporal aspects of action:

Temporal structure has a major role in human understanding of everyday events. Observers are able to segment ongoing activity into temporal parts and sub-parts that are reliable, meaningful and correlated with ecologically relevant features of the action. [...] a network of brain regions is tuned to perceptually salient event boundaries [...]. Activity within this network may provide a basis for parsing the temporally evolving environment into meaningful units. (Zacks et al., 2001, p. 651)

Psychological experiments indicate that "there are different retrieval patterns between object and event concepts. The differences in the retrieval pattern [...] reflect significant differences in the underlying mental organizations and cognitive processes" (Chen, 2003, p. 966). This is also supported by neurophysiological findings:

Indeed, there are a broad range of behavioural, neuropsychological, and electrophysiological findings that support a dissociation between object knowledge in temporal cortex and action knowledge in parieto-frontal cortex. (Plaut, 2002, p. 631)

Greenfield quotes the neurophysiologist Ad Aertson:

[We] should distinguish between structural and anatomical connections on the one hand and functional or effective connectivity on the other. The former can be described as quasi-stationery, whereas the latter may be highly dynamic [...]. It appears that dynamic co-operativity is an emergent property of

neuronal assembly organisation in the brain which could not be inferred from single neuron observation. (Greenfield, 1998, p. 217)[6]

In the biological realm we will find several examples of stabilization. For example, the DNA molecule is the basic building block in all living organisms. Although its basic structure is simple (there are only four types of nitrogenous bases in the DNA molecule), it gives rise to an overwhelming abundance of organisms. Very simple rules like "do what your neighbor does" may lead to a variety of collective behaviors of social insects like honey-bees or army ants.

Greenfield (1998) describes how a group of neurons with relative long-lasting connections between them, may engage neuronal assemblies when a particular stimulus is reaching the brain. Depending on the size of the neuronal assembly, different forms of consciousness may arise. Small neuronal assemblies may result in a kind of consciousness that would just be reactive to whatever crossed one's path without a lot of reflection and memories and thoughts. This form of consciousness appears in a child's perspective and in a schizophrenic person. On the other hand abnormally large neuronal assemblies would have the opposite effect. The outside world would appear remote, grey, and distant. This form of consciousness can be found in clinical depression. Thus it appears that a balance between the two extremes of chaos and petrification exist in a developed and healthy brain.

Integrating Construct

In order to move closer to the biological constitution of humankind, I have grounded my arguments in the structure of language and its connections to the conceptual structure as suggested by Jackendoff. The evidences provided from this analysis for cross-modal integration of sensory modalities are strengthened by findings in neurobiology (Shimojo & Shams, 2001). Vision, auditory/hearing, somatosensory/touch, gustatory/taste, and olfactory/smell sensations are integrated into a coherent perception of the environment, enabling an organism to evaluate a situation and act upon it.

In general, different modalities are engaged depending on what modality dominates the perception in the context of a task. Vision has a higher spatial resolution, hence its dominance in spatial tasks whereas audition has a higher temporal resolution, hence its dominance in temporal tasks (ibid, p. 506). However, there is plenty of evidences of integration across modalities. For example, vision may alter speech perception: the sound 'ba' tends to be perceived as 'da' when coupled with a visual lip movement associated with 'ga' (ibid, p. 506). Other examples of sound altering temporal aspects of vision, and vision altered by auditory or tactile sensations have also been reported (ibid, pp. 506-507).

Exactly how this integrating mechanism works is still being investigated. However, it is clear that some kind of mental structure exists where sensory modalities are integrated:

[It is] understandable that evolution has provided multimodal sensory convergence in order to perform space-time integration on the multi-sensory percept, a Gestalt, not on its components prior to their assembly. (Freeman, 2004, p. 525)

This means that there exist, in addition to the linguistic arguments, neurophysiological and cognitive indications that sensory modalities are integrated into spatialization and temporalization as basic modes of action. It is likely that this integration also includes the other activity modalities.

Congruence

Since the neuropsychological stratum is tightly related to the phylogenic evolution of humans, it is not to be expected that we can find substantial evidence in this stratum of congruence between the mind and the environment. However, one aspect is evident: that an organism is only capable of acting on what can be sensed in the environment. Every organism lives in its particular "Umwelt" (von Uexküll, 1983). This is also valid for humans:

As living systems, we have a closed (autopoietic) organization but are interactively open to our environment. The nature and limitations of such interactions are determined primarily by our own physical structure (structure-determined) rather than by the environment. [It] is the structure that determines what can or cannot be a stimulus for the organism – organisms without eyes or the equivalent cannot be triggered by light. (Mingers, 2001, p. 119)

Strata Transition

The cognitive elements in the neuropsychological stratum consist of a large number of simple but highly connected neurons. These receive excitatory or inhibitory inputs and transmit activity to other units according to some function of the inputs. The processing is done in parallel, and the strength of the connections between the units changes according to experiences made from the interaction with the environment. The result of the organism's interaction with the environment is the formation of patterns of connected neurons. The high dimensionality of the neuropsychological stratum makes it implausible that the strengths between connections are independent, since the learning curve would become prohibitively long. It is therefore likely that the connection strengths are correlated according to high-level structures in the conceptual stratum, which has a substantially lower dimensionality.

This raises the interesting issue of where the control of the brain functions lies. The neural patterns and thus the control of its formation cannot be explained in terms of single neurons in the biological stratum. Rather, the control is an emergent phenomenon which appears in the interaction with the environment through the higher strata (e.g. Sperry, 1976). The formation of the control and our consciousness is a two-way movement: an upward causation from the connectionist level and a downward causation from the environment and higher strata (Lemke, 2000). Thus, the formation of the connectionist level is dialectical in character, since it provides the means of apprehending and possibly changing the environment and at the same time is formed by that very environment.

REFERENCES

Chalmers, A. F. (1976). *What is this thing called science?* St Lucia, Australia: University of Queensland Press.

Chen, X. (2003, December). Object and Event Concepts: A Cognitive Mechanism of Incommensurability. *Philosophy of Science*, *70*, 962–974. doi:10.1086/377381

Engeström, Y. (2001). Expansive Learning at Work: toward an activity theoretical reconceptualization. *Journal of Education and Work, 14*(1), 133–156. doi:10.1080/13639080123238

Freeman, W. J. (2004). How and why Brains Create Meaning from Sensory Information. *International Journal of Bifurcation and Chaos in Applied Sciences and Engineering, 14*(2), 515–530. doi:10.1142/S0218127404009405

Gärdenfors, P. (1992). *Blotta tanken.* [The mere thought (in Swedish)]. Nora, Sweden: Nya Doxa.

Gärdenfors, P. (2000). *Conceptual Spaces: The geometry of thought.* Cambridge, MA: MIT Press.

Gärdenfors, P. (2000b). *Hur Homo blev Sapiens. Om tänkandets evolution.* [How Homo became Sapiens. About the evolution of thinking (in Swedish)]. Nora, Sweden: Nya Doxa.

Greenfield, S. (1998). How Might the Brain Generate Consciousness? In S. Rose (Ed.) *From Brain to Consciousness - Essays on the new science of the mind* (pp. 210-227). London: Penguin Books.

Hampe, B. (2005). Image Schemas in Cognitive Linguistics. In B. Hampe (Ed.), *From Perception to Meaning, CLR 29* (pp. 1–13). New York: Mouton de Gruyter.

Jackendoff, R. (1983). *Semantics and Cognition.* Cambridge, MA: MIT Press.

Jackendoff, R. (1990). *Semantic Structures.* Cambridge, MA: The MIT Press.

Lakoff, G. (1987). *Women, Fire, and Dangerous Things: What Categories Reveal about the Mind.* Chicago: University of Chicago Press.

Langacker, R. W. (1988). The nature of grammatical valence. In B. Rudzka-Ostyn (Ed.), *Topics in Cognitive Linguistics* (pp. 91–125). Amsterdam: John Benjamins.

Lemke, J. (2000). Material Sign Processes and Ecosocial Organization. In P.B. Andersen, C. Emmeche, & N.O. Finnemann-Nielsen (Eds.), *Downward Causation: Self-organization in Biology, Psychology, and Society* (pp. 181–213). Aarhus, Denmark: Aarhus University Press. Retrieved Oct 5, 2008, from http://www-personal.umich.edu/~jaylemke/jll-new.htm

Mingers, J. (2001). Embodying information systems: the contribution of phenomenology. *Information and Organization, 11*(2), 103–128. doi:10.1016/S1471-7727(00)00005-1

Parsons, J. (1996). An Information Model Bases on Classification Theory. *Management Science, 42*(10), 1437–1453. doi:10.1287/mnsc.42.10.1437

Plaut, D. C. (2002). Graded modality-specific specialization in semantics: A computational account of optic aphasia. *Cognitive Neuropsychology, 19*(7), 603–639. doi:10.1080/02643290244000112

Rogers, Y., & Ellis, J. (1994). Distributed Cognition: an alternative framework for analysing and explaining collaborative working. *Journal of Information Technology, 9*(2), 119–128. doi:10.1057/jit.1994.12

Rumbaugh, J., Blaha, M., Premerlani, W., Eddy, F., & Lorensen, W. (1991). *Object-Oriented Modeling and Design.* Upper Saddle River, NJ: Prentice-Hall International, Inc.

Shimojo, S., & Shams, L. (2001). Sensory modalities are not separate modalities: plasticity and interactions. *Current Opinion in Neurobiology, 11*(4), 505–509. doi:10.1016/S0959-4388(00)00241-5

Sperry, R. W. (1976). Mental phenomena as causal determinants in brain function. In G. Globus, G. Maxwell, & I. Savodnik (Eds.), *Consciousness and the Brain* (pp. 163–177). New York: Plenum Press.

Vološinov, V. N. (1986). *Marxism and the Language of Philosophy*. London: Harvard University Press.

Von Uexküll, J. (1983) An introduction to Umwelt. *Semiotica, 2001*(134), 107–110. From J. von Uexküll (1936). Niegeschaute Welten, 11–17. Berlin: S. Fischer Verlag. Excerpt selected and translated by Gösta Brunow.

Wertsch, J. V. (1991). *Voices of the mind: a sociocultural approach to mediated action*. Cambridge, MA: Harvard University Press.

Wittgenstein, L. (1953). *Philosophical investigations*. Oxford, UK: Blackwell.

Zacks, J. M., Braver, T. S., Sheridan, M. A., Donaldson, D. I., Snyder, A. Z., & Ollinger, J. M. (2001). Human brain activity time-locked to perceptual event boundaries. *Nature Neuroscience*, *4*(6), 651–655. doi:10.1038/88486

ENDNOTES

1. Some branches of cognitive science have started to move away from this position (see for example Rogers & Ellis (1994) and Wertsch (1991).
2. In Lemke (2000) a distinction is made between "topological and typological" levels. These correspond roughly to the connectionist and linguistic levels. However, like Gärdenfors I believe that there is a need for the intermediate, conceptual level.
3. Gärdenfors notes that the symbolic paradigm in cognitive sciences has by and large been a failure due to several reasons, one of them being the lack of notion of contextuality.
4. An interesting observation is that the conceptual representations (concepts, properties, domains, objects, regions, relations, etc.) are similar to the constructs of object-oriented relational models (types, type hierarchies, attributes, objects, relations, etc.). This implies that an object-oriented model (e.g. Rumbaugh, Blaha, Premerlani, Eddy, & Lorensen, 1991) may be well aligned with the conceptual spaces in our brains. Thus, the emergence of object-oriented modeling paradigm may be seen as an indication of congruence between our cognitive spatial capabilities and a preferred way of modeling systems.
5. For example, a cat can "simulate" a mouse behind a hole in the wall and sit waiting for hours for it. A snake, on the other hand, does not have this capability. As soon as the prey disappears out of the snake's sensory range, the attention of the snake ceases. If the mouse suddenly appears again, the snake senses this as a new prey (Gärdenfors, 2000b, p. 45).
6. It can be noted that this quotation is in line with the stance of dialectical interaction as a fundamental epistemological category. Dynamic cooperation brings out new qualities even in the connectivist stratum.

Chapter 7
Operationalizing the Theory

By operationalization, I refer to the transformation of the theoretical elements in ADT into mediational means that can be manipulated, measured, or observed in order to construct an activity domain. In line with the correspondence between mind and the socio-cultural environment, such means should be aligned with the activity modalities as much as possible to alleviate the construction of communal meaning. Moreover, the operationalized elements must be treated as dialectically related to each other; a change in one will impact all the others. The proposed set of operationalization means is:

- An entity-relationship information model operationalizing spatialization.
- An information interaction model operationalizing temporalization.
- A stabilizing core operationalizing stabilization.
- An elaborated specification-based data model operationalizing transition.
- The domain construction strategy operationalizing enactment.
- A coordination information system (cIS) as a mediational means for managing coordination items.

DOI: 10.4018/978-1-60566-192-6.ch007

THE INFORMATION MODEL

Spatialization is operationalized by an *information model,* which signifies what information entities are subject to coordination and how these entities are related to each other[1]. The information model frames the spatial context of the domain; thus, it is also manifests contextualization. Moreover, since the information model captures the structure of informative things, the model expresses the *information architecture* of the domain.

Information models are usually conceptualized as some variant of entity-relationship models (Chen, 1976), such as the Object Modeling Technique (OMT: Rumbaugh, Blaha, Premerlani, Eddy, & Lorensen, 1991) and the Universal Modeling Language (UML: Jacobson, Christerson, Jonsson, & Övergaard, 1992). In order to alleviate the construction of communal meaning, the nomenclature should be easy to comprehend. UML might be advocated since it is a de facto standard for modeling software systems. However, the formalism of UML is hard to grasp for non-experts, which is a strong argument for using a simpler nomenclature such as OMT.

The information model needs to be defined at various levels of granularity depending on the purpose of the model. In Table 1, an overview of these levels is shown, derived from practical experience at Ericsson. The term "items" refers to anything needed to define the model, such as type definitions, relations, attributes, state sets, cardinalities, rule for revising elements, etc.

The first level focuses on achieving common understanding about the model across cooperating activity domains. A fairly small set of items is used, around 20. The next level, which comprises about 50 items, concerns common understanding about what should be subject to coordination within a specific domain. In Figure 1 an example of a level II information model is given.

The third level is detailed enough to specify how the model should be implemented in an IS. The number of items at this level is in the order of several hundreds. In Figure 2, a level III model corresponding to the requirement context in Figure 2 (encircled) is shown. This model is directly implementable in the coordination information system (cIS) Matrix.

It is at level III that the mathetic phase of meaning construction is emphasized by a constant iteration between the model and its implementation in the IS. The level IV, finally, comprises IS internal items needed for making the IS fully operational such as scripts, report generators, triggers, queries, etc. Here, the pragmatic phase of meaning construction is accentuated. More than 1000 items are engaged at this level.

Somewhere between the third and fourth level, the momentum driving the implementation turns from the information model to the IS. The model loses gradually its role as a means for achieving common understanding, and is replaced by the implementation in the IS. It is quite easy to realize that the effort of achieving common understanding about all items, at all levels across all domains is prohibitive.

Table 1. Various levels of granularity in an information model

Level	Purpose	Characteristics	Number of items
I. Cross-activity domain	Common understanding	Mandatory items	~20
II. Activity domain	Common understanding	Domain specific	~50 / activity domain
III. Activity domain	IS specification	Detailed rules	~100 / activity domain
IV. Activity domain	IS implementation	IS specific items	> 1000 / activity domain

Figure 1. A level II information model for the A-domain at Ericsson (2001) (Taxén, 2005. ©Wiley Inter Science. Used with permission)

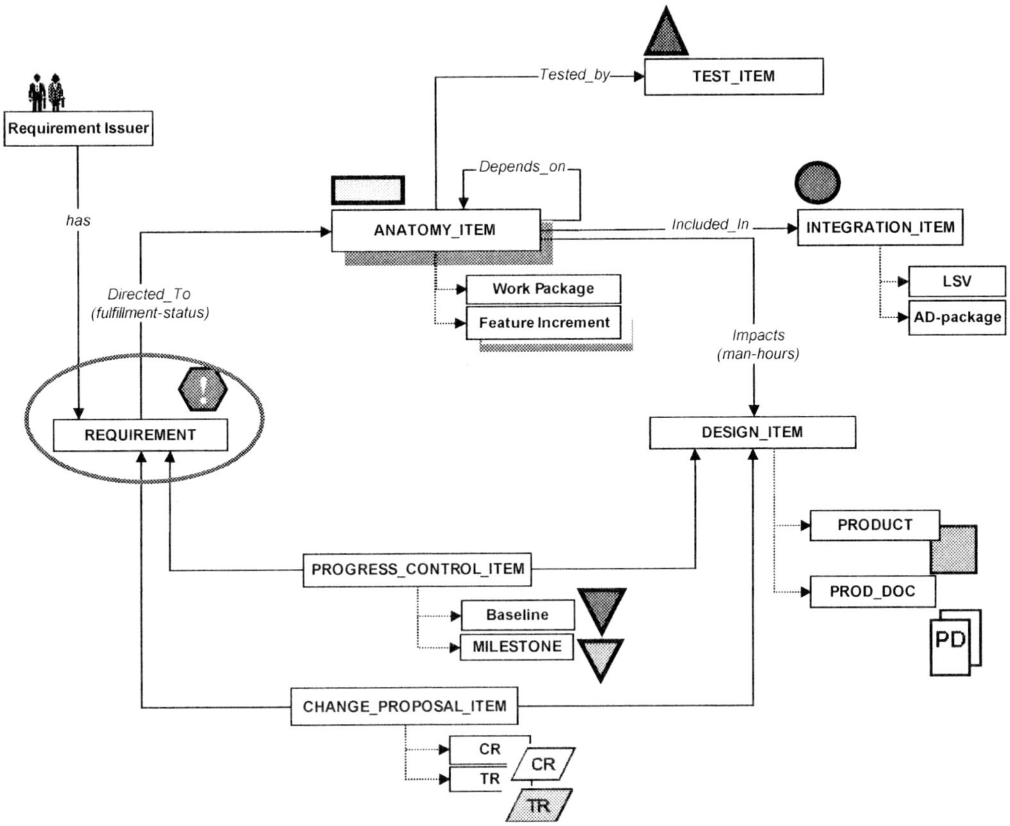

Figure 2. An information model at level III

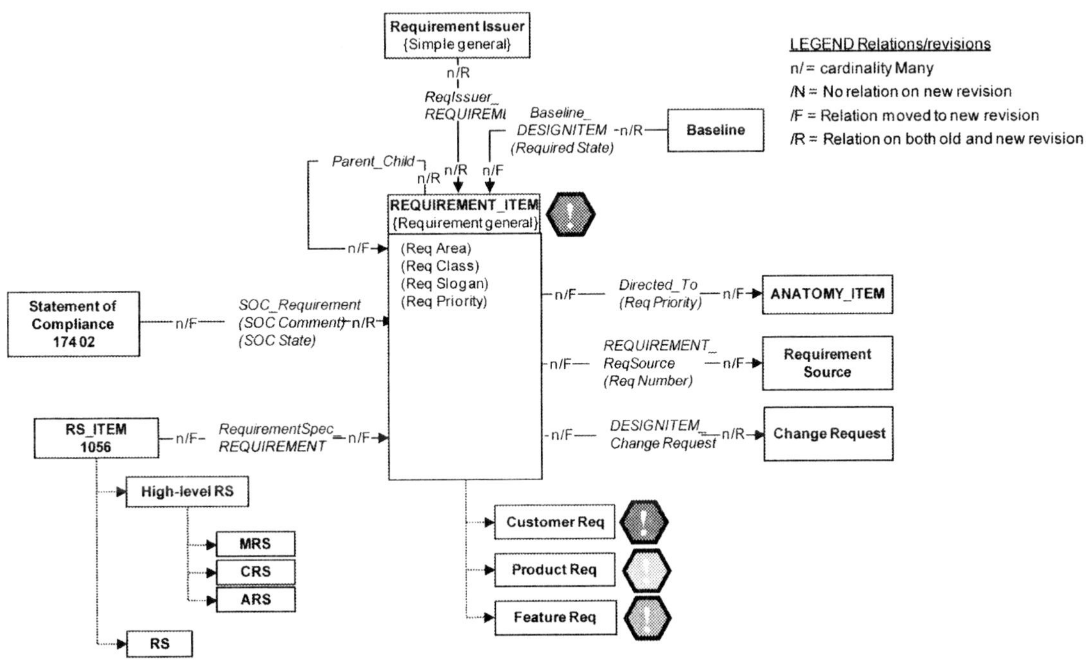

For example, at Ericsson, three domains were constructed: the S-, A-, and L domains (see Section 1). Consolidating these into one domain (the C domain) would imply achieving common understanding of about 3000 items. This is simply not achievable, which is one reason why several activity domains cannot possibly attain commonality down to every single detail, spanning all levels. Commonality needs to be constructed at the proper level but not beyond that.

THE PROCESS MODEL

Temporalization is operationalized as a *process model*. Such models are commonly based on an activity-centric view of processes that emphasizes dependencies between activities (see Figure 3). In this sense, the process model manifests coordination as defined by Malone & Crowston (1994). The progression of the process is usually signified by statuses of activities such as "Idle", "Started", "Done", etc.

A disadvantage with activity-based models is the difficulty of seeing the progression of entities. For example, the chain of events causing the entity A in Figure 3 to change its state from α' to α'' is almost impenetrable in spite of the simplicity of the example. One reason for this is that the entities are depicted in several places in the model. From an activity modality point of view, manifestations of spatialization (the entities) are entangled with manifestations of temporalization (the activities and the state changes).

Alternatively, processes may be expressed as *entity-based* models, where progress is signified by entity state rather than activity state (Humphrey & Kellner, 1989). An example of such a model is the information interaction model (IIM) (see Figure 4).

As with activity-based models, IIMs consist of entities, lifecycle statuses, activities, and dependencies between entities and activities. Entities are shown as boxes piled up vertically to the left. Lifecycles of entities are indicated by horizontal lines. At the bottom of the diagram, activities are outlined. The

Figure 3. An activity-based process model

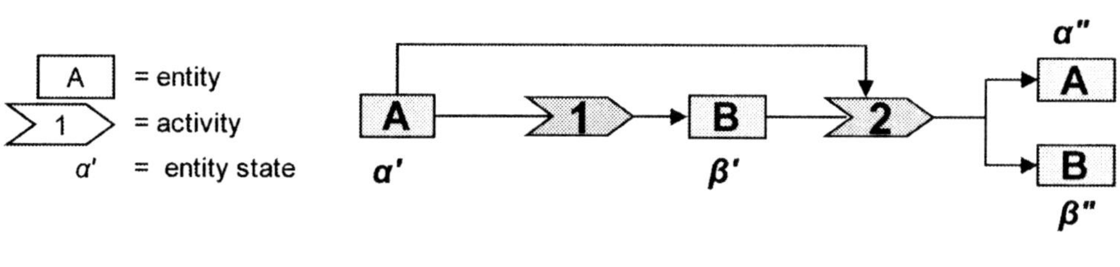

Figure 4. The structure of information interaction models

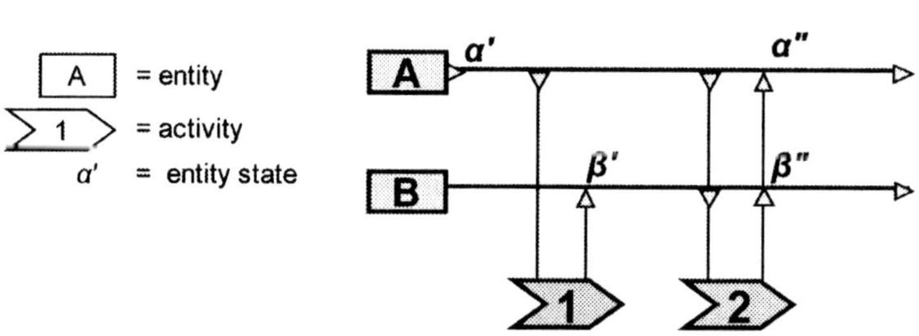

dependencies between activities and entities are signified by vertical arrows showing inputs to (downwards pointing) and outputs from (upwards pointing) activities. The junction of an upwards pointing arrow and an entity line implies a state change of the entity.

The origin of IIMs is obscure, but they were frequently used at Ericsson in the 1990s. IIMs have several appealing qualities. First, the progress of an entity is easy to follow. Second, entity statuses provide a more stable control mechanism than the activity statuses in volatile situations. When things change rapidly, it is more important to manage information rather than activities. At the end of the day, it is the state of the information that counts, not the activities that have been carried out. Third, the dependencies between entities and activities are easier to comprehend. From the activity modality perspective, spatialization and temporalization, as well as their interdependencies, are clearly separated, which facilitates the construction of communal meaning about the process. Thus, the preferred modeling notation in operationalizing temporalization is the IIM.

THE STABILIZING CORE

Operationalization of stabilization implies the definition of elements pertinent to the stabilization of the activity domain. Such elements may take many forms: regulatory documents; business rules implemented in ISs; common IT platforms; naming conventions; and so on. Often, there are specific organizational units dedicated to defining and maintaining mandatory business rules. In ADT, the ensemble of stabilizing elements is called the *stabilizing core*.

THE TRANSITION MODEL

Transition is operationalized by a *transition model*. As a basis for such a model, the specification-based data model (SBDM) suggested by Gandhi & Robertson (1992) may be used. The basic feature of this model is a recursive structure, which makes the SBDM particularly suitable for modeling transition.

In Figure 5 the SBDM is illustrated. A specification component expressed as description, behavior, and interface objectives is implemented by an implementation component that in turn needs other specifications in order to be realized. Thus, the model "provides alternating layers of specifications and implementations. Each specification may have many alternative implementations. A specification, by itself, does not describe the internal structure for the entity it specifies. It is the implementation that details the internal structure" (ibid, p. 6).

The primary purpose of the SBDM is to model configurations of engineering products implemented in many technologies. However, the SBDP can be adapted to model the transition between two domains (see Figure 6):

The recursive core of SBDM is kept, and the specification/implementation components are replaced with the external/internal facets of the activity domain. This requires that:

- The specification component is apprehended as the external side of the domain providing a resource needed in another domain. This resource is expressed in the ideology of the external domain.

Figure 5. The Specification-based Data Model (after Gandhi & Robertsson, 1992)

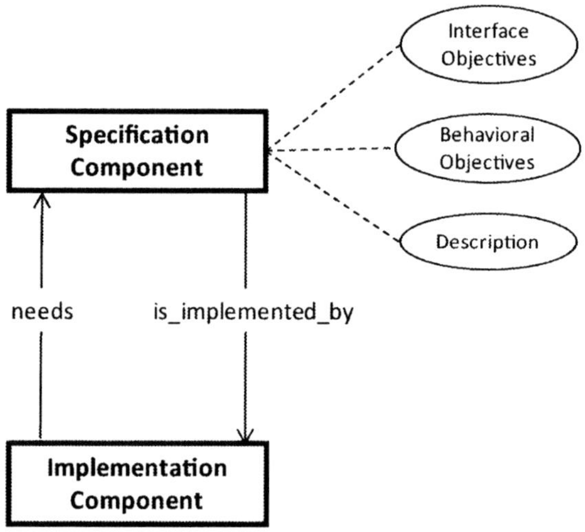

Figure 6. Transition between two activity domains

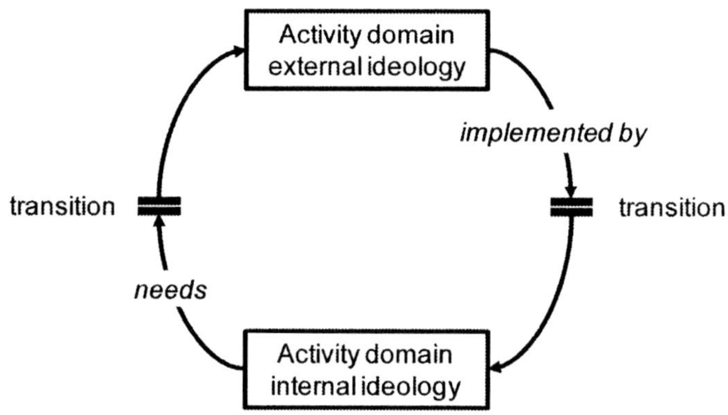

- The implementation component is apprehended as the same resource expressed in the internal ideology of the providing domain.
- A mapping can be defined between the internal and external ideologies.

In this way, a recursive structure of cooperating activity domains can be modeled. As pointed out in Section 1, however, the transition model was never put to use in the Ericsson practice since the transition between the S-, A-, and L domains remained unattended.

THE COORDINATION INFORMATION SYSTEM

The items in the three models described are type categories that signify persistent spatial, temporal, and transitional coordination patterns in the activity domain. These patterns are recurrently employed in the execution of coordination tasks. During each execution of a particular task, instances of the types

Figure 7. A view of instances managed in Matrix

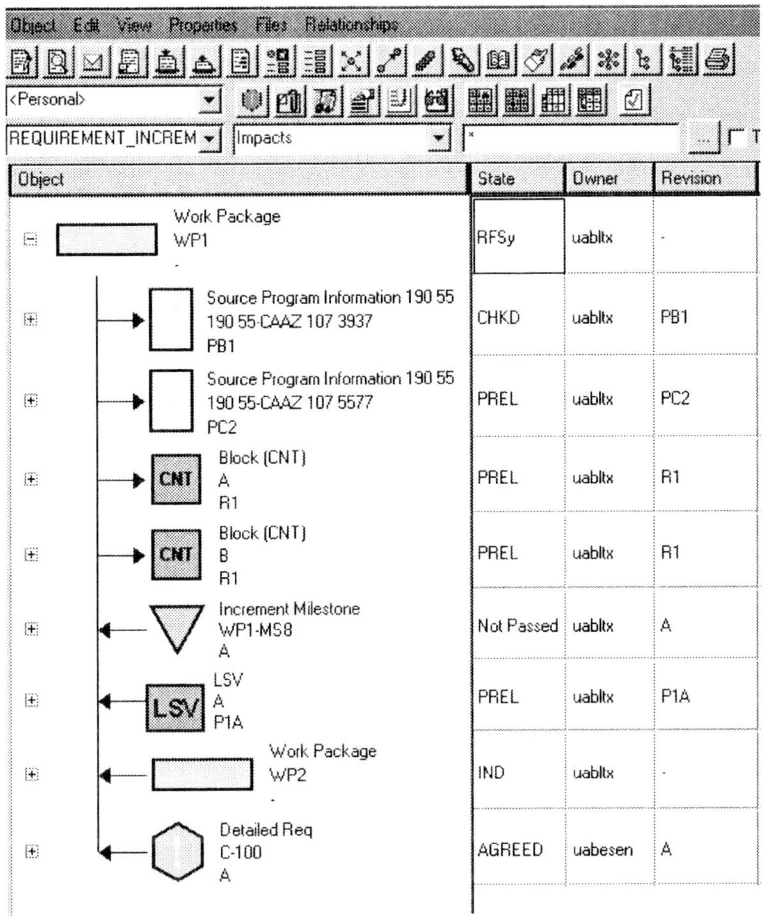

are managed in the *coordination information system* (cIS). This system must have the same qualities as Matrix in order to enable the construction of the domain. Above all, the implementation of the information model must be extremely easy to change. In Figure 7 a view of instances and their relations corresponding to the information model in Figure 1 is shown.

THE DOMAIN CONSTRUCTION STRATEGY

The domain construction strategy (DCS) is a systematic way to construct the activity domain. A central target of this strategy is to achieve common understanding among actors. In essence, the DCS is an enactment process, meaning that mediational means turn into coordinative resources as the strategy is implemented. In other words, the outcomes of the DCS are three models manifesting the spatial, temporal and transitional modalities, a stabilizing core manifesting the stabilization modality, IS support for coordination, and common understanding about how to employ these means. In essence, the DCS is an experiential learning approach (Kolb, 1984), which can be illustrated as in Figure 8 (for readability, only the information model and the cIS are shown):

Figure 8. The experimental leaning approach in DCS

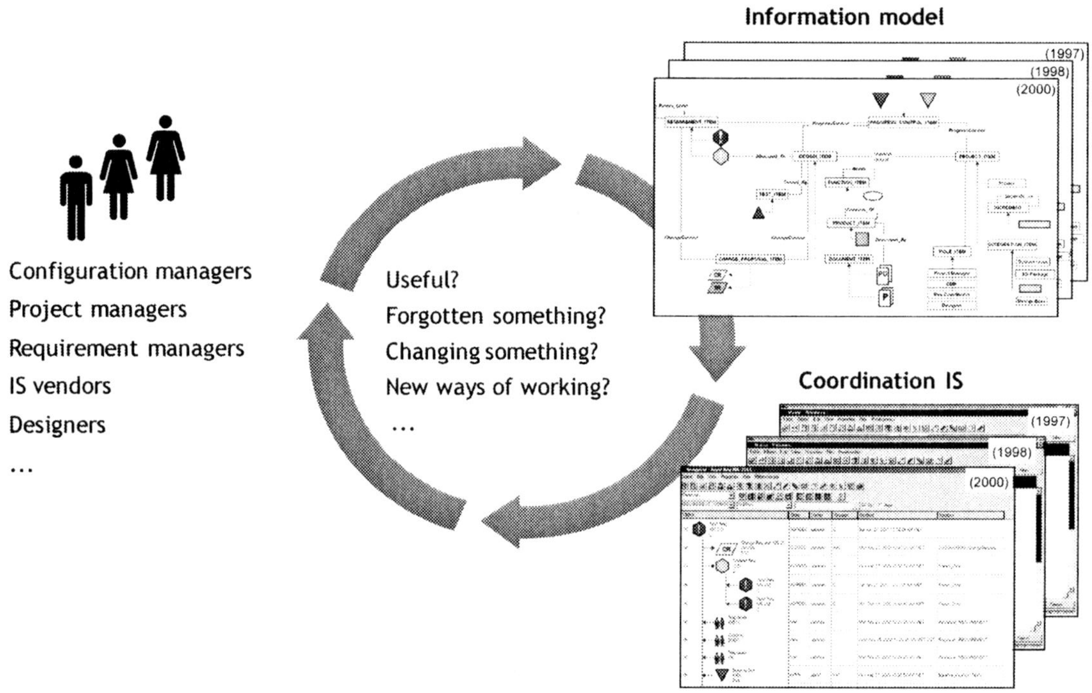

Figure 9. The domain construction strategy

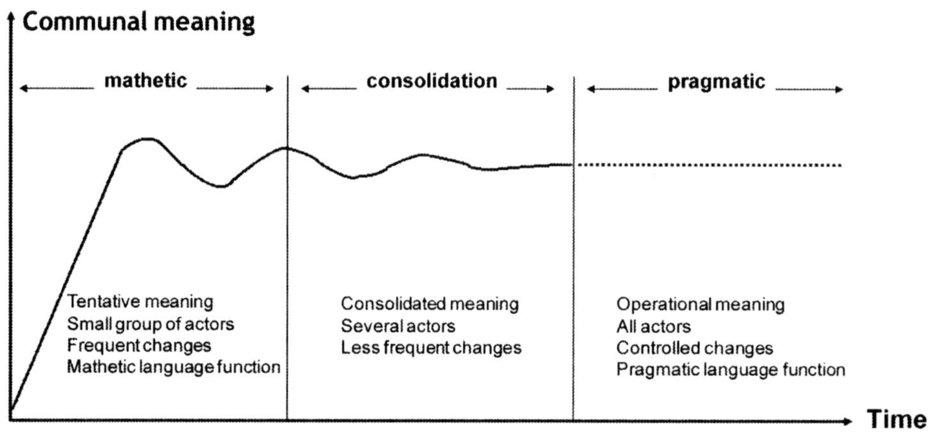

The evolution of the DCS proceeds in three phases that I have called the *mathetic*, *consolidation*, and *pragmatic* ones. The terms mathetic and pragmatic are adopted from Halliday (1975), who distinguishes between pragmatic and mathetic functions of language. Pragmatic functions involve coordination of actions when common understanding is established, while mathetic functions have to do with the construction of communal meaning. In Figure 9 the evolution of the DCS is illustrated.

During the evolution, a gradual shift is made from the mathetic to the pragmatic functions of language. Based on experiences from Ericsson, each phase proceeds as follows:

- *Mathetic*: In this phase, the initial construction of the domain is carried out. The focus is on the mathetic function of language. The main purpose is to achieve a tentative common understanding. The work is carried out in a "daily build" manner by a small "task force". Provisional models, rules, etc., are suggested and implemented in the cIS. The results are discussed and evaluated with respect to usefulness. Changes are suggested to the models and implemented anew. This iterative process is continued until a working consensus is achieved.
- *Consolidation:* The purpose of this phase is to boost the trust of the feasibility of the domain as constructed in the mathetic phase. Key issues are getting all actors to trust the data in the IS, and to validate the performance of the system. This may be done in a "sharp" development project, that is, a project that develops a product for a customer. The task force is still driving the construction. Additional user roles around the project are involved and immediate, personalized support is provided. The construction of the domain in the consolidation phase progresses by controlled changes. No major reconstruction is allowed at this stage.
- *Pragmatic*: In this phase, actors in several projects are included. As in the consolidation phase, the construction is done by controlled changes, however now in a formalized way. The domain may also be expanded to include new types of information entities. The focus in this phase is on the pragmatic function of language.

In summary, the DCS aims at the construction of the entire activity domain. The intention is that communal meaning construction and construction of artifacts should proceed simultaneously. Since the activity domain is considered to be in constant evolution, the DCS should be conceived as a continuous activity domain development process. A similar approach for IS development has been suggested by Truex, Baskerville, & Klein (1999).

REFERENCES

Chen, P. (1976). The Entity Relationship Model - Towards a Unified View of Data. *ACM Transactions on Database Systems*, *1*(1), 9–36. doi:10.1145/320434.320440

Gandhi, M., & Robertson, E. L. (1992). A Specification-based Data Model. *Indiana University Computer Science Department Technical Report TR344*. Bloomington, IN: Indiana University. Retrieved Sept 30, 2008, from http://www.cs.indiana.edu/ftp/techreports/index.html

Halliday, M. (1975). *Learning How To Mean*. London: Edward Arnold.

Humphrey, W., & Kellner, M. (1989). Software Process Modelling: Principles of Entity Process Models. *CMU/SEI-89-TR-2*. Pittsburgh, PA: Carnegie Mellon University, Software Engineering Institute.

Jacobson, I., Christerson, M., Jonsson, P., & Övergaard, G. (1992). *Object-oriented software engineering: a use case driven approach.* Wokingham, UK: ACM Press, Addison-Wesley.

Kolb, D. A. (1984). *Experiential Learning: Experience as the Source of Learning and Development.* Englewood Cliffs, NJ: Prentice Hall.

Malone, T., & Crowston, K. (1994). The Interdisciplinary Study of Coordination. *ACM Computing Surveys*, *26*(1), 87–119. doi:10.1145/174666.174668

Rumbaugh, J., Blaha, M., Premerlani, W., Eddy, F., & Lorensen, W. (1991). *Object-Oriented Modeling and Design*. Englewood Cliffs, NJ: Prentice-Hall International, Inc.

Taxén, L. (2005). A Socio-technical Approach Towards Alignment. *Software Process Improvement and Practice*, *10*(4), 427–439. doi:10.1002/spip.245

Truex, D. P., Baskerville, R., & Klein, H. (1999). Growing Systems in Emergent Organizations. *Communications of the ACM*, *42*(8), 117–123. doi:10.1145/310930.310984

ENDNOTE

1 In information modeling it is customary to distinguish between conceptual models, information models, and data models. The purpose of these models are: the conceptual model – to achieve common understanding about terms; the information model – to specify what items to manage; the data model – to specify how to implement the information model in an IS. This classification is not used in the operationalization of the ADT. The information model fulfils all these purposes at the same time; however at different levels of granularity.

Chapter 8
Positioning Against Other Theories

In this chapter, the ADT is compared to two closely related theoretical strands: complex systems' theory and practice-based theories.

COMPLEX SYSTEMS THEORY AND ADT

Complex systems theory (e.g. Bar-Yam, 1997) is a rich and vital research area that has a huge potential for organizational research (Lewin, 1999). In particular, the study of complex adaptive systems (CASs) provides many insightful results concerning the adaptability and strategic management of organizations in turbulent environments (ibid).

There is no clear definition of complexity that is generally accepted (see e.g. Morel & Ramanujam, 1999). However, some characteristics can be identified. Complex systems are self-organizational; order emerges naturally in open systems that exchange resources with its environment (Lewin, 1999). They consist of interconnected parts that are often complex systems themselves. The behavior of complex systems is an emergent property "contained in the behavior of the parts if they are studied in the context in which they are found" (Bar-Yam, 1997, p. 10). A complex organization consists of "a set of interdependent parts, which together make up a whole that is interdependent with some larger environment" (Anderson, 1999, p. 216). More specifically, complexity in organizations has been defined with respect to the number of levels in the organizational hierarchy, the number of departments in the organization, and the number of geographical locations (ibid, p. 216).

CASs are special cases of complex systems. A general definition is given by COSI (2009): "Macroscopic collections of simple (and typically nonlinearly) interacting units that are endowed with the ability to evolve and adapt to a changing environment." In the literature, the following characteristics of CASs can be identified:

DOI: 10.4018/978-1-60566-192-6.ch008

- *Agents with schemata*: A CAS is a "dynamic network of many agents (which may represent cells, species, individuals, firms, nations) acting in parallel, constantly acting and reacting to what the other agents are doing" (Wikipedia, 2009c).
- *Self-organizing networks sustained by importing energy*: Agents are "partially connected to one another, so that the behavior of a particular agent depends on the behavior (or state) of some subset of all the agents in the system [...].Maintaining a self-organized state requires importing energy into the system" (Anderson, 1999, p. 219-220).
- *Coevolution to the edge of chaos*: Agents coevolve with one another. Each agent adapts to its environment by striving to increase a payoff or fitness function over time. The equilibrium that results from such coevolution lies at the edge of chaos (ibid, p. 220).
- *System evolution based on recombination*: CASs "evolve over time [...]. New agents may be formed by recombining elements of previously successful agents. Furthermore, the linkages between agents may evolve over time [...]. CASs can contain other complex adaptive systems, as, for example, organisms have immune systems" (ibid, p. 220).
- *Self-similarity*: Self-similarity indicates that a CAS system is invariant under a change of scale (Morel & Ramanujam, 1999). Self-similarity is closely related to fractals, where the structure of a system on a coarse scale is repeated on finer scales.

In the following, I will take this characterization of CASs as a starting point for exploring possible connections between CAS and ADT. This is done against the background of the previously reported complexity inherent in both telecom systems and projects developing these systems.

Agents with Schemata

In CAS, agents are usually characterized as following a fixed set of simple rules. A nice example is the collective behavior of a flock of birds, where each bird follows simple rules such as (Reynolds, 2009):

- Separation: steer to avoid crowding local flock mates
- Alignment: steer towards the average heading of local flock mates
- Cohesion: steer to move towards the average position of local flock mates

When applying this feature of CASs to organizations, it is clear that the rule concept has to be refined. Human actions cannot be reduced to automatons following a few simple rules. First, rules may be elaborated to *schemata*:

Each agent's behavior is dictated by a schema, a cognitive structure that determines what action the agent takes at time t, given its perception of the environment (at time t, or at time t - k if theoretical considerations suggest applying a lag structure). Different agents may or may not have different schemata (depending on one's theory), and schemata may or may not evolve over time. (Anderson, 1999, p. 219)

Second, an element of hierarchy can be introduced: "For example, in a model of an organization, agents might be individuals, groups, or coalitions of groups" (Anderson, 1999, p. 219). This enables agents to evolve different schemata depending on the level of hierarchy in the organization. A further elaboration is to consider the evolution of schemata as a result of interactions among agents, resulting in a social order based on shared interpretations and meanings (ibid, p. 221).

The ADT Connection

The agents in ADT are actors working on the work object of the activity domain. In this activity, the capabilities of means and actors are enacted into resources that make sense in transforming the work object. Enactment is a dialectical process that brings about objectivated manifestations in the minds of the actors and objectified manifestations in the domain along the activity modalities. Consequently, in the ADT perspective, "schemata" correspond to objectivated manifestations.

In ADT, "individuals, groups and coalitions of groups" are separate kinds of entities. Objectivations/schemata are intrinsically bound to individuals. Groups and coalitions of groups cannot have schemata. Instead, these entities would be considered as activity domains, provided that a clear motive and work object can be associated with them. The hierarchical element is conceptualized in ADT as nested activity domains; a domain may use other domains to achieve its outcome.

Interactions in ADT are taken place both between agents in an activity domain, and between activity domains. Different objectivations/schemata will be enacted in different domains, providing domain specific communal meanings. However, in order for activity domains to cooperate, communal meaning must be enacted across domains as well.

Multilevel and Integrative

CAS models of organizations are inherently multilevel, since "order is considered an emergent property that depends on how lower-level behaviors are aggregated. Accordingly, they respond well to contemporary calls for more integrative, cross-level research in organization science" (Anderson, 1999, p. 220). Evolution occurs through a "nested hierarchy of selective systems" (ibid, p. 222). Order requires that the components of systems are partly connected:

Systems in which every element is connected to each other in a feedback loop are hopelessly unstable (Simon 1996). Instead, CASs tend to form a decompositional hierarchy, in which elements are loosely coupled with one another (Simon 1996). Most components receive inputs from only a few of the system's other components, so change can be isolated to local neighborhoods. (Anderson, 1999, p. 222)

Agents are connected to one another by feedback loops where "each agent observes and acts on local information only, derived from those other agents to which it is connected" (Anderson, 1999, p. 220). Organization and order emerges naturally in systems where the actions of actors are governed by rules and local information. Such systems are self-organizing as long as energy is imported to the systems; there is no need for central control. Transferring these characteristics to organizations, social entities self-organize "as long as their members contribute work; this is why informal structures emerge and persist in a way that is remarkably robust to changes in the formal organizational structure" (ibid, p. 222).

The emphasis on multilevel and local aspects of CASs makes it interesting to elaborate a contextual and interpretative approach to organizations:

One promising avenue of inquiry that the CAS perspective opens up is exploration into how ideas, initiatives, and interpretations form an internal ecology within an organization. As McKelvey (1997) points out, organizational scholars have emphasized macroevolution (within organizational populations) at the expense of microevolution (within organizations). (Anderson, 1999, p. 221)

Another striking property of CASs is that they are self-similar in the same sense as fractals: "by magnifying one part of an object, we find the same structure as of the original object" (Bar-Yam, 1997, p. 259). An approach based on a fractal perspective is interesting in organizational theory building:

[This] kind of approach could consist in detecting or suspecting similarities between different scales and building a mathematical object to represent it, which can be used to establish in a quantitative way that fractals can be used to model this self-similarity. This is an area of great significance to [organization theory] as the issue of crosslevel analysis is a recurring theme (for example, research on the relationships between individual learning, group level learning, and organizational level learning. (Cohen 1996)) (Morel & Ramanujam, 1999, p. 281)

The ADT Connection

The activity domain provides a conceptualization of locality since it frames the context in which an "internal ecology" emerges as determined by the work object of the domain. The "internal ecology" is equal to the ideology enacted in the domain. Moreover, since the activity domain is recursive (activity domains may employ other domains), the multi-level aspect of the organization can be expressed in terms of a constellation of activity domains. Furthermore, since the internal structure of each domain is the same in every domain (see Figure 25, p. 88), the activity has a fractal structure (if you "open up" a domain, you will see the same structure as for the original domain). Altogether, this means that the activity domain may be one way to approach organizations from a CAS perspective, where multilevel, locality and self-similarity aspects are in focus.

Co-Evolution on the Edge of Chaos

An interesting result from applying CAS ideas to organizations is that adaptative organizations may be conceptualized as having evolved structures that poise the organization on the edge of order and chaos (Brown & Eisenhardt, 1997, p. 31). The most effective organizations implement strategies that maintain equilibrium between order and disorder. The structure of such organizations is "sufficiently rigid so that change can be organized to happen, but not so rigid that it cannot occur. Too little structure makes it difficult to coordinate change. Too much structure makes it hard to move" (ibid, p. 29). Brown & Eisenhardt (ibid) also point out that sustaining a balanced state is a challenging managerial task, since constant supervision is needed to avoid the organization from slipping into pure chaos or pure structure.

In their investigation of what characterizes successful, continuously changing organizations Brown & Eisenhardt (ibid, p. 3) found three key properties:

- "Semistructures" that balance between order and disorder.
- "Links in time" that direct attention simultaneously to different time frames and the ties between them.

- "Sequenced steps", which is the recipe by which these organizations are created over time.

In semistructured organizations some features, such as responsibilities, key meetings, project priorities, and time intervals between projects are strictly regulated. Other aspects are given complete freedom, such as the design process used. Plans for the future are not rigid, nor reactive; rather a strategy of "probing" the future using experiments, strategic alliances, and the like, is effectuated.

Another hallmark of successfully adapting organizations is choreographed transitions at predicted time intervals from current to future projects. These transitions or "links in time are neither haphazard nor rigid connections between present and future" (ibid, p. 28). The choreographed transition can be likened with extensively routinized procedures, where each actor has a specific rule; much like the pitstop in car racing. The established time interval creates a rhythm that synchronizes people's actions, maintaining a sense of urgency, and eventually giving rise to a "flow" that focus attention on the task at hand.

"Sequenced steps" refers to the fact that continuously changing organizations rely on "growing" the organization rather than adhering to road-maps of the future. Starting with some stable parts, the near future is embarked upon, evaluated, and reacted on. Linkages across time is established to take the organization from past to future. Strategic equilibrium over time is a combination of frequent small changes that occasionally cumulate into radical strategic innovations, changing the terms of competition fundamentally (Anderson, 1999, p. 224). In contrast, organizations that rely heavily on planning (making "to-be" scenarios) are often rammed by unforeseen events. The gist of the "sequenced step" property is that adaptation must be evolved, not planned.

The ADT Connection

One immediate connection to ADT is the possibility to articulate "semistructures" as activity modalities. The modalities provide structure to the activity domain in the sense that these must be enacted into both objectivated manifestations in the minds of the actors, and as objectified manifestations in the domain. Driving the organization towards equilibrium between chaos and order implies that some expressions of the modalities are regulated across the organization, while other expressions remain the responsibility of each domain. Examples of regulated expressions are rules for naming and revising products. These should be the same for all domains. In contrast, the ideology and capabilities enacted in each domain as determined by the work object should not be regulated. For example, enforcing a common development process (the temporalization modality) in all domains is most certainly counterproductive. In essence, the drive towards equilibrium concerns the stabilization modality, as illustrated by Figure 31, p. 101. Stabilization influences all modalities, and together these form targets for establishing and maintaining equilibrium.

A particular interesting connection to ADT is the "links in time" for transition between the past and the future. The transition activity modality was primarily focused on the transition between the ideologies in different activity domains at a certain point in time. This kind of transition enables the coordination of activities. However, the property of "links in time" in successfully adapting organizations indicates that transition can be envisaged also in the temporal modality. Thus, transition has both a synchronic aspect (concerned with events existing in a limited time period and ignoring historical antecedents) and a diachronic aspect (of, relating to, or dealing with phenomena (as of language or culture) as they occur or change over a period of time).

An additional observation is that the rhythm introduced by the predicted time intervals is, for example, explicitly introduced in the One-Track agile software development process currently being used at Ericsson (described in the Section *Anatomy-centric agile development of software*, p. 180). A strict planning is enforced to release a complete build of an entire radio network every fifth week. However, the actual development of the software is not controlled centrally. The One-Track process is a nice example of balancing stable and loose elements in an organization.

The "sequenced steps" property has, as far as I can tell, no direct connection to ADT. However, it can be noted that this property aligns well with the epistemology of praxis as described in Section 2. The embarking on a mission to change the organization will inevitably change any conception of the future by the mere actions taken to implement the change. Thus, the conception of a pre-planned, steady progression from an "as-is" state to a "to-be" state is based on an illusion.

Complexity and Simplicity

Complexity might be a disguise for more simplistic structures. The central task of natural science is to "make the wonderful commonplace; to show that complexity, correctly viewed, is only a mask for simplicity; to find patterns hidden in apparent chaos" (Simon, 1996, p. 1). When organizations face complex, highly interdependent patterns of activity, there is a need for efficient communication in order to coordinate actions. Efficient communication can be provided by means that convey large amounts of information with relatively few symbols (March & Simon, 1958, p. 184). An example of such a means is the blueprint, whose symbols have definite and common meanings in organizations where blueprints make sense.

Another point is that limited and simplified structures enable people to comprehend fast-changing environments in which it is easy to be confused and make mistakes (Brown & Eisenhardt, 1997, p. 15). Such structures provide the overarching framework without which there are too many degrees of freedom (ibid, p. 16). Thus, efforts to manage complex situations might be better grounded by laying bare the essence of complexity.

The ADT Connection

The structure of the activity domain can be seen as a simplified blueprint of all kinds of organizational manifestations such as units, departments, teams, groups or any other social constellation of actors working together on a work object. As such, the activity domain makes it possible to comprehend the essence of the multitude of organizational constructs. Another example of a "blueprint" artifact is the anatomy, which shows a simplified image – an architecture – of a complex system, by which actors can orient themselves and take proper actions.

Gaussian Adaptation and Complexity

An interesting example of the order-disorder problematic is found in quite another area. For many years, I worked with a stochastic optimization algorithm which was used extensively at Ericsson to optimize the performance of telecommunication systems. For example, the filtering function of a transmission system could be optimized by randomly adjusting filter parameters of system elements according to the algorithm.

The algorithm is based on a Gaussian search process (called Gaussian Adaptation) derived from the natural evolution of species (Kjellström & Taxén, 1981; Kjellström & Taxén, 1992). Samples of independent variables are randomly generated from a Gaussian n-dimensional distribution and evaluated for "success" or "failure". The outcome is regarded as a success if the sample belongs to a region R_A in R^n which is determined by an n-dimensional criterion function of the independent variables. Depending on the outcome of successive samples, the first and second order moments of the distribution is gradually aligned to R_A.

The interesting point here is that the optimal convergence rate of the search algorithm is achieved if the probability of a sample to hit R_A is in the vicinity of 0.3 (actually 1/e) (Kjellström, 1991). At both extremes the convergence is zero. A hitting probability of 0 means that each sample falls outside R_A, while a hitting probability of 1 means that every sample falls inside R_A. In both cases, no information of how to adapt the Gaussian distribution is achieved. Thus, in this application the order-disorder balance is crucial. In Kjellström & Taxén (1981, p. 707) the convergence rate as a function of the hitting probability is shown in a figure, which has the same structure as in Figure 31, p. 101.

It is near at hand to interpret these findings from an organizational point of view. An organization that has a "hitting probability" near 1 would correspond to a stable organization with a long tradition to supply basically the same product or service to a stable and regular customer for many years. Such organizations are extremely ill prepared for a sudden change in market, technology or whatever other circumstances. At the other extreme, a hitting probability near 0, we find organizations that fail to deliver results, either because the products/services are never realized or are not accepted by the market for various reasons such as quality, price, etc. These organizations are bound to go bankrupt.

Somewhere in between 0 and 1 lies the optimum hitting probability where the organization manages to adapt to changing circumstances with maximum efficiency. Sometimes its products/services fail and sometimes they succeed. This observation resonates with the strategies that successful, continuously changing organizations apply (Brown & Eisenhardt, 1997). Such strategies blend stable elements with freedom for local initiatives. The future is "probed" using experiments, strategic alliances, and other means. This can be interpreted as "looking outside" the organization in order to lower the hitting probability.

However, the analogy between stochastic optimization and CAS models of organizations should not be driven too far. In CAS theories, the conjecture is that agents coevolve on a fitness landscape (Anderson, 1999, p. 224). This landscape can be interpreted as the region of acceptability R_A in the Gaussian Adaptation process. The coevolvement conjecture has been criticized by McKelvey (1999) for being too simplistic. From the ADT view, the assumption of a given fitness landscape fails to take the dialectical relation between actors and the environment into consideration. The environment and the actors mutually influence each other. Nevertheless, the striking similarities between the Gaussian Adaptation process and CASs are well worth to explore further.

Summary

As Anderson (1999, p. 220) point out, CAS models represent a new way of modeling organizations based on properties exhibited by natural systems. Complex outcomes are derived from simple agent schemata and interconnections among agents. Rather than examining complex systems in a holistic way, "CAS models allow investigators to focus on an agent in its local environment. It becomes possible to grasp complex behavior by varying assumptions about the schemata, connections, fitness functions, or population dynamics that characterize the agents" (ibid, p. 220).

In several aspects, ADT provides an elaborated and experience based connection between complexity theory in the form of CASs and organizational theory. The central notion is that of the activity domain, which can be interpreted as the context within which "locality" in CAS can be framed in terms of organized human activity. For example, McKelvey (1999) has studied the coevolution of firms using Kauffman's "NK[C] model" (1993) as a basis. McKelvey employs the rather nebulous notion of "value chain competencies" for the internal "parts" of the firm. Replacing value chain competencies with the activity domain might produce more realistic and interesting results. In activity domains, the schemata of agents/ actors are enacted along with mediational means in purposeful activity. At the same time, the activity domain provides an intermediate construct between the whole, the organization, and the individual.

Expressed differently, the activity domain is a meso construct mediating between the micro and macro perspectives (Rousseau & House, 1994). The domain shields all elements in the organization from being connected to each other. By conceptualizing the organization as a set of interrelated activity domains, the organization may be seen as a loosely coupled system. This in turn makes it possible to devise means to drive the organization to equilibrium at the edge of chaos. Maintaining the equilibrium entails balancing between commonality and variability in all activity modalities. The organization must make strategic decisions about what rules (stabilization), process elements (temporalization), information (spatialization) and transitions between domains and in time (transition) shall be common, and which of these elements are concerns for local domains only.

Moreover, the activity domain renders a fractal structure (domain may utilize other domains) by which the multilevel aspect of CASs can be modeled. Thus, the complexity of organizations can be reduced to a canonical, simple core construct. Altogether, the ADT offers an alternative towards conceptualizing CASs in organizational contexts.

PRACTICE-BASED THEORIES AND ADT

As with the activity domain in ADT, an increasingly number of research approaches take some kind of practiceconstruct as the basic Unit of Analysis for addressing the problems associated with the "macro" – "micro" levels of conceptualization. Examples of such constructs are activity (Leont'ev, 1981), social world (Strauss, 1985); activity system (Engeström, 1987); communities of practice (Lave & Wegner, 1991); thought world (Dougherty, 1992); functional department (Schmidt & Bannon, 1992); regionalized communication space (Clement & Wagner, 1995); knowledge domain (Boland & Ramkrishnan,1995); work domain (Schmidt & Simone, 1996); work situation (Bannon & Bødker, 1997); work arrangement (Fjuk, Smordal, & Nurminen, 1997); communities of interest (Fisher, 2001); organizational function (Carlile, 2002); workpractice (Goldkuhl & Röstlinger, 2002); work context (Bechky, 2003); work system (Alter, 2006); ecology (Bergman, Lyytinen, & Mark, 2007).

The underlying belief in all these practice related constructs is that goal-oriented, situated human activity is the genesis of individual and constructional aspects of the society in which the individual is immersed. One may ask why there is a need to introduce yet another practice construct – the activity domain – into this already entangled discourse. The reason, I claim, is that there are some characteristics of the ADT approach that distinguishes it from other approaches.

The evolutionary path of the ADT started in practice and progressed towards theoretical grounding in existing theories. By being conceived and shaped by the very practical circumstances found in the industrial development of complex telecom systems, ADT has incorporated a deep understanding about the nature of the phenomena to be explained and influenced. Most other theoretical approaches have progressed in the opposite direction; conceived in the academic world and applied in practice. Hence, they often lack an appreciation of the complexity to be explained and informed by the theory.

Theories in applied sciences are often appropriated from some original target domain into the application domain. For example, there are numerous attempts to use Structuration Theory – a grand theory of society – in IS development research (e.g. Devadoss & Pan, 2007; Jones, 1999; Jones & Karsten, 2008; Orlikowski, 1992; Orlikowski, 2000). Although such maneuvers are certainly valid, they bring about new problems of interpretations and adaptations of the original theory. In contrast, ADT originated in the target domain and sought theoretical enforcement in other theories such as Activity Theory.

Most alternative approaches have been concerned with *explaining* and *understanding* rather than *informing*; that is, providing practical and demonstrable guidelines how to improve practice:

[Structuration Theory] has been used to theorize the field of IS and to analyze empirical situations involving IS, but little attempt has been made to 'operationalise' the theory – that is, to use it in an attempt to directly influence IS practice. In an applied field, it should be taken as axiomatic that useful theory should lead to improvements in the capacity for effective action. (Rose & Scheepers, 2001, p. 218)

To the best of my knowledge, none of the alternative approaches has been used for informing the configuration and application development of large, complex ISs like ERP (Enterprise Resource Planning) or PLM (Product Lifecycle Management) systems in industrial settings. ADT has demonstrated its capability of improving the "capacity for effective action" in at least one industrial setting, that of Ericsson.

A particular feature, which I believe is unique for ADT, is the integration of individual cognition, technology, and social structures from a coordination perspective. Although other approaches include these elements to a varying degree, they emphasize some of them at the expense of the others. For example, the focus of Structuration Theory is on the dualism between social structures and individual agency. Technology, which certainly is structuring a practice, is clamped together with other structures, giving rise to "memory traces" in the individual (Giddens, 1984, p. 17). The technological focus has, on the other hand, been emphasized by Orlikowski (1992; 2000) using Structuration Theory as a point of departure.

Although all of the practice oriented approaches deal extensively with the phenomenon of "context", few of them, if any, treat the *transition* between contexts, that is, what happens when practices need to coordinate their efforts. This aspect, which is one of the cornerstones in ADT, is of course necessary to include in any theory inquiring into the nature of coordination.

Based on these considerations, it might be interesting to position the ADT in relation to other theoretical approaches that have been proposed as socio-technical frameworks for practical purposes, in particular the development of ISs. The main reason for this exercise is that I want to emphasize the practical dimensions of ADT, while still making it plausible that many issues in socio-technical approaches are covered by ADT. The approaches I will discuss are:

- Structuration Theory (ST)
- Actor Network Theory (ANT)
- Organisational Semiotics (OS)
- Work System Framework (WSF)
- Cultural-Historical Activity Theory (CHAT)

Since each of these approaches is enormously rich in detail and scope, it is of course not possible to provide a decent coverage of them in this book. By necessity, I have to select features that I regard as essential for the comparison, which means that the analysis is by necessity superficial and certainly arguable. However, the intention is to provide a broad illumination of main features related to the ADT.

The analysis is carried out in two steps. First, I provide a short description of each approach and evaluate the approach against ADT. In this step each separate approach is in the foreground. Next, I change focus and put the ADT in the foreground. Each approach is valuated with respect to the following set of prominent characteristics in the ADT:

- Basic categories (meaning, historicity & change, human mind, cognition, technology & artifacts).
- Integrative construct.
- Activity modalities.
- Communicative function (mathetic, pragmatic).
- Practical results (analytical, constructive).

EVALUATION OF EACH APPROACH

Structuration Theory

Structuration Theory (ST: Giddens, 1979, 1984; Jones, 1999; Rose, 1998; Rose & Lewis, 2001; Rose & Scheepers, 2001) was developed by Giddens[1]. ST is a comprehensive theory which seeks to understand how human agency in the sense "capacity for accomplishing changes" is related to social structures. In ST, the focus of social theory is social *practices,* not individual action, nor the existence and constraints of some kind of societal totality. Social practices constitute both individuals and society.

In traditional social sciences there has been a sharp division between those who regard social phenomena to be products of the actions of human "agents" (e.g. Weber, 1983), and others who regard them as caused by the influence of social structures (e.g. Durkheim, 1972). Instead of regarding them in a dualistic way as separate and opposite things, Giddens proposes that structure and agency should be viewed as a mutually interacting duality. The notion of *duality of structure* is central in ST, and refers to the fact that structure is not only constraining, but is also enabling human action. The actions of human agents discursively and recursively form the sets of rules, practices and routines which, over time and space constitute structure. This process is called *structuration.*

When structure is being continuously produced and reproduced through action, this leads to another significant aspect of structuration: *routinization.* Routine is fundamental to the continuity of the personality of agents and to the institutions of society. Individuals acquire ontological security through their daily engagement in familiar routines and encounters. These encounters are also constitutive of social institutions, which mean that they enable the continuity of social life and the creation of order over time and space. From this follows that time and space are important elements in ST. Moreover, ST claims that social structures, which Giddens refers to as "memory traces" in the human brain, do not exist independent of human action. However, the emphasis in ST on the mental nature of structures has been criticized for disregarding material aspects of structure.

The insight that ISs are closely associated with social and human issues has inspired a number of researchers to use ST as a starting point for theoretical development, analysis and operational purposes (see e.g. Jones, 1999; Rose, 1998). Since ST is a very comprehensive theory it has lead to a number of different appropriations and interpretations. Mostly ST has been used for analytical purposes, i.e., as an analytical framework for understanding empirical situations or cases. The usage of ST for operational or constructive intents has been limited. One obvious difficulty is that ISs have a technological, material character which is hard to conciliate with the non-material stance of ST. Attempts to adapt structuration to incorporate the material aspects of IS have encountered a number of problems which have not been resolved.

Evaluation

Obviously, ST has much in common with the praxis-based approach taken in this book. Contextualization, spatialization, temporalization, and stabilization are all important elements of ST as well as enactment in the form of recursive constitution of routines and structure. The notion of "dualism" recognizes that a dialectical attitude is important.

However, in ADT there is an objective, material aspect which is not present in ST. The objectivation – objectification aspects of enactment appears to have been reduced to the objectivation aspect in ST in the form of "memory traces". Thus, ST recognizes objectivation and ignores objectification. This is a weak point in ST when it comes to constructive purposes of artifacts like IS. This is also confirmed by the research community so far. The ADT approach on the other hand has resulted in practical, concrete implementations of IS.

Actor Network Theory

Actor Network Theory (ANT: Latour, 1992; Walsham, 1997) is one strand of research dealing with the social construction of technology where social and technical phenomena are considered together. ANT is concerned with the creation and maintenance of coexisting networks of human and non-human elements. Instead of apprehending the social and technical phenomena as separate elements, ANT treats them as inseparable. Latour (ibid) argues that people and artifacts should be analyzed with the same apparatus. There is a need for new methodological and theoretical devices that we can use to analyze hybrids of people and artifacts such as for example ISs.

ANT examines the motives and actions of groups of actors who form networks of aligned interests. These networks include both humans and non-humans as actors. This standpoint has caused considerable controversy since it can be claimed that human responsibility and intentions are left out of the theory. Non-human resources can "stand in or speak for" or be delegates for particular viewpoints or truth-statements which help to maintain a particular network of alliances.

The empirical focus of ANT is to trace and explain the processes of creating and maintaining stable networks of aligned interests, or to examine why such networks fail to establish themselves. Thus, ANT is both a theory and method combined. It provides theoretical concepts as ways of viewing the world and suggests that these elements should be traced in the empirical work. The researcher is led to investigate and document *network elements*, both human and non-human, processes of *translation* and *inscription*, the creation of *black boxes* or *immutable mobiles* and the degree of stability and irreversibility of networks and elements.

Walsham (1997) has reviewed the status of ANT in IS research. Many researchers use concepts from ANT to analyze their findings, and the main point seems to be that ANT does not privilege either social aspects or technology. Some critique has been issued against ANT, for example that the theory addresses the local and contingent but pays little attention to broader social structures. However, Walsham's overall conclusion is that ANT has much to offer in both theoretical and methodological terms. A combination of, for example, ANT and the Structuration Theory of Giddens might be fruitful.

Evaluation

As with Structuration Theory, ANT has a lot in common with the ADT approach, among all, the dialectical view of the interaction between human agents and artifacts. There is also a strong notion of contextualization, spatialization and temporalization in ANT. Moreover, the constructs in ANT are obviously very suitable for analysis of the complicated interplay between networks of actors (human or non-human) over time and space. In short, ANT seems to be an interesting and useful methodology based on a somewhat extreme ontology. However, the view on non-humans as actors having motives, goals, intentions, etc., is alien to the ADT approach.

Results concerning the usage of ANT for constructive purposes seem to be limited so far. A combination of the constructive power of the ADT approach with the analytical power of ANT might be an interesting perspective to elaborate.

Organisational Semiotics

Organisational Semiotics (OS: Stamper, 2001; Liu, 2000) is defined by Stamper as informatics from a social angle (Stamper, 2001). This discipline deals with technological, human and social aspects of information, and will according to Stamper play a major role in the future. The reason why Stamper somewhat provocatively characterizes this discipline as "informatics without the computer" is that he wants to base his approach in a precise philosophical understanding of the information concept. Traditional computer informatics went wrong because failure in building upon primitive concepts whose meaning must be defined ostensively, i.e., by explaining the meaning of a word by ostension, that is, by pointing to something to which the word applies.

Stamper claims that semiotics leads to a more precise definition of *information as various properties of signs*. This will cross the bridge between language and reality. Semiotics can solve this problem since it builds on an ostensively definable concept: the notion of a sign. Stamper wants to advance semiotics to study sign products: shared knowledge, mutual commitments and institutions. The key problem is linking linguistic signs to the reality we are talking about. This is a problem both of ontology and epistemology.

Figure 1. The work system (Wikipedia, 2008b, Work systems theory, 2009)

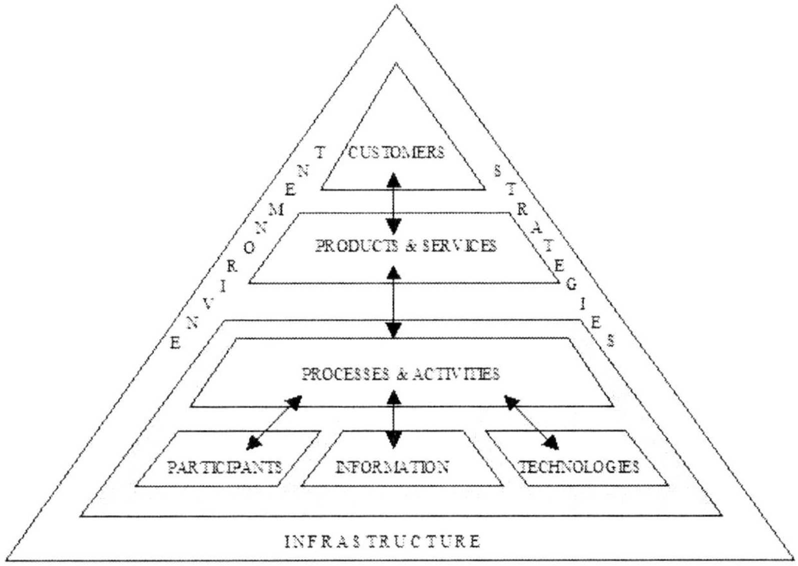

Stamper means that we live in two more or less orthogonal worlds, the material and non material worlds. Information creates the understanding and relationships from which organizations and communities are built. Information is the "material of our social world". This leads Stamper to define a new ontology: *actualism*. The actualist ontology entails a profound shift in the way we talk about the world. The only reality we can know directly consists of here and now. Signs exist here and now. Past and future exist only through the signs that represent them. The only knowable reality consists of *affordances* – invariant repertoires of behavior, either substantial affordances or social norms. Some agent bears the responsibility for every feature of the world as we know it. *Society* is always the root agent for human knowledge and behavior. This means that society is regarded as a responsible agent. Other agents can be individuals or groups.

In order to capture socially established meanings, Stamper defines a specific formalism to model invariant repertoires of behavior. This formalism, Norma, is a language of norms and affordances. A particular set of repertoires valid for an organization can be illustrated in a schema, where the nodes do not represent concepts in anyone's head but socially shared physical or social affordances. According to Stamper, the Norma approach has been applied in a large number of organizations with large cost savings.

Evaluation

The importance placed on the role of signs is a feature that Organisational Semiotics share with the ADT approach. From this follows that intersubjectivity is regarded in both approaches as a major characteristic of human activity. Intersubjectivity is the basis for the Norma language. Since this method has been used with good results in practice, it demonstrates the importance of considering intersubjectivity in constructive situations. In more traditional methods, common sense often passes unnoticed which might create huge maintenance costs of IS when redesign becomes necessary. Another similarity with the praxis approach is the importance placed on contextualization.

A major difference between OS and the ADT is the importance placed on the individual. In OS perceptions and conceptions cannot be used, since they are private to the individual in whose mind they occur. Thus "informatics has no need for perceptions and conceptions" (Stamper, 2001, p. 123). This means that OS can be regarded as the extreme opposite to ST that acknowledges only the existence of "memory traces". The ADT on the other hand, considers both individual cognition and communal meaning.

Another difference is that human agency is downplayed in OS. "Society" cannot be an actor in ADT, not even in the ANT. Enactment, temporalization, spatialization and transition do not seem to be salient in the OS approach.

Work System Framework

The notion of work systems has been used in several publications since the 1970s. More recently, Alter has defined a work system as

A work system is a system in which human participants and/or machines perform work using information, technology, and other resources to produce products and/or services for internal or external customers. (Alter, 2002, p. 92)

Business organizations comprise work systems that "obtain materials from suppliers, produce and deliver end products, find customers, create financial reports, hire employees, coordinate work across departments, and perform many other functions" (Alter, 2002, p. 92). From a business viewpoint, organizations are best understood as work systems, regardless of whether they use IT extensively or not. A work system can be conceived as the Work System Framework (WSF) in Figure 1.

The following description of a work system is collected from Wikipedia (2008b). The work system itself consists of four elements: the processes and activities, participants, information, and technologies. Five other elements must be included in even a rudimentary understanding of a work system's operation, context, and significance:

- *Work practices, processes and activities* include everything that happens within the work system. The term processes and activities is used instead of the term business process because many work systems do not contain highly structured business processes involving a prescribed sequence of steps, each of which is triggered in a pre-defined manner. In effect, business process is but one of a number of different perspectives for analyzing the activities within a work system. Other perspectives with their own valuable concepts and terminology include decision-making, communication, coordination, control, and information processing.
- *Participants* are people who perform the work. Some may use computers and IT extensively, whereas others may use little or no technology.
- *Information* includes codified and non-codified information used and created as participants perform their work. Information may or may not be computerized. Data not related to the work system is not directly relevant, making the distinction between data and information secondary when describing or analyzing a work system. Knowledge can be viewed as a special case of information.
- *Technologies* include tools (such as cell phones, projectors, spreadsheet software, and automobiles) and techniques (such as management by objectives, optimization, and remote tracking) that work system participants use while doing their work.

- *Products and services* are the combination of physical things, information, and services that the work system produces. This may include physical products, information products, services, intangibles such as enjoyment and peace of mind, and social products such as arrangements, agreements, and organizations.
- *Customers* are people who receive direct benefit from products and services the work system produces. They include external customers who receive the organization's products and/or services and internal customers who are employees or contractors working inside the organization.
- *Environment* includes the organizational, cultural, competitive, technical, and regulatory environment within which the work system operates. These factors affect system performance even though the system does not rely on them directly in order to operate. The organization's general norms of behavior are part of its culture, whereas more specific behavioral norms and expectations about specific activities within the work system are considered part of its processes and activities.
- *Infrastructure* includes human, informational, and technical resources that the work system relies on even though these resources exist and are managed outside of it and are shared with other work systems. For example, technical infrastructure includes computer networks, programming languages, and other technologies shared by other work systems and often hidden or invisible to work system participants.
- *Strategies* include the strategies of the work system and of the department(s) and enterprise(s) within which the work system exists. Strategies at the department and enterprise level may help in explaining why the work system operates as it does and whether it is operating properly.

There is also a work system method (Alter, 2002; 2006) that business professionals (and/or IT professionals) can use for understanding and analyzing a work system at whatever level of depth is appropriate for their particular concerns. This method is iterative in nature and consists of the main steps: initiation, development, implementation, and operation and maintenance.

Evaluation

The work systems approach has much in common with ADT. It places the focus on the elements of a system whose purpose it is to produce products/services for some client/customer. Thus, elements like processes, information, actors/participants, technology, etc., are subordinated the overarching construct "work system". In this sense the works system is directed towards the same Unit of Analysis as the activity domain. It is also clear that the aim of the work systems approach is to be a guiding framework for analyzing and informing practice.

However, there are several decisive differences between the work systems approach and ADT. The work object, the products/services that is the outcome of the work, is placed outside the work system: "...products and services are not part of the work system, but should be considered when attempting to understand or analyze a work system" (Alter, 2006b, p. 304). Taken literally, this means that the work object is not directly influencing the structure of the work system; contrary to the activity/activity domain where the work object together with the motive is determinant for the structure of the domain.

Moreover, the definition of the work system is from an ontological point of view unclear. For example, an IS is sometimes included in the work system: "The concept of a 'work system' is a general case that encompasses ISs, projects, value chains, supply chains, and other special cases" (Alter, 2002, p. 91).

Figure 2. The structure of the activity system according to CHAT

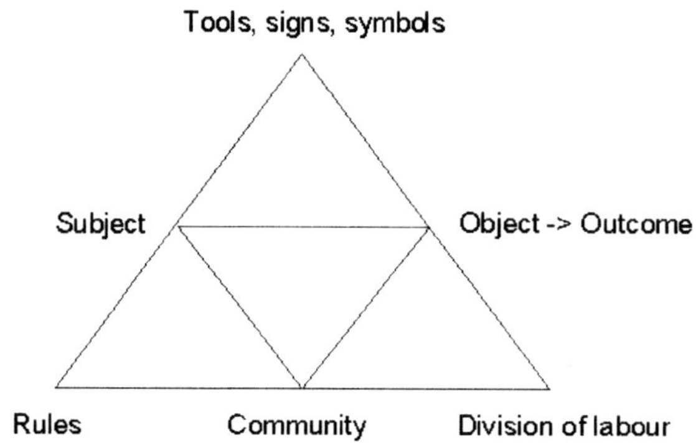

Elsewhere, the IS is a work system: "Information systems constitute a special case of work systems in which the business processes performed and products and services produced are devoted to information" (Alter, 2002, p. 95). This ambiguity makes it hard to conceptualize the essence of a work system. For example, if ISs are work systems, this would mean that a participant, which is one element of a work system, is included in an IS.

Many of the activity modalities are traceable in the work system: temporalization (business processes, activities, and work practices), spatialization (information), stabilization (strategies), and enactment (the work system method). However, these elements are not structured as a dialectical whole in which they mutually influence each other. The elements seem to be more or less independent in the work system. The most striking difference is that there is no transitional element in the WSF. Consequently, the work system is not recursive. The issue of work systems utilizing other work systems is not addressed other than indirectly. This is also clear from the fact that the environment of the work system is structured quite differently from the work system itself.

The issue of sense-making or communal meaning is not up front in the work systems approach. Cognitive aspects are addressed only indirectly[2]. Concerning practical applications, the WSF approach has been mainly used for analytical purposes (Alter, 2007). A final difference is the absence of references to a theoretical fundament for work system in line with praxis and Activity Theory for ADT.

Cultural-Historical Activity Theory

The original Russian theory of Activity has undergone several transformations since its instigation by Vygotsky in the early 1920s. A comprehensive account for this is given in Kaptelinin & Nardi (2006). In this Section I, will focus on the elaboration of Activity Theory in the form of Cultural-Historical Activity Theory (CHAT) proposed by Engeström (1987).

The first generation of AT focused on the individual in her social environment. Engeström expanded AT by adding elements to it in order to incorporate cooperative and transformative aspects (see Figure 2). *Community* is added to form a triangle with three mediating relationships between subject, community and object: *tools, signs and symbols* mediate between subject and object, *rules* mediate between subject and community, and *division of labor* mediates between community and object. Taken together, these elements form an *activity system* (see Figure 2).

The extension of AT to activity systems focuses on production of artifacts fulfilling needs. This makes it necessary to consider the interactions between several activity systems, i.e., the same scope that ADT has. According to Engeström, such an elaboration should adhere to five basic principles (Engeström, 2001):

- *Unit of analysis*: The activity system in relation to other activity systems is the primary unit of analysis. Goal-directed individual actions as well as operations are relative subordinate analysis units. In ADT the unit of analysis has been is the activity domain, and the relations between several domains have been the focus in ADT from the outset.
- *Multi-voicedness*: Activity systems are multi-voiced. There are always multiple points of view, traditions and interests in an activity system. This is valid also in networks of activity systems. In ADT, the coordination point of view of the activity system is emphasized. Other views are submerged.
- *Historicity*: Activity systems are historically shaped and transformed over lengthy periods of time. This is shared by ADT.
- *Contradictions*: Contradictions are the source of change in activity systems. This is an area, which is not attended explicitly in ADT. The principle of contradictions as the source of change is however inherent in ADT as well due to its roots in the praxis philosophy.
- *Expansive transformations*: Activity systems may undergo qualitative transformations where historically new forms of activity emerge as the result of contradiction aggravation and resolution. In principle, this is shared by ADT. However, this line of inquiry has so far not been pursued in ADT.

Evaluation

CHAT and ADT have common ontological and epistemological roots. As with praxis, CHAT tries to capture the totality of human activity. The elements in activity system (object, outcome, subject, community, tools, signs, symbols, division of labor, rules, etc.) are found in ADT as well, however differently structured. Enactment and continuous change are also common to both approaches.

However, there are several differences between CHAT and ADT. CHAT downplays the role of the individual since the subject can be both individuals and groups. Furthermore, signs appear to be under-researched in CHAT (Leiman, 1999). Another difference is that transition is not treated in CHAT until recently (e.g. Virkkunen, 2004). This means that the cooperation between activity systems is less elaborated. Finally, the use of CHAT for constructive purposes seems to be limited. Although CHAT has prompted a tremendous variety of research in IS, the results appear to be mainly of analytical nature (Bertelsen, 2001; Bertelsen, Korpela, & Mursu, 2004). In contrast to this, ADT has proven to be operational in very demanding practices of coordinating the development of complex systems.

EVALUATION AGAINST THE ACTIVITY DOMAIN THEORY

Basic Categories

Integrative Construct

In ADT, the activity domain is a central category which integrates elements of human activity into a coherent whole. CHAT has a corresponding construct in the activity system. In OS, the concept of organization is inherent: "Organisational semiotics" is defined as informatics from a social angle. However, Stamper does not define "organisation" very precisely. Some traces of an integrative construct can be found in "sub-cultures", which are described as "overlapping groups of people subject to shared norms that enable them to behave in an organized fashion." (Stamper, 2001, p. 154). In ST, the practice is a salient element. However, the internal structure of it is only vaguely articulated. In ANT, the actor networks of aligned interests, including people, organisations, and standards have some resemblance with activity domains. Again, these networks are only vaguely defined. In the WSF approach, the work system is an integrating element, which however is unclear on how the elements are integrating into a coherent whole.

Meaning

OS is particularly expressive in its focus on meaning. Signs are inherent in OS and articulated in the form of the semiotic ladder. Signs can be dealt with in the intersubjective domain, and a language community can bridge the sign - reality gap. Moreover, OS emphasizes strongly communal meanings and mutual commitments. In the early period of AT, meaning was a central theme. This has however become less attended in the CHAT elaboration of AT. In CHAT, signs are put on equal terms as instruments, machines, methods, laws, etc., in mediating between the subject and object. The object in CHAT can be material or intangible, and shared for manipulation into the outcome. Thus, there are some, albeit vague, notions of communal meaning in CHAT. Concerning ST, the interpretative schemes are considered to be embedded in social structures as signification or meaning. In ANT, meaning is not a salient theme. This is understandable since signs mark a clear distinction between humans and technology, something which is subdued in ANT. In WSF, meaning is not in focus.

Historicity & Change

Historicity and change are inherent in ST, CHAT, ANT, and WSF. In contrast, OS means that the only reality we can know directly consists of here and now. The past and future exist only through signs. Thus, in OS historicity and change are weak.

Human Mind, Cognition

The nature of the human mind was the prime knowledge interest to the Russian Theory of Activity as conceived by Vygotsky, Leont'ev, Luria, and others. In CHAT, mediation can take place both at the individual and the societal levels (Engeström, 1999). However, it seems that the main influence in the IS community has been to re-focus the unit of analysis from an individual, cognitive oriented focus to a focus on social oriented activity (Virkkunen & Kuutti, 2000; Kuutti, 1996). The subject can be both individuals and groups, which makes the individual less salient in CHAT. In ST, the structures of society are seen as traces in the mind, which are instantiated only through action. The individual as an actor is not emphasized in ANT. Concerning OS "... informatics has no need for perceptions and conceptions" (Stamper, 2001, p. 123). By this, Stamper means that a theory of information cannot take individual perceptions and conceptions as its starting point. In WSF, cognition is only implicit (Alter, 2006b).

Mediation

In both ADT and CHAT, technology and artifacts are inherent elements. In ANT artifacts are agents and meanings can be inscribed in artifacts. OS considers the only reality we can know as consisting of the here-and-now and being composed of actual, performable repertoires of behavior. In this context, technology is perceived only as "affordances" of agents. In spite of this, there is an outspoken purpose in OS to provide methods and tools for working with organisational semiotics, and thus making organisations more effective and improving the use of technology. ST regards technology only as "resources" to be drawn upon in the structuration. Technology is one of the elements in WSF.

Activity Modalities

Contextualization

Contextuality is present in all approaches. In ST, the study of context is inherent in the investigation of social reproduction. Context is salient also in ANT on an abstract level as the substitution (paradigm) construct in the program – anti-program trajectory of an invention. In CHAT, cognitive processes "are processes occurring in concrete, practical activity and are formed within the limits of this activity" (Kuutti, 1996, p. 33). In OS, contextuality is inherent since the affordances of certain agent are context dependent. The work system in WSF delimits a context in which the production of products and services take place. However, there is no indication that contextualization is a constituting element in the same sense as in ADT. In other words, there is no discussion in WFS of how the context impacts the elements in the work system.

Spatialization

Spatialization is salient in all approaches to some extent. However, with the exceptions of CHAT and OS, this category is not particularly articulated. Especially OS emphasizes spatialization in the form of the semantic model, which signifies socially shared affordances and their relations.

Temporalization

Temporalization is also salient in all approaches albeit not particularly articulated. The emphasis in CHAT on actions has a strong temporal character. In OS every universal and particular affordance has its start and finish for which some agent is responsible. However, this is a rather weak notion of temporalization. The element "work practices" in WSF has a temporal character.

Table 1. Comparison between different approaches

		ST	ANT	CHAT	OS	WSF	ADT
Basic categories	Integrative construct	*	*	***	**	***	***
	Meaning	**	-	**	***	-	***
	Historicity & change	***	***	***	-	**	***
	Human mind, cognition	**	-	**	-	-	***
	Mediation	*	***	***	***	**	***
Activity modalities	Contextualization	***	***	***	***	***	***
	Spatialization	*	*	**	***	***	***
	Temporalization	*	*	**	*	***	***
	Stabilization	***	***	***	***	**	***
	Transition	*	*	*	-	-	***
Communicative function	Mathetic	*	-	**	**	-	***
	Pragmatic	*	-	*	**	-	**
Results achieved	Analytical	***	***	***	***	***	**
	Constructive	*	-	*	*	*	***
		ST	ANT	CHAT	OS	WS	ADT

Stabilization

Stabilization is salient in all approaches. In ST, for example, routinization is considered fundamental. In CHAT, rules are mediating between the subject and the community. In OS, norms play a major role. In WSF, the most apparent sign of stabilization is the "strategies" element.

Transition

Only ADT appears to emphasize transition. In CHAT, the construct "division of labor" mediates between community and object in a particular activity system. However, there is no indication of a mapping and interpretation between activity systems. A transitional element is traceable in the discussion of "boundary objects" (Bertelsen, 1999). These are objects that can be interpreted differently by different groups, say users and designers, but still maintain some commonly recognized feature that ties different praxes together. In ANT the "black box" concept indicates something relevant to context. However, the internal structure of the black box is not accessible or relevant in that particular context. The black box construct can possibly be apprehended as an activity domain, the outcome of which is used in another domain. In ST, the notion of "bracketing" an area of investigation in the analysis is akin to transition (Giddens, 1979, p. 248 ff.).

The work system in WSF lends itself easily to a discussion of how several work systems may coordinate their efforts in an organization. This is indicated by the following statement: "All businesses and organizations consist of multiple work systems that perform essential functions such as hiring employees, producing products, finding customers, selling to customers, providing customer service, and planning for the future" (Alter, 2009). In Alter (2006), there is a discussion about how several work systems interact. However, this discussion is on a general and superficial level and does not consider the transition between, for example, different communal meanings in the work systems.

Communicative Function

Mathetic

The mathetic function of communication is concerned with the construction of communal meaning. In OS, there is an iterative element in the Semantic Analysis method when eliciting requirements put on an IS. Communal meaning is gradually achieved by refining the semantic model (Liu, 2000). However, this model serves as a specification for further design, for example, using object oriented methods. Thus, the method does not include feedback from the usage of the IS. In essence, OS argues for a linear development model where specification and design are two distinct phases. Concerning CHAT, Engeström describes an "expansive learning" cycle consisting of seven steps which has an experiential learning resemblance (Engeström, 1999b). The experiential mode of learning is present also in ST as routinized and recursive social practices. ANT does not appear to have this element, neither does WSF.

Pragmatic

The pragmatic function of communication is concerned with the coordination of acts when communal meaning has been established. This is to some extent present in OS. The other approaches do not emphasize this aspect.

Results Pertinent to Practice

The comparison in this section concerns analytical and constructive results related to practice. I am interested in the footprint each approach has left in complex organizational and industrial settings. This means that the type of ISs involved are large, globally accessible enterprise systems like, for example, ERP or PLM systems. Thus, we are not concerned with applications developed in academy for demonstrating a particular approach or advancing the theoretical development of the approach.

There is substantial evidence that all the approaches are viable for analytical purposes. Concerning the constructive aspect, ADT and OS approaches appear to be the strongest ones. OS has been applied in a number of organisations with large cost savings (Liu, 2000). However, it seems that the applications reported are of less complexity compared to ADT. Concerning ST, some recent constructive results have been reported (Rose & Lewis, 2001). The same is valid for CHAT (Korpela, 2004). There are no examples of ANT being used in industrial settings. This also seems to be the case for WSF.

Comparative Analysis

In the table below, the analysis is summarized. A hyphen (-) indicates no presence. The stars indicate different emphasis given to the element, from weak (*) over modest (**) to strong (***). Naturally, these judgments are subjective and should be taken as fairly coarse indicators of the relative weight of each category in relation to the same category in ADT.

Some observations can be drawn from the analysis. The approach that has the closest affinity with ADT is not surprisingly CHAT, since ADT and CHAT are based on similar principles. WSF is close to ADT, except for issues that have to do with cognition and meaning. No approach is strong with respect to all categories characterizing ADT.

A most striking observation is that the transition modality is not salient in any approach except ADT. This is even more surprising since contextualization is strong in all approaches. It would seem close at hand to consider not only contextual aspects but also the interaction between different contexts. This is possibly due to the lack of integrative constructs that can leverage the attention to context interdependencies. An awakening interest in this matter can be traced in the discussion of the so called 3rd generation of Activity Theory (Engeström, 2001).

Most approaches seem weak on the *construction* of communal meaning, that is, the mathetic function of language. The focus appears to be on the pragmatic, communicative function. In contrast, meaning construction was one of the main drivers in the development of ADT. This is possibly due to the fact that controversies concerning the meaning of things surge when concrete and detailed IS implementation decisions are imminent (Bititci & Muir, 1997). This is particularly proliferated in, for example, industrial settings. Since most IS applications developed in academia do not make an inroad into industry, it might be that the issue of meaning construction is less attended to.

Finally, it can be seen that only ADT has had a substantial influence on the particular type of IS practice I am concerned with in this book. The reason for this might be that ADT, in contrast to the other approaches, was devised in the very same practice it influenced. Thus, it was geared from its outset to the specific needs found in complex, industrial settings.

REFERENCES

Alter, S. (2002). The Work System Method for Understanding Information Systems and Information Systems Research. [from http://www.stevenalter.com/newsletter.htm]. *Communications of the Association for Information Systems*, *9*(9), 90–104. Retrieved April 2, 2008.

Alter, S. (2006). *The Work System Method: Connecting People, Processes, and IT for Business Results.* Larkspur, CA: Work System Press.

Alter, S. (2006b). Work Systems and IT Artifacts: Does the Definition Matter? *Communications of the Association for Information Systems, 17*(14), 299–313. Retrieved April 21, 2008, from http://cais.aisnet.org/articles/17–14/default.asp?View=Journal&x=45&y=11

Alter, S. (2007). Making the Making the Work System Method More Useful, In *Proceedings of the 6th Annual ISOnEworld Conference*, April 11-13, 2007, Las Vegas, NV.

Alter, S. (2009) *The Work System Method.* Retrieved March 16, 2009, from http://www.worksystem-method.com/works.htm

Anderson, P. (1999). Complexity Theory and Organization Science. *Organization Science*, *10*(3), 216–232. doi:10.1287/orsc.10.3.216

Bannon, L., & Bødker, S. (1997). Constructing common information spaces. In J. A. Hughes, W. Prinz, T. Rodden, & K. Schmidt (Eds.), *Proceedings of the Fifth Conference on European Conference on Computer-Supported Cooperative Work (Lancaster, UK, September 07 - 11, 1997) ECSCW* (pp. 81–96). Norwell, MA: Kluwer Academic Publishers.

Bar-Yam, Y. (1997). *Dynamics of complex systems.* Reading, MA: Perseus Books.

Bechky, B. A. (2003). Sharing meaning across occupational communities: The transformation of understanding on a production floor. *Organization Science, 14*(3), 312–330. doi:10.1287/orsc.14.3.312.15162

Bergman, M., Lyytinen, K., & Mark, G. (2007). Boundary Objects in Design: An Ecological View of Design Artifacts. [JAIS]. *Journal of the Association for Information Systems, 8*(11), 546–568.

Bertelsen, O. (1999). Mediation and Heterogeneity in Design. In Y. Dittrich, C. Floyd, N. Jayaratna, F. Kensing, R. Klischewski (Eds.), *Social Thinking - Software Practice* (pp. 16–20). Dagstuhl-Seminar-Report; 250, 05.09.1999 - 10.09.1999 (99361). Wadern, Germany: IBFI gem. GmbH.

Bertelsen, O. (2001). Operationalizing Activity Theory. IRIS 24 workshop, *24th Information Systems Research Seminar in Scandinavia*, Ulvik, Norway, August 11th–14th 2001.

Bertelsen, O., Korpela, M., & Mursu, A. (2004) Preface. In O. Bertelsen, M. Korpela, A. Mursu, (Eds.), *ATIT 2004, Proceedings of the First International Workshop on Activity Theory Based Practical Methods for IT-Design* (p. 3). September 2004, Copenhagen, Denmark. Report PB-574, Daimi, Denmark: University of Aarhus. Retrieved Oct 6, 2008, from http://www.daimi.au.dk/research/papers-reports/reports

Bititci, U., & Muir, D. (1997). Business process definition: a bottom-up approach. *International Journal of Operations & Production Management, 17*(4), 365–374. doi:10.1108/01443579710159950

Boland, R. J., & Ramkrishnan, V. T. (1995). Perspective Making and Perspective Taking in Communities of Knowing. *Organization Science, 6*(4), 350–372. doi:10.1287/orsc.6.4.350

Brown, S. L., & Eisenhardt, K. M.(199). The Art of Continuous Change: Linking Complexity Theory and Time-Paced Evolution in Relentlessly Shifting Organizations. *Administrative Science Quarterly, 42*(1), 1–34. doi:10.2307/2393807

Carlile, P. (2002). A Pragmatic View of Knowledge and Boundaries: Boundary Objects in New Product Development. *Organization Science, 13*(4), 442–455. doi:10.1287/orsc.13.4.442.2953

Clement, A., & Wagner, I. (1995). Fragmented Exchange: Disarticulation and the Need for Regionalized Communication Spaces. In H. Marmolin, Y. Sundblad, & K. Schmidt (Eds.), *Proceedings of the Fourth Conference on Computer-Supported Cooperative Work, ECSCW '95* (pp. 33–50). Dordrecht: Kluwer Academic Papers.

Cohen, M. D. (1996). Individual learning and organizational routine. In Cohen, M. D., Sproull L. S. (Eds.) *Organizational Learning* (pp. 188–194). Thousand Oaks, CA: Sage.

COSI. (2009). *Complexity in Social Science.* Retrieved February 19, 2009, from http://www.irit.fr/COSI/index.php

Devadoss, P., & Pan, S. L. (2007). Enterprise Systems Use: Towards a Structurational Analysis of Enterprise Systems Induced Organizational Transformation. *Communications of the Association for Information Systems, 19*, 352–385.

Dougherty, D. (1992). Interpretive barriers to successful product innovation in large firms. *Organization Science, 3*(2), 179–202. doi:10.1287/orsc.3.2.179

Durkheim, É. (1972). *Selected writings.* A. Giddens (Trans.). Cambridge, UK: Cambridge University Press.

Engeström, Y. (1987). *Learning by expanding: An activity-theoretical approach to developmental research.* Helsinki: Orienta-Konsultit.

Engeström, Y. (1999). Activity theory and individual and social transformation. In Y. Engeström, R. Miettinen, R.L. Punamäki (Eds.) *Perspectives on Activity Theory* (pp. 19–38). Cambridge UK: Cambridge University Press.

Engeström, Y. (1999b). Innovative learning in work teams: Analyzing cycles of knowledge creations in practice. In Y. Engeström, R. Miettinen, R.L. Punamäki (Eds.) *Perspectives on Activity Theory* (pp. 377–404). Cambridge UK: Cambridge University Press.

Engeström, Y. (2001). Expansive Learning at Work: toward an activity theoretical reconceptualization. *Journal of Education and Work, 14*(1), 133–156. doi:10.1080/13639080123238

Fischer, G. (2001) Communities of Interest: Learning through the Interaction of Multiple Knowledge Systems. In S. Bjornestad, R. Moe, A. Morch, A. Opdahl (Eds.), *Proceedings of the 24th IRIS Conference* (pp. 1–14). Bergen, Norway: Department of Information Science.

Fjuk, A., Smordal, O., & Nurminen, M. (1997). Taking Articulation Work Seriously—an activity theoretical approach. *TUCS TR 120 (Technical Report).* Turku, Finland: Turku Centre for Computer Science.

Giddens, A. (1979). *Central problems in social theory: action, structure and contradiction in social analysis.* Berkeley, CA: Univ. of California Press.

Giddens, A. (1984). *The Constitution of Society: Outline of the Theory of Structuration.* Los Angeles: University of California Press.

Goldkuhl, G., & Röstlinger, A. (2002). The practices of knowledge - investigating functions and sources. *The 3rd European Conference on Knowledge Management (3ECKM)*, Dublin. Retrieved June 2, 2008, from http://www.ida.liu.se/~gorgo/engpub.html

Jasperson, J., Carter, P. E., & Zmud, R. W. (2005). A Comprehensive Conceptualization of Post-Adoptive Behaviors Associated with Information Technology Enabled Work Systems. *MIS Quarterly, 29*(3), 525–557.

Jones, M. (1999). Structuration Theory. In W. Currie, R. Galliers (Eds.), *Rethinking management information systems: an interdisciplinary perspective* (pp. 103–135). Oxford, UK: Oxford Univ. Press.

Jones, M., & Karsten, H. (2008). Giddens's Structuration Theory and Information Systems Research. *MIS Quarterly, 32*(1), 127–157.

Kaptelinin, V. (1996). Computer-Mediated Activity: Functional Organs in Social and Development Contexts, In B. A. Nardi (Ed.) *Context and consciousness. Activity theory and human-computer interaction* (pp. 45–68). Cambridge, MA: MIT Press.

Kaptelinin, V., & Nardi, B. (2006). *Acting with Technology - Activity Theory and Interaction Design.* Cambridge, MA: The MIT Press.

Kauffman, S. (1993). *The Origins of Order: Self-Organization and Selection in Evolution.* New York: Oxford University Press.

Kjellström, G. (1991). On the Efficiency of Gaussian Adaptation. *Journal of Optimization Theory and Applications, 71*(3). doi:10.1007/BF00941405

Kjellström, G., & Taxén, L. (1981). Stochastic Optimization in System Design. *IEEE Transactions on Circuits and Systems, CAS-28*(7), 702–715. doi:10.1109/TCS.1981.1085030

Kjellström, G., & Taxén, L. (1992). Gaussian Adaptation, an evolution-based efficient global optimizer. In I. C. Brezinsk, & U. Kulish (Eds.), *Computational and Applied Mathematics* (pp. 267–276). Thousand Oaks, CA: Elsevier Science Publishers.

Korpela, M. (2004). ActAD in Information Systems Analysis and Design. In O. Bertelsen, M. Korpela, A. Mursu, (Eds.), *ATIT 2004, Proceedings of the First International Workshop on Activity Theory Based Practical Methods for IT-Design* (pp. 134–136), September, Copenhagen, Denmark. Report PB-574. Daimi, Denmark: University of Aarhus. Retrieved Oct 6, 2008, from http://www.daimi.au.dk/research/papers-reports/reports

Kuutti, K. (1996). Activity theory as a potential framework for human-computer interaction research. In B. A. Nardi (Ed.) *Context and consciousness. Activity theory and human-computer interaction* (pp. 17–44). Cambridge, MA: MIT Press.

Latour, B. (1992). Technology is society made durable. In J. Law (Ed.), *A Sociology of Monsters. Essays on Power, Technology and Domination* (pp. 103–130). London: Routledge.

Lave, J., & Wenger, E. (1991). *Situated Learning: Legitimate Peripheral Participation.* Cambridge, UK: Cambridge University Press.

Leiman, M. (1999). The concept of sign in the work of Vygotsky, Winnicott, and Bakhtin: Further integration of object relations theory and activity theory. In Y. Engeström, R. Miettinen, R.L. Punamäki (Eds.) *Perspectives on Activity Theory* (pp. 419–434). Cambridge, UK: Cambridge University Press.

Leont'ev, A. N. (1981). *Problems in the development of the mind.* Moscow: Progress Publ.

Lewin, A. Y. (1999). Application of Complexity Theory to Organization Science. *Organization Science, 10*(3), 215. doi:10.1287/orsc.10.3.215

Liu, K. (2000). *Semiotics in information systems engineering.* Cambridge, UK: Cambridge University Press.

March, J. G., & Simon, H. A. (1958). *Organizations.* Cambridge, MA: Blackwell Publishers.

McKelvey, B. (1997). Quasi-Natural Organization Science. *Organization Science*, *8*(4), 352–380. doi:10.1287/orsc.8.4.351

McKelvey, B. (1999). Avoiding Complexity Catastrophe in Coevolutionary Pockets: Strategies for Rugged Landscapes. *Organization Science*, *10*(3), 94–321. doi:10.1287/orsc.10.3.294

Morel, B., & Ramanujam, R. (1999). Through the Looking Glass of Complexity: The Dynamics of Organizations as Adaptive and Evolving Systems. *Organization Science*, *10*(3), 278–293. doi:10.1287/orsc.10.3.278

Orlikowski, W. (1992). The Duality of Technology: Rethinking the Concept of Technology in Organizations. *Organization Science*, *3*(3), 398–427. doi:10.1287/orsc.3.3.398

Orlikowski, W. (2000). Using Technology and Constituting Structures: A Practice Lens for Studying Technology in Organizations. *Organization Science*, *11*(4), 404–428. doi:10.1287/orsc.11.4.404.14600

Reynolds, C. (2009) *Boids. Background and Update*. Retrieved February 20, 2009, from http://www.red3d.com/cwr/boids/

Rose, J. (1998). Evaluating the contribution of structuration theory to the information systems discipline. In W. R. J. Baets (Ed.), *6th European Conference on Information Systems* (pp. 910–924). Aix-en-Provence, France: Euro-Arab Management School.

Rose, J., & Lewis, P. J. (2001). Using structuration theory in action research: an intranet development project. In L. Russo, B. Fitzgerald, & J.I. DeGross (Eds.), *Realigning Research and Practice in Information Systems Development: the Social and Organisational Perspective* (pp. 273–296). Dordrecht: Kluwer Academic Publishers.

Rose, J., & Scheepers, R. (2001). Structuration theory and information systems development; frameworks for practice. In S. Smithson, & S. Avgerinou (Eds.), *The 9th European Conference on Information Systems* (pp. 217–231), Bled, Slovenia, June 27–29.

Rousseau, D. M., & House, R. J. (1994). Meso organizational behavior: avoiding three fundamental biases. *Trends in Organizational Behavior*, *1*, 13–30.

Schmidt, K., & Bannon, L. (1992). Taking CSCW Seriously: Supporting articulation work. *Computer Supported Cooperative Work, 1*(1–2), 7–40. Retrieved April 4, 2008, from http://citeseer.ist.psu.edu/schmidt92taking.html

Schmidt, K., & Simone, C. (1996). Coordination Mechanisms: Towards a Conceptual Foundation of CSCW Systems Design. *Computer Supported Cooperative Work, 5*(2/3), 155–200. Retrieved April 2, 2008, from http://citeseer.ist.psu.edu/schmidt96coordination.html

Simon, H. A. (1996). *The Sciences of the Artificial,* (3rd Ed.). Cambridge, MA: MIT Press.

Stamper, R. K. (2001). Organisational semiotics. Informatics without the computer. In K. Liu, R. J. Clare, P. B. Andersen, R. K. Stamper (Eds.), *Information, organisation and technology. Studies in organisational semiotics* (pp. 115–171). Boston: Kluwer Academic Press.

Strauss, A. (1985). Work and the Division of Labor. *The Sociological Quarterly*, *26*(1), 1–19. doi:10.1111/j.1533-8525.1985.tb00212.x

Virkkunen, J. (2004). Hybrid agency in co-configuration work. In *3rd Nordic conference on Cultural and Activity Research*, 3–5 September, Copenhagen, Denmark.

Virkkunen, J., & Kuutti, K. (2000). Understanding organizational learning by focusing on "activity systems." *Accounting . Management and Information Technologies*, *10*(4), 291–319. doi:10.1016/S0959-8022(00)00005-9

Walsham, G. (1997). Actor-network theory: Current status and future prospects. In A. Lee, J. Liebenau, J. I. DeGross (Eds.), *Information systems and qualitative research* (pp. 466–480). London: Chapman & Hall.

Weber, M. (1983). *Ekonomi och samhälle. Förståendesociologins grunder*. [Economy and society: an outline of interpretive sociology] (pp. 3–22). Lund, Sweden: Argos (in Swedish).

Wikipedia (2008b). Retrieved Aug 10, 2008, from http://en.wikipedia.org/wiki/Work_system

Wikipedia (2009c). Retrieved Jan 30, 2009, from http://en.wikipedia.org/wiki/Complex_adaptive_system

Work systems theory (2009). Retrieved from http://www.fsc.yorku.ca/york/istheory/wiki/index.php/Work_systems_theory

ENDNOTES

1 The survey of ST is mainly collected from Jones (1999) and Rose (1998).

2 An elaborate discussion of sense-making in relation to an alternative work systems approach suggested by Jasperson, Carter, & Zmud (2005) is found in Alter (2006b).

Chapter 9
The Practical and Theoretical Trails in Hindsight

In Section 1 and Section 2, I described the practical and theoretical background of the ADT. In Section 3, I have given an account of the ADT itself. At this point, it is time to look back and see how these three parts are related to each other.

RECAPITULATION OF THE PRACTICAL TRAIL

The insights and needs from the empirical experience at Ericsson were summarized in Section 1 as follows:

- **Common understanding***:* Achieving common understanding is a difficult task that is easily overlooked.
- **Integrating construct:** Many observations pointed to the need of some organizational construct that could serve as a whole; integrating elements like IS/IT, business processes, information structures, corporate business rules and norms. Usually, the business process is assigned this role, which leads to a dilution of the process concept, and makes it less pregnant for analytical purposes.

DOI: 10.4018/978-1-60566-192-6.ch009

- **Contextualization***:* Many observations indicated that situational and contextual aspects must be given more attention and importance in the organizational discourse.
- **Recurrent patterns***:* Certain patterns appeared over and over again – temporal ordering of things, how things were related to each other, how to do things, transitions between contexts. This indicated the existence of an underlying structure whose existence could be perceived in the everyday workings of the organization.
- **Enactment***:* the importance of enactment – the emergence of new structures by constant engagements of people with the technology at hand. Enactment also emphasizes the importance of being sensitive to the history of organizational artifacts like information models, process models, etc.

RECAPITULATION OF THE THEORETICAL TRAIL

From another direction, a collection of theoretical instruments were outlined in Section 2:

- *Praxis*: The central concept in Marx's philosophy; the nexus of human activity in which mankind creates the social existence of itself.
- The *dialectical relation*: Indicating that the relationship between pairs is *internal*, in which opposites are different but mutually depending on and impacting each other within a totality.
- The *epistemology of praxis*: Issues such as "truth", "reality", "facts" cannot be treated as something external to praxis. On the contrary, they lose their meaning outside praxis.
- The *dialectical method*: The object of inquiry is a whole where the parts are related internally to each other and mutually constitute the whole and the parts – a *concrete totality*. The structure of the whole is based on the diversity of its parts, and this diversity gives rise to the dynamic evolution of the whole and its parts. In every concrete totality there is a particular component that has a unique position in the sense that it determines the other. This component is called the *concrete universal* of the totality, and can be thought of as an elementary form of the whole. The concrete universal is the germ, the central category which determines the evolution and structure of its own progress and of the whole. The task of the dialectical method is to reproduce the concrete totality and its "seed", the concrete universal.
- The concept of *activity* in Activity Theory: The social phenomenon in which a group of humans join their forces in order to fulfill social needs. The activity itself is oriented towards a motive, which actors ultimately need to attain. The actions needed to achieve the motive are directed towards work objects. The activity is the social fabric in which individual actions are meaningful.
- *Mediation*: Refers to the idea that humans always put something else between themselves and their object of work.
- *Meaning*: Fundamental for understanding the human mind.

THE ACTIVITY DOMAIN THEORY: SUMMARY

The main features of ADT are as follows (see Figure 1):

The *activity domain* is a work setting whose existence is *motivated by* its capability to produce some *outcome* that fulfils some social needs. The result is achieved through the actions of socially organized

Figure 1. A conceptual model of the activity domain

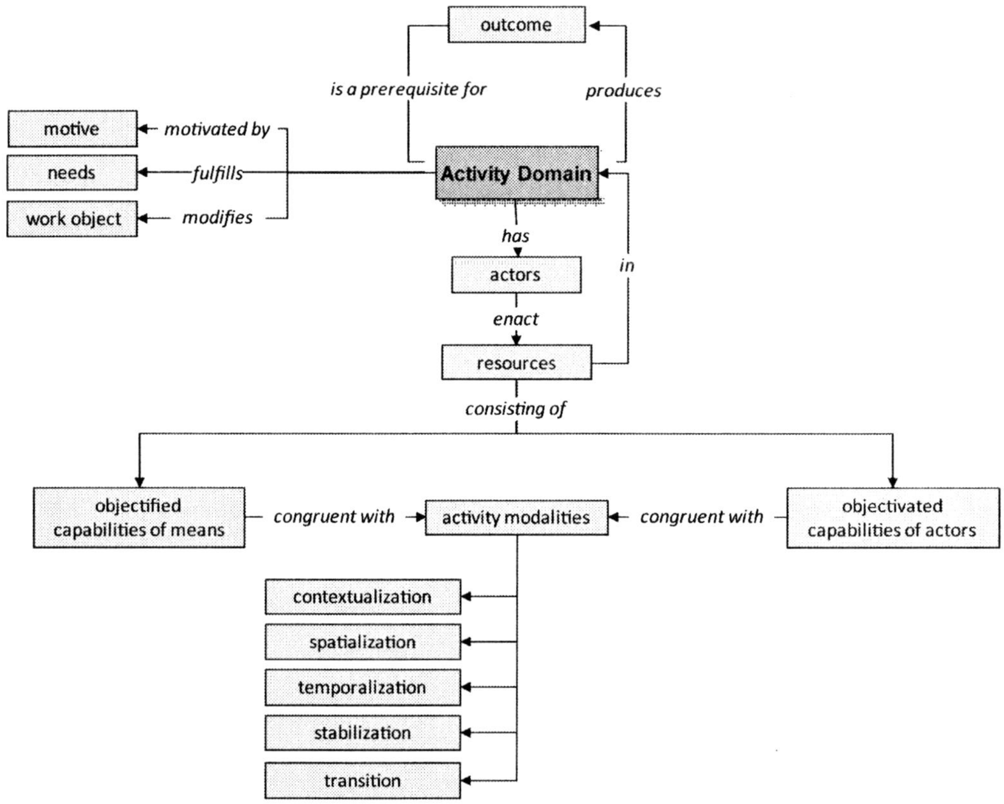

actors, which modify a *work object* into an outcome. In this activity, actors *enact* useful resources consisting of *objectified capabilities of means* and *objectivated capabilities of actors*. Potential resources do not become resources until the actors have drawn these into the social fabric of the activity domain, and collectively learnt how to use them for the benefit of the domain.

The capabilities are enacted along the dimensions of the *activity modalities contextualization*, *spatialization*, *temporalization*, *stabilization* and *transition*. The objectified and objectivated forms of enactment are *congruent* in the sense that both are derived from the activity modalities; humans draw on the given physical world to enact modality specific capabilities. We draw maps because our cognitive apparatus can cognize spatial structures; we print calendars because our cognitive apparatus can cognize temporal structures, and so on.

The outcome of one domain may be the *prerequisite* for another domain, which means that the activity domain construct is recursive and scalable. The same structure is found in every domain. Teams, projects, functional units, divisions, subsidiaries, the entire organisation, networks of organisations, etc., can all be apprehended as different activity domains. In this way, the problematic issue of "levels" in organizational discourse can be circumvented; the key issue is whether these work contexts have a motive and a work object being transformed, and what constellations of domains are necessary for an organization or cluster of organizations to develop a product or a service.

TRACING THE ACTIVITY DOMAIN THEORY TO ITS EMPIRICAL AND THEORETICAL ROOTS

One of the first aspects that captured my interest in the Ericsson practice was the glaring difficulties in arriving at communal meaning about how coordination was to be conceived at the S-domain in Stockholm where I worked. It was quite obvious from the daily practice of implementing the Matrix support that meaning somehow had to be "managed" along with more tangible things like products, requirements, base-lines, milestones, and so on. It was near at hand to regard the construction of meaning as something that had to be done systematically and subjected to engineering methods, similar to the engineering of systems in general. In order to achieve this, the essence of meaning had to be articulated. What is meaning? How does it come about? Is meaning individually or collectively based? Is there some way that the construction of meaning can be accelerated in practical settings? Which are the prerequisites for the emergence of communal meaning? Some answers to such questions I found in Activity Theory, where the concept of meaning has been extensively investigated.

The next unanticipated experience at Ericsson was the emergence of three separate activity domains – the S-domain in Stockholm, the A-domain in Aachen and the L-domain in Linköping. All these domains were heavily involved in the 3G projects, and it made good sense to assume that they should have evolved in a similar way. However, although not intentional, these domains quickly developed into more or less incompatible ones. The information models were quite dissimilar; the entities differed in type and characterization, the relations were different, the revision rules were different, etc. Subsequent efforts to consolidate these domains into one (the C-domain) were arduous to say the least. Even at the time of writing this book (2008) the A-domain still remains as an independent domain, and the C-domain has evolved into a prodigious, inert domain serving large parts of Ericsson.

It is natural to see the evolution of separate, incompatible domains serving the same projects as a failure, and of course, in one sense it is. There were obvious needs for some commonality between the domains. For example, the central entities delivered to the integration activity such as the "work package" needed to be characterized in the same way in all domains. This was never achieved and the necessary coordination between domains had to be maintained by other means.

However, from another point of view the separation into disparate domains, each evolving according to its own logic, was a complete success. Each domain provided coordination support that was previously unmatched in the history of Ericsson. The management of requirements, engineering change orders, error reports, support for the anatomy-centric development method, base-lines, milestones, products, documents, etc., in one and the same IS had never before been achieved. At that time (1999-2002) the functionality implemented in each domain was quite unique in the PDM/PLM area, and still is to a large extent.

Since the platforms that these applications were built on (Matrix and its PDM/PLM competitors) are widespread and established in many organizations, there had to be other reasons than the platform technology for the successful implementations at the different Ericsson sites. It is my firm conviction that the main explanation has to do with the way the construction of communal meaning was carried out. It appeared that the effort of arriving at communal meaning, with the myriads of distinct implementation decisions to be taken, simply became insurmountable when too many stakeholders were involved and too large a number of conflicting requirements had to be considered. Also, it was clear that most of the functionality finally implemented was enacted along the way as new insights, needs and possibilities unveiled in action. The sites were in constant evolution, manifested both as tangible imprints in the implementation in Matrix, in the models, and as intangible imprints of communal meaning among actors. In this process, the Matrix tool and the models were mediational means in achieving the desired coordinative capabilities.

A consequence of the different paths taken by the various domains was that opinions differed between them about what things were relevant; what the world consisted of, so to say. What made perfect sense in one domain was nonsense in the other domains. This was especially evident when the construction of the domains became detailed enough for the models to be implemented in Matrix. The different world-views were striking illustrations of the "epistemology of praxis". Issues about "truth" and "facts" can be settled only in relation to the activity domain. It makes no sense to argue if a construct in domain is "true" or not. Usefulness is what matters.

The ideas of Activity Theory also made me realize the profundity of coordination of human activity. It goes without saying that coordination is a fundamental issue in all organizations, even if the nature of this phenomenon is less well understood. Coordination is often conceived in a superficial way as managing the dependencies between actions. According to Activity Theory, however, coordination is a cardinal element of socially organized human activity due to the division of labor.

Even if many of the experiences from the Ericsson practice can be addressed by the praxis approach, there is one major issue that I found insufficiently treated in Activity Theory or, for that matter, in other theories addressing the nature of ISs and IS development[1]. This issue concerns the nature and relevance of the recurrent patterns that I experienced in the daily work at Ericsson. The responsibilities for tasks such as improving business processes, running information modeling projects, deciding business rules, outline organization structures, and defining collaboration between units, were assigned to different business units. This allocation of responsibility always seemed to introduce glitches in the interaction across the responsibility borders. It was quite obvious that the patterns, which I subsequently coined activity modalities, were much more interrelated than the organizational responsibility allocation indicated. Thus, I could sense an underlying structure behind these patterns; a structure that I felt had to be related to meaning and to the cognitive capabilities of humans.

In order to place the modalities on a firm ground, I needed to collect evidences for these in cognitive sciences. On the surface, it is self evident that the modalities play a key role in the coordination of human actions; otherwise we would not see traces of them all around us in everyday life. However, the step from the self evident to the underlying mechanisms is not by any means trivial. An even more intriguing proposal is the congruence between activity modalities and the social fabric that we erect in our societies. In other words, we construct not only physical artifacts according to "the measures of humankind", but also meaningful, symbolic artifacts. Again, on the surface this is no mystery. Why should we construct symbols that are meaningless to us; symbols that we cannot make any sense of? Whatever the outcome of future research will be, the activity modalities are at the core of the ADT. In a broad sense, this construct should be seen as the main contribution of the book.

The dialectical method captivated me as an inspiring approach towards understanding what was happening at Ericsson. The notion of a concrete totality provided me with a theoretical apparatus to approach this problem. In the language of praxis, activity and its subsequent elaboration into the activity domain in ADT can be understood as a *concrete totality* composed of individual phenomena; a concretization of previously isolated, abstract elements. These elements are of course the activity modalities, and the activity domain is the *concrete universal*, the irreducible unit of analysis around which everything revolves. The dialectical interrelationships between the domain and the modalities imply that none of these elements can be treated in isolation; they all mutually constitute each other. Like the DNA is the core building block of living organisms, the activity domain can be seen as the basic building block of organizations; the "DNA of the organization", so to say.

So, to conclude Section 3, the works of early Marx and Engels and their followers have provided me with "sensitizing devices" for comprehending the complex phenomena that I was confronted with in the Ericsson practice. Once I had conceived the activity domain as a concrete totality, it followed that the domain was an excellent candidate for the integrative construct that the practical experiences had called for. The path from the original praxis philosophy, over the "activity" concept in Activity Theory to the activity domain and the activity modalities, seemed to provide a thorough theoretical ground for further inquiries. The implications of this line of reasoning are what Section 4 is all about.

REFERENCES

IS theories (2008). *Theories Used in IS Research.* [Wiki]. Retrieved Sept 21, 2008, from http://www.fsc.yorku.ca/york/istheory/wiki/index.php/Main_Page

ENDNOTE

1 For a comprehensive overview of theories used in IS Development, see IS theories (2008).

Section 4
Implications

The implications of the Activity Domain Theory for managing complex projects, enterprise architectures, product lifecycle management, and alignment of business strategies, knowledge and IS/IT. Business process reconceptualization. From information system development to domain construction. The activity domain as an integration construct. Concluding remarks.

INTRODUCTION

As with any good theory, it should be possible to apply it in many areas. In this concluding Section 4 of the book the implications, or consequences, of taking the Activity Domain Theory (ADT) as the point of departure for organizational inquiries is discussed. The main focus is on the coordination of complex projects, which is quite natural since this area is the origin of ADT. With the maturing of the theory, I came to realize that it had broader implications. I have concentrated on some areas that are highly relevant for organizations today: the development of enterprise architectures, managing product related data throughout the lifecycle of a product, and aligning business strategies, knowledge and IS/IT. Other areas could have been included, for example, knowledge integration (e.g. Grant, 1996b).

The book is concluded with the message that new insights into organizational inquiry may be opened up by taking the activity domain as the fundamental organizational construct. The simple reason for this is that the activity domain is where socially organized work becomes meaningful to actors. Through the activity modalities, individual cognition and coordination of actions are bridged in a way that makes it possible to devise mediational means that can indeed make a change in practice.

Chapter 10
The Anatomy-Centric Approach Towards Managing Complex Projects

The purpose of this chapter is to describe and discuss the anatomy-centric approach towards coordinating complex development projects. The content of the chapter is mainly based on my empirical experience from Ericsson.

The development of a telecom system must deal with a multitude of technical, market related, cultural and organizational interdependencies. Most often, this requires mutual adjustment across many types of boundaries (Adler, 1999). For example, telecom systems are to a great extent implemented in software and developed on a global basis. The interest in distributed software development[1] has increased due to factors such as reduced costs, the access to well-educated labor pools, the possibility of 24-hour development, global presence, and proximity to customers (e.g. Battin, Crocker, Kreidler, & Subramanian, 2001; Damian, 2003; Ebert & De Neve, 2001; Gorton & Motwani, 1996; Heeks, Krishna, Nicholson, & Sahay, 2001; Herbsleb & Moitra, 2001; Karlsson, Andersson, & Leion, 2000; Komi-Sirviö & Tihinen, 2005; Kraut & Streeter, 1995; McChesney & Gallagher, 2004; Ovaska, Rossi, & Marttiin, 2003; Prikladnicki, Audy, & Evaristo, 2003; Sakthivel, 2005).

Many challenges already present in centralized software development are aggravated by the distribution. At the core of these challenges lies the issue of coordination. "While there is no single cause of the software crisis, a major contributor is the problem of coordinating activities while developing large software systems" (Kraut & Streeter, 1995, p. 69). The anatomy-centric approach is one possible answer to these challenges.

DOI: 10.4018/978-1-60566-192-6.ch010

FROM WATERFALL TO INCREMENTS

During the late 1980s and early 1990s many software producing organizations were trying to find alternatives to the then prevalent "waterfall" model for software development (e.g. Brooks, 1995). Examples of such alternatives are evolutionary delivery models, spiral models, iterative development, etc. The traditional waterfall model consisted of phases performed in sequence: analysis, implementation, and integration & test. In the analysis phase, the total functionality of the system was distributed to subsystems and modules, which were developed and tested in the implementation phase. In the integration phase, modules were gradually integrated and system tested until the entire system functionality was achieved.

However, the waterfall model had several intrinsic problems (ERI-1996-04-04; Karlsson, 2002):

- The time span between specification of the system and feed-back to customers was wide. After providing the requirements at the beginning of the project, the customer had to wait for a long time before seeing any results.
- Requirement changes were hard and expensive to accommodate for. Changes in functionality might lead to the re-opening and re-design of already completed modules.
- The delayed integration of modules (called "Big Bang" integration) meant that potential interface problems between the modules were discovered late in the project. Moreover, the concentration on testing and integration towards the end of the project put a heavy burden on test personnel and system test equipments.
- The early distribution of the total functionality on modules brought about a focus on module quality rather than system quality. There was a tendency for designers to concentrate on their specific module, and paying less attention to interactions between modules.
- The focus on modules resulted in an impaired project control. Since the progress of the project was measured from how many modules were ready, problems caused by potential integration problems and/or lack of customer focus appeared late in the project, which might have cause unforeseen delays.
- Since the project participants executed each phase only once, the potential for learning was low. Good and bad experiences collected in one phase were hard to apply anew as the first opportunity to do so did not come until a new project was started. By that time, most lessons learnt had been forgotten.

These and other problems stimulated Ericsson to look for alternative development models. Various more or less isolated initiatives were taken at different development units during the early 1990s. All these initiatives had in common some kind of *incremental* developmentapproach, where the total functionality of the system was built in several steps in such a way that each step provided a testable part of the overall system. This is illustrated in Figure 1.

Several advantages were anticipated from the transition from the waterfall to the incremental model:

- Since the increments can be developed to a certain extent in parallel, the lead-time of the project can be reduced. Although each phase in fact stretches over a longer period of time (see Figure 1), the overall time span is shorter. However, this reduction comes at the expense of a more

Figure 1. The difference between the waterfall and the incremental model

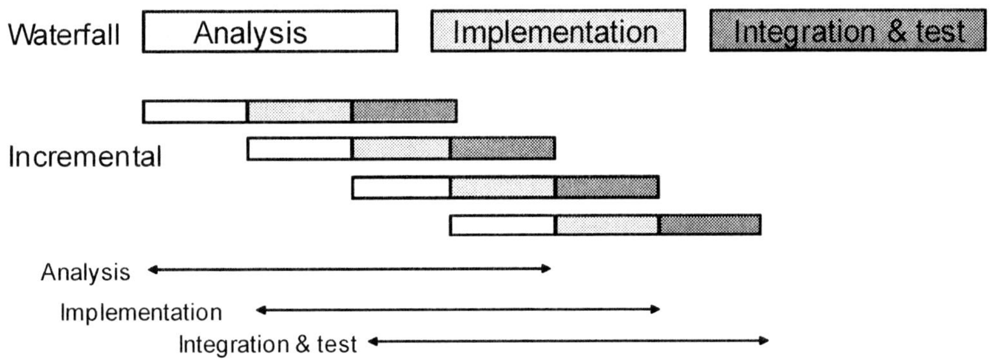

complicated procedure for integration and testing the increments, since each increment may build on previous ones. For example, late detection of errors implies that earlier, already tested and integrated increments have to be tested anew (so called regression testing).

- The incremental model is based on iterating analysis, implementation, and integrating & testing several times. This means that the model is, to a certain extent, better prepared to cope with unstable requirements. A late change in a requirement can in some cases be accommodated for in a later increment. The project can start with implementing those requirements that are believed to be most stable.
- Each increment can be seen as a "mini"-version of the entire system. This means that it is easier to keep a focus on the whole picture.
- Since increments can be defined in such a way that each increment delivers an executable part of the overall functionality, the feedback to customers and designers is much quicker. This provides a better control of the overall project since possible misunderstandings of what is to be delivered can be dealt with early on.
- Integration problems can be solved earlier since the "Big Bang" integration towards the end of the project is avoided. In addition, more time can be spent on system testing since the entire testing period can start earlier. Also, this means that the load on test resources can be relieved. An important aspect concerning testing is that all tests are performed on the "real" system. By avoiding testing against emulators and prototypes, more comprehensive fault coverage is achieved. Furthermore, platforms can be tested earlier. Possible bottlenecks in performance can be remedied early.
- The incremental development model enforces planning, coordination and monitoring of the project to a greater extent than the waterfall model. Since a particular module may be involved in several increments, there is a greater need to understand how the functionalities of different increments depend on each other. However, the planning needs not to be performed centrally; this can be done by teams working with the increments. Moreover, constant re-planning is rather the norm, mainly due to two reasons: customers may change their requirements, and new insights are acquired as the project proceeds.

Given the problems of the waterfall method and the anticipated promises of incremental development to cope with these, the question was how to make the incremental approach operational. The first real test was the previously mentioned project delivering a mobile system to the Japanese market. The problems faced were enormous. The product development task was very complex, and involved coordinating deliveries from several hundreds of participants. In particular, three basic problems were looming large (Järkvik, Berggren, & Söderlund, 2007). First, the project members needed to understand interdependencies, and acquire a common view of how these were to evolve during the project. The second problem can be denoted "the fog of errors", which refers to unknown errors that have to be identified and corrected in a particular sequence without introducing new errors on the way. The third problem was the dynamic nature of the technical development and the corresponding volatile expectations of the customer. Thus, the challenge was, wryly summarized, to deliver "anything at all on the right day, with properties we do know anything about" (ibid, p. 11).

A key insight, which gradually emerged in dealing with these problems, was that the most important thing when working with complex systems is to *work from how things depend on each other*. In order for this to be possible, there is a need for some conceptualization of dependencies that people can agree upon and use for taking actions. Now, several different types of such conceptualizations are conceivable. The question was, however, which one would suit the project needs the best? The answer turned out to be the *anatomy* of the system. This was an insight that had far reaching consequences. Over time, the anatomy became the cornerstone in the anatomy-centric approach towards coordinating complex projects at Ericsson.

THE ANATOMY

The anatomy is an illustration – preferably on one page – that shows *the dependencies between capabilities* in the system from start-up to an operational system (Adler, 1999; Anderstedt, Anderstedt, Karlsson, & Klasson, 2002; Jönsson, 2006; Lilliesköld, Taxén, Karlsson, & Klasson, 2005; Taxén & Lilliesköld, 2005). Here, "capability" shall be understood as the capability of a certain system element to provide a utility that other system elements need.

In order to explain the anatomy construct, we may use a variant of the specification based data model suggested by Gandhi & Robertson (1992; 1995). Suppose we have a system as illustrated in Figure 2. A certain user capability 1 ("text processing") is implemented by module 1 ("Word 2007") in the system. In order for module 1 to provide capability 1, two other capabilities are needed: "executing capability" and "Internet access" (for accessing clip arts, program updates, etc.). These in turn are implemented by two more modules: "Windows OS" and "Internet Explorer". It should be noted that capabilities can in general be implemented in several ways. For example, "executing capability" may be implemented by the "Mac OS Leopard" and "Internet access" by "Google Chrome".

Various alternatives to the basic structure in Figure 2 are possible. For example, a module may implement several capabilities, a capability may be needed by several modules, and a capability may be implemented by several modules.

An important aspect is that *a capability can only be defined in a context*. To take another example, suppose a capability is "containing water for drinking purposes", which can be implemented by a bucket. This capability of the bucket needs other capabilities; the capability of "water access" (might be implemented by a tap), and "water supply" (implemented, for example, by a well). It is only in the context of "containing water" that this capability of the bucket is engaged. One might conceive other contexts where other capabilities are brought to the fore. For example, the bucket may be turned upside-down and climbed on in order to reach something.

Figure 2. An example of a system

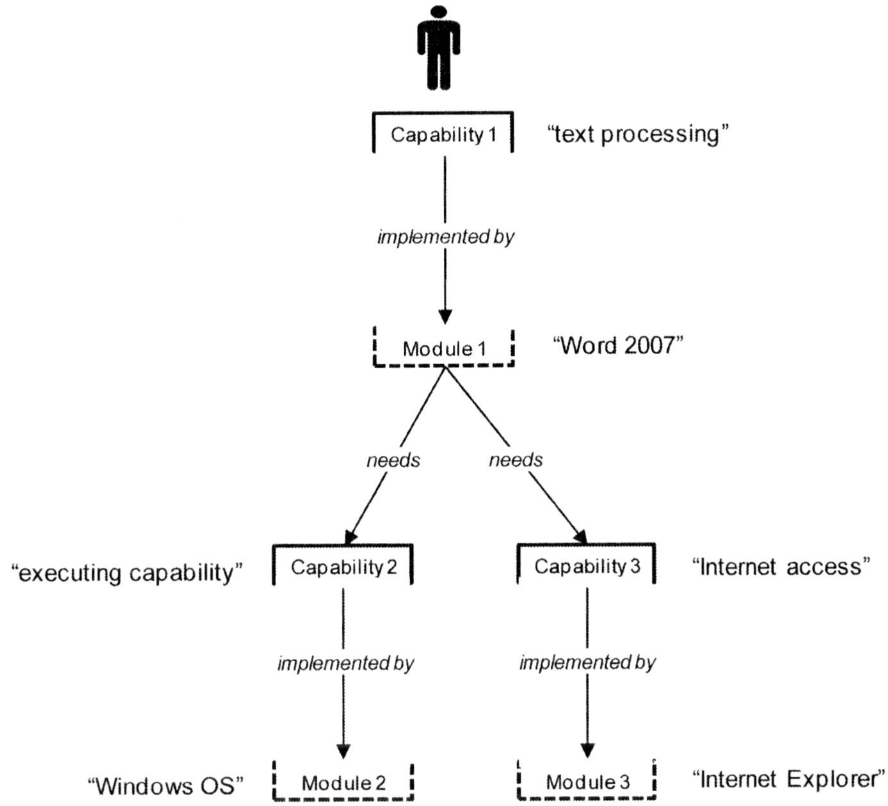

Now, the anatomy is achieved by *suppressing the implementing modules* (see Figure 3).

Thus, with the anatomy the system is, so to say, reduced to its "canonical" form; it is simplified as much as possible without losing its basic systemic feature: the dependencies between capabilities. In a way, the anatomy can be seen as an example of the principle attributed to Einstein: "Everything should be made as simple as possible, but not simpler."

An example of a realistic anatomy from the Ericsson practice was given in Figure 7, p. 25. Another example is shown in Figure 4.

The boxes (the details of which are less important here) should be read as a capability provided by one or several modules (subdued in the figure). The dependencies (lines) proceed from the bottom to the top of the anatomy. For example, the capability "EMRPS Start" (an extension module regional processor with a speech bus interface) is a basic capability. If this capability fails, the whole system will fail. In line with this, the gist of the anatomy-centric approach is to design and test the system *in the same order as the capabilities* are invoked. In a metaphorical sense this can be seen as the order in which the system "comes alive", hence the term "anatomy".

Since the anatomy is a simplified illustration of the system, it is of course necessary to attend which modules implement which capabilities. This can, for example, be done in an appropriate IS such as the Matrix system used at Ericsson (see the example in Figure 5).

Figure 3. The anatomy of the same system as in Figure 2

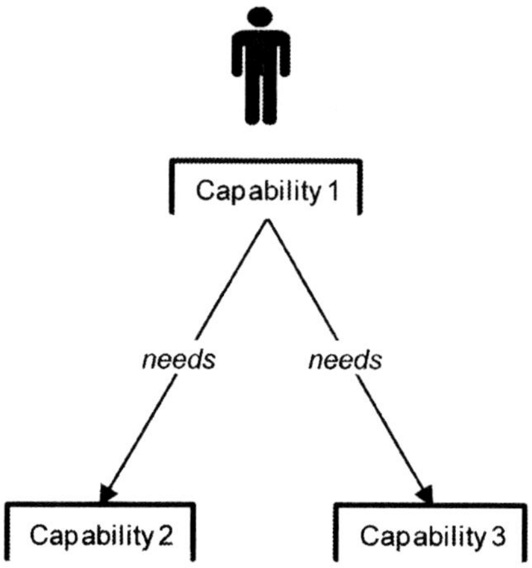

Figure 4. An example of an anatomy for a radio-base station

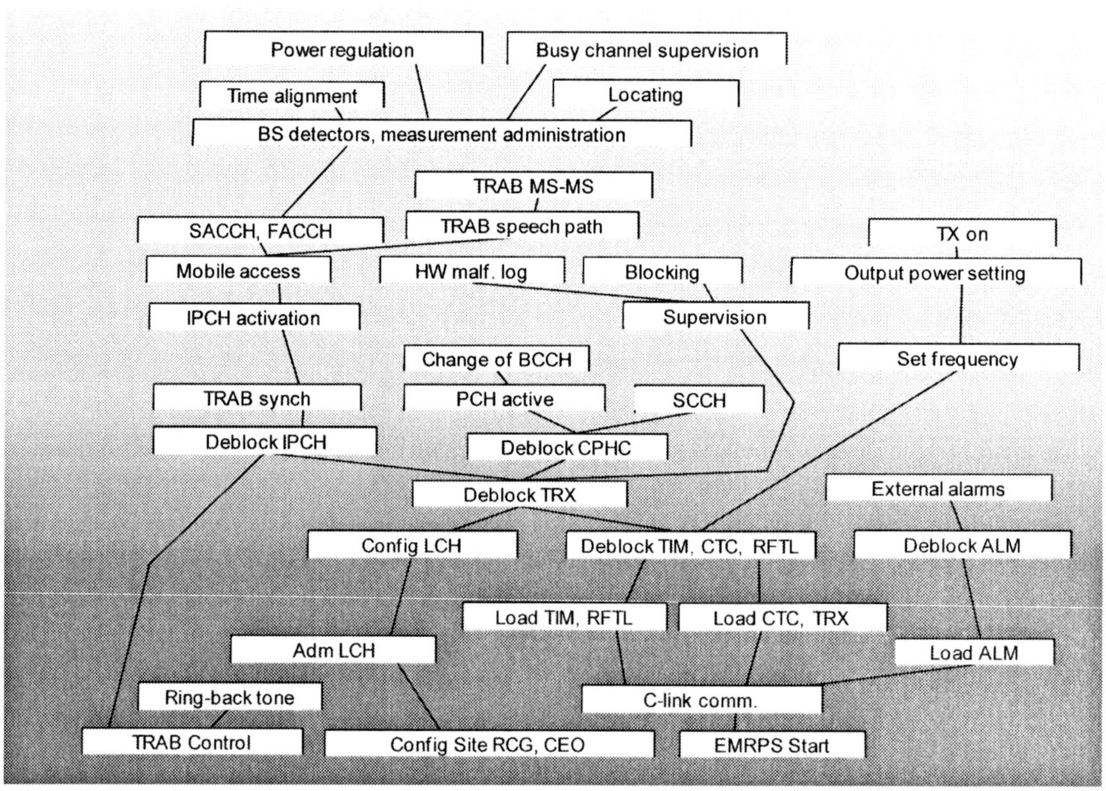

The importance of the anatomy lies in the fact that it provides a clean and easy-to-apprehend illustration of the fundamental dependencies in the system. The consequences of a missing or delayed capability in a project can easily be inferred from the anatomy.

Figure 5. Capabilities (work packages/columns) versus implementing modules (blocks/rows). The state "RFLSV" (Ready for latest system verification) indicates that the work package is ready for inclusion in the next system build

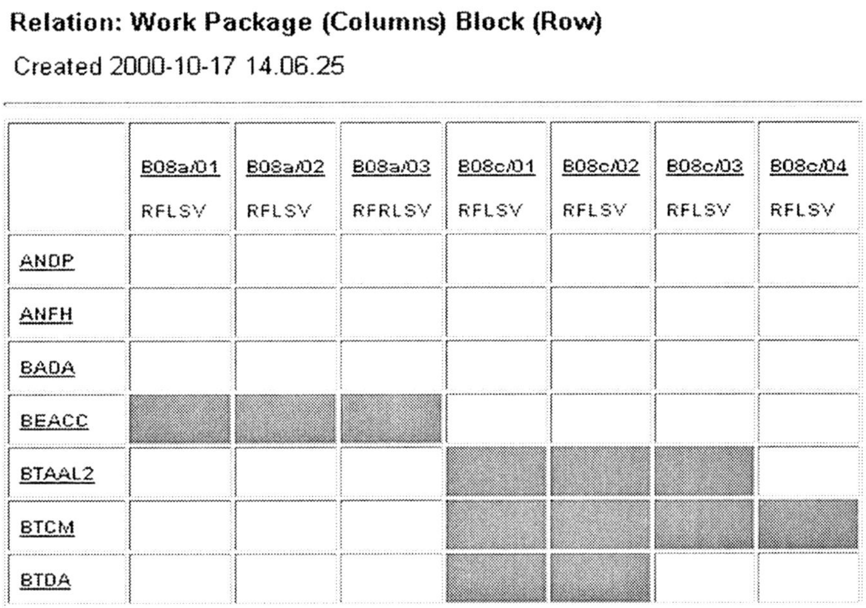

Relation: Work Package (Columns) Block (Row)

Created 2000-10-17 14.06.25

	B08a/01 RFLSV	B08a/02 RFLSV	B08a/03 RFRLSV	B08c/01 RFLSV	B08c/02 RFLSV	B08c/03 RFLSV	B08c/04 RFLSV
ANDP							
ANFH							
BADA							
BEACC							
BTAAL2							
BTCM							
BTDA							

Since the anatomy is an image of the way the system is built up, it is near at hand to conceive the anatomy as an *architecture* of the system. So, let's see how architectures can be related to the anatomy.

ON ARCHITECTURES

It has long been recognized that the architecture is vital for quality and performance of a system:

An explicit understanding of the underlying architecture is a prerequisite for the design, evolution and maintenance of modern information systems that must complement today's complex business processes spread across internal divisions and external partners. (Smolander, Rossi, & Purao, 2008, p. 575)

According to Booch (2006, p. 16) an engineering discipline "shows signs of maturity when we can name, study, and apply the patterns relevant to that domain." Another reason for attending architectures becomes evident as the size of the project grows. The necessity to divide the work into manageable work packages allocated to different teams, inevitably leads to the emergence of separate "thought worlds" formed by the character of the work. Consequently, the communication between teams is aggravated, and to ameliorate the necessary communication, an easy to comprehend architecture is indispensable (e.g. Ovaska, Rossi, & Marttiin, 2003).

Although the motives for attending architectures are indisputable, there is less agreement of what architecture is (Booch, 2006). Some suggested, general properties of architectures are *simplicity*, *appropriate* for the intended audience, *communicative*, and *minimal* (Jensen, Møller, Sönder, & Tjørnehøj,

2006). The architecture should emphasize those things that are *difficult to change;* the *foundation* on which the rest must be built (Fowler, 2004).

Many contributions in the literature make use of the IEEE Standard 1471-2000, which defines architecture as

[The] fundamental organization of a system embodied in its components, their relationships to each other and to the environment and the principles guiding its design and evolution. (Maier, Emery, & Hilliard, 2001, p. 108)

As can be seen, this definition conceptualizes architecture in a value-free, objective manner. The architecture can be "extracted" from the system, illustrated, and documented as an objective fact. Quite a different view on architecture is given by Johnston:

In most successful software projects, the expert developers working on that project have a shared understanding of the system design. This shared understanding is called "architecture". (related by Fowler, 2003, p. 11)

This definition, to which I adhere, emphasizes that the architecture is a social accomplishment, unavoidably marked by "footprints" from those defining the architecture. This view implies that the architecture, when operationalized in a particular setting, ultimately will be idiosyncratic, at least in parts.

Software Architectures

One of the most widely architectures used in connection with software development is the "4+1" view model proposed by Krutchen (1995), consisting of five *viewpoints*, each representing a certain perspective of the architecture (see Figure 6).

Figure 6. The "4+1" view model of the architecture (adapted after Krutchen, 1995, p. 43)

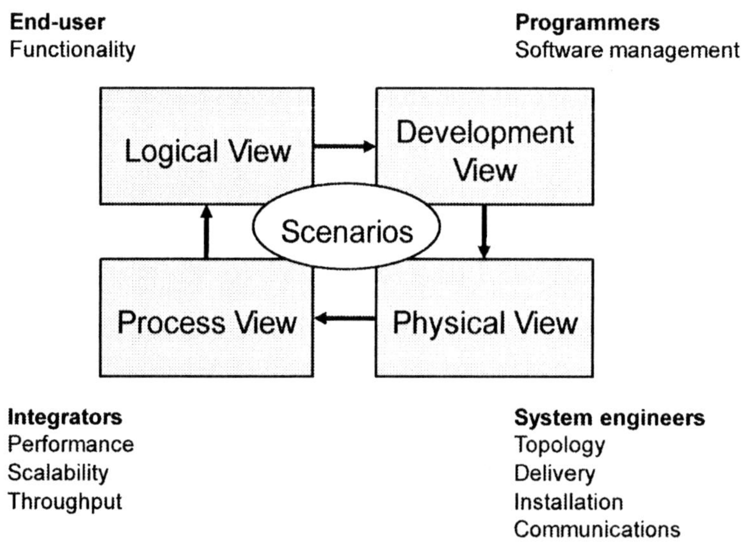

The purpose of the *logical* view is to show what services/features the system provides to its users. For object-oriented systems, this view is usually represented by class diagrams, communication diagrams, and sequence diagrams. The *development* view focuses on the organization of software modules, and is represented by module diagrams showing "export" and "import" relationships. Thus, the development view shows the physical components out of which the system is to be constructed (e.g. source code, executables, physical databases, documentation, code libraries). The *process* view deals with dynamic aspects such as communication between system processes, and the run-time behavior of the system. It takes into account non-functional requirements such as performance and availability. The *physical* view maps the software to the hardware in terms of processing nodes such as a network of computers. Finally, the *scenarios* view puts it all together by integrating the content of the other four views. Usually, the scenario view is represented by use-case diagrams. In Booch, Rumbaugh, & Jacobson (1999) the five views are called *design* view, *implementation* view, *process* view, *deployment* view and *use case* view respectively.

The "4+1" architecture can be related to the anatomy through alternating the "design views" and "implementation views" as illustrated by the example in Figure 7. The first design view corresponds to the uppermost capability in the anatomy. The first implementation view is related to the module(s) that implement the feature (Module 1 in the figure). This in turn needs other capabilities that can be considered as system internal "design views" implemented by system-internal "implementation views".

Figure 7. An example of anatomy-centric feature development

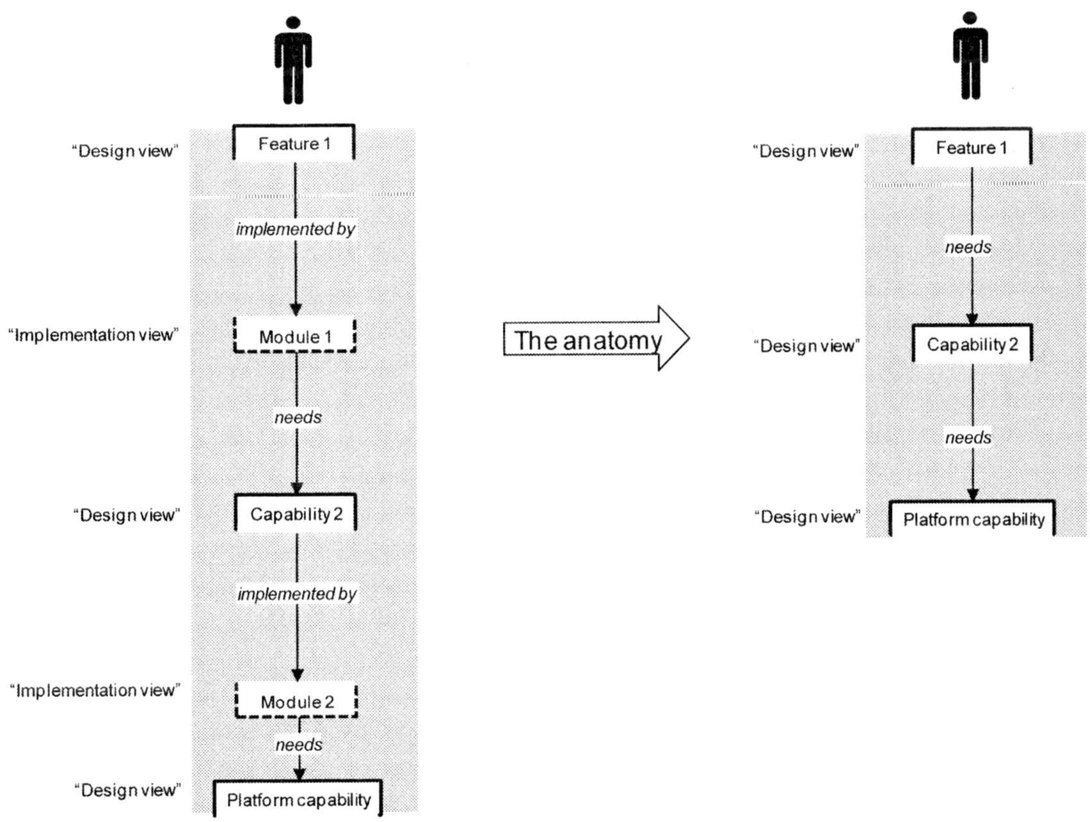

By suppressing the implementations the anatomy can be seen as illustrating the dependencies between "design views", interpreted as capabilities.

Summing up, it can be seen that the anatomy has many of the properties associated with viable architectures: it is simple, appropriate for its intended audience, communicative, and minimal in its expression. It can be defined in such a way that the "hardest to change" elements of the architecture are in focus. For software architectures, the anatomy can be related to the design view in the "4+1" architecture. The other views can be seen as complements to the anatomy. Above all, the anatomy is central in the sense that it centers on the most essential property of a system: the dependencies between capabilities. As we shall see in the following, the anatomy is also explicitly conceptualized as a social construct, which makes it possible to deliberately include social aspects when defining the anatomy.

Recently, there has been a growing interest in so called "architecture-centric methods" (e.g. Nord & Tomayko, 2006; Henry, 2007) in connection with agile development of software. In line with this, I propose the anatomy as the architecture in system development; hence, the notion of the "anatomy-centric" approach, which is the subject of the next section.

THE ANATOMY-CENTRIC APPROACH

The anatomy-centric approach is built around the anatomy as the central construct, and is executed in three phases: *anatomy definition, increment planning,* and *integration planning.* These phases, which are tightly interrelated, have different purposes as described in the following sections.

Anatomy Definition

The purpose of the anatomy definition phase is to achieve common understanding of how the system works in terms of dependencies between capabilities. In Figure 8, an example of a realistic anatomy taken from the Ericsson practice is shown. The boxes and lines indicate capabilities and dependencies respectively. The nature of the dependency is not important, only that there is some kind of dependency that must be attended during the development. Equally important as the presence of dependencies is the *absence* of dependencies.

If the dependencies are under control, a firm ground for coordination has been laid. For example, in the center of the anatomy in Figure 8 is a critical capability (encircled) that many other capabilities depend on. If this capability fails for some reason, the other capabilities will also fail. Thus, it is intuitively clear that it is important to closely watch the progress of the work implementing this particular capability in the project.

The anatomy is created in several stakeholder meetings where the mindset should be: "what is the first capability needed to make the system work?" For most electronic equipments this is access to the electrical power provided by the capability of the "power-on" button. In order to follow the logic of the anatomy-centric approach, the first thing to develop and test should be this button. The anatomy is gradually defined by asking "if you have 'powered-on', what happens then? And then?"And so on. This question is repeated until the full capabilities of the system are reached.

The main stakeholders in the anatomy definition phase are system architects. In addition, other key persons such as customers, project managers, requirement managers, etc., may be included whenever relevant. The final anatomy can be interpreted as the expression of common understanding about the

Figure 8. The anatomy of a processor in a telecom system (Taxén & Lilliesk, 2008. ©Elsevier. Used with permission)

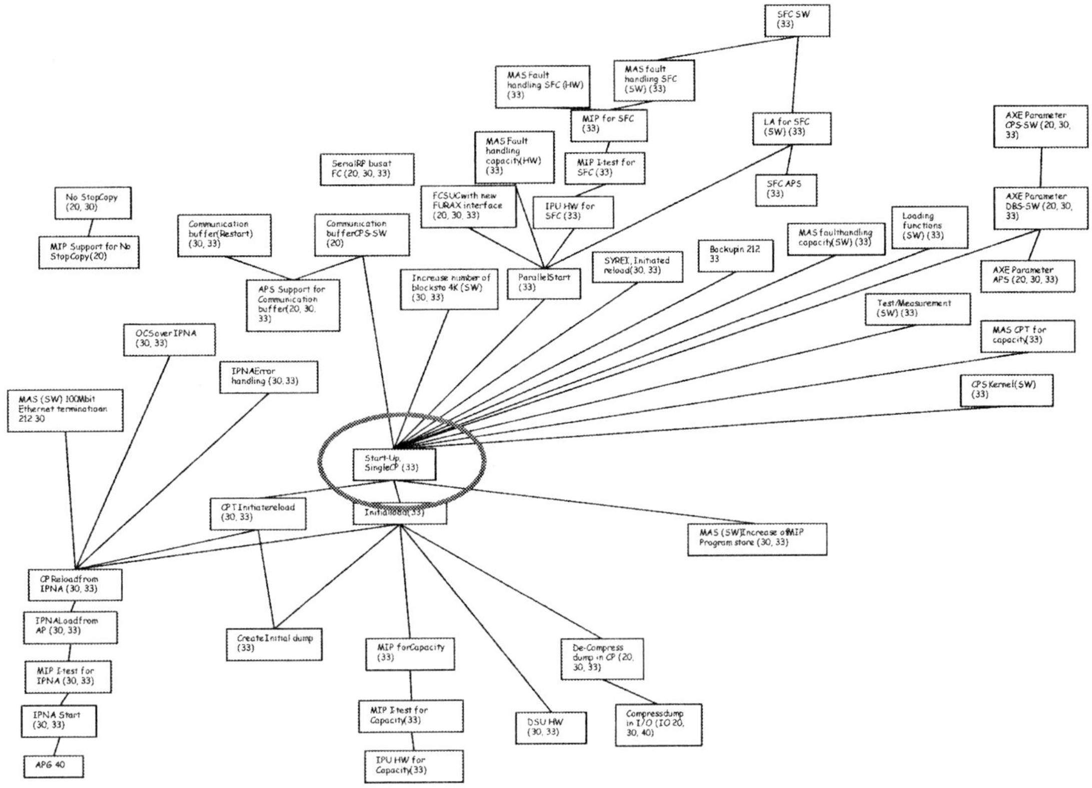

architecture of the system among those participating in the anatomy definition work. This means that the anatomy is dependent on both the system and the participants; thus, it is a social construct. Other participants might have included different capabilities and chosen a different level of detail.

Increment Planning

The purpose of the second phase is to outline the implementation of the system. The capabilities are grouped into development and integration steps — increments — in such a way that the additional capability after each increment is executable and verifiable. The intention is to parallelize design and testing as much as possible. The distribution of the development work is determined by a number of circumstances such as available resources, customer feedback, complicated or simple capabilities, geographical proximity between resources, capabilities that can be tested jointly, etc. The main stakeholders in increment planning are system integrators, system testers and project managers. In this step, the dependencies between increments are brought to the fore. Capabilities and dependencies within each increment are subdued.

Two main ways of laying out the increments can be identified: stacking increments "laterally/horizontally" on top of each other in much the same way as when a brick wall is built, or slicing increments "longitudinally/vertically" based on a common, established and proven platform. The lateral way has

Figure 9. An increment plan of the processor (Taxén & Lilliesköld, 2008. ©Elsevier. Used with permission)

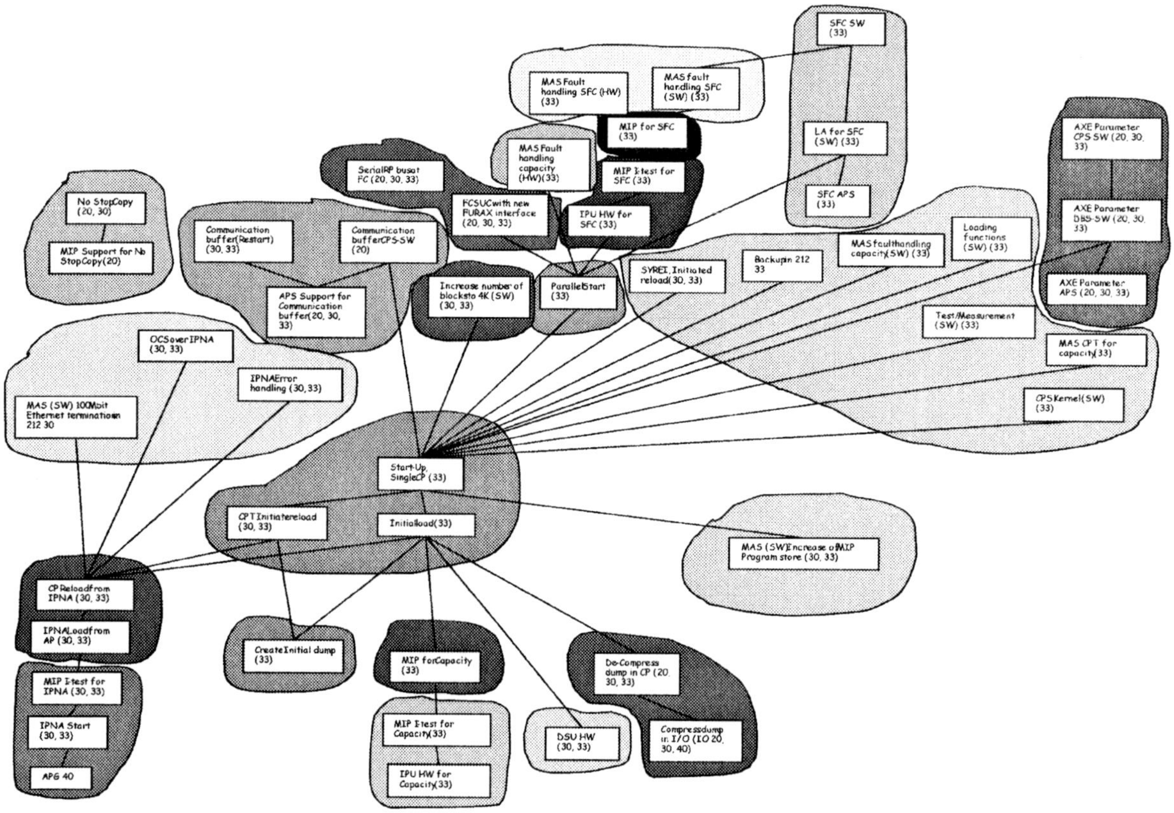

mainly been used in the telecom industry for core architecture development of systems containing a mixture of hardware and software modules. In Figure 9, a possible laterally oriented increment plan for developing the system in Figure 8 is shown.

The longitudinal way is more frequent in connection with software based, agile development of customer features, which will be treated in more detail later on.

The lateral and longitudinal ways of ordering increments represent two extremes. In most cases, in particular in new product development projects, a mixture between these extremes may be utilized. Regardless of which, there is a need to make sure that the architecture of the system is upheld and communicated to all stakeholders in the project.

Integration Planning

In the third phase, the purpose is to divide the work between subprojects and establish common understanding of what is delivered from whom and when. The main stakeholders are project managers and test managers. Resources are assigned and dates for deliveries of the increments to system integration & test are negotiated. For each increment, traditional time and resource plans are made. The dependencies in focus are those between subprojects.

Figure 10. Integration plan of the processor (Taxén & Lillieskóld, 2008. ©Elsevier. Used with permission)

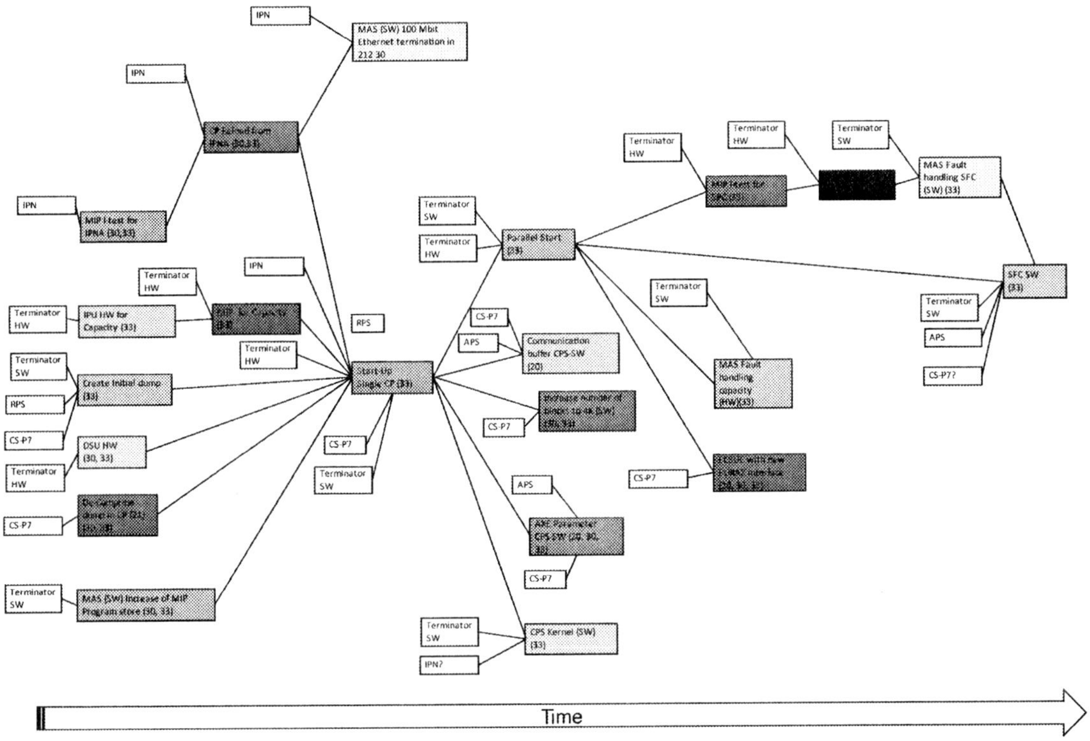

During the project, the integration plan is used as a means for communicating the progress of the project. The state of each increment may be visualized by traffic-light cues such as Green – On Plan, Yellow – Warning, Red – Off Track. Impacts of delays are clearly shown, which give the project management time to react and take corrective actions. In Figure 10, an integration plan of the processor is shown.

At a quick glance, the integration plan appears similar to traditional PERT (Program or Project Evaluation and Review Technique) diagrams in the sense that it shows dependencies between tasks. However, the way the integration plan is constructed is completely different from the traditional construction of PERT diagrams. Moreover, the focus of the integration plan is on expressiveness and ease of use, the ultimate purpose of which is to manage complexity. A comparison of various project-planning techniques from this point of view is given in Ekstedt, Johnson, Lillieskóld, & Jonsson (2003).

In summary, the anatomy-centric approach uses a simple illustration – the anatomy – as a means to achieve common understanding among different stakeholders about how the project should be coordinated. This is of outmost importance when frequent re-planning of the project is necessary. In doing so, all the phases may be affected. Thus, these phases should be seen as interrelated perspectives that make it possible for various stakeholders to control a complex situation.

The Domain Construction Strategy

The anatomy is one way of conceiving the architecture of the system; an architecture that provides a solid ground for managing the coordination of complex projects. However, having a clear picture of the dependencies between capabilities in the system is not enough for coordinating the project. In addition, the coordinative capabilities of the activity domain have to be enacted, i.e., the capabilities that are used in coordinating the transformative actions.

The purpose of the domain construction strategy (DCS) is to operationalize the coordination of the anatomy-centric approach. The main means employed are the information model and the coordination information system (cIS). The information model shows what entities actors consider relevant to manage, and how these entities should be characterized and related to each other. Examples of such entities in product development organizations are products, articles, requirements, documents, engineering change orders, error reports, and the like. For example, in order to trace requirements to be fulfilled by a product, the information model must include a requirement type, a product type and a relation type linking these types to each other. If the anatomy-centric approach is used, the information model must also include elements for managing the anatomy and dependencies between capabilities.

The information model is defined in terms of types, type hierarchies, relationships, attributes, cardinalities, revision stepping rules, state sets, etc. From this model, data for particular projects are entered into the cIS. By continually iterating between modifying the information model, implementing it in the cIS and evaluating the result, coordination is gradually operationalized in terms of the information model, the cIS support and common understanding among the actors about what constitutes coordination (see the Section *The domain construction strategy*, p. 131, for more details). The result is a constructed and enacted coordination domain, meaning that capabilities of actors and mediational means have been established as resources for managing transformative actions in the project.

In Figure 11, an example from Ericsson of an information model from 1999 is shown. The encircled part shows entities relevant for managing the anatomy.

A detail of the requirement context in Figure 11 is shown in Figure 12.

The entire model in Figure 11 must be defined to the level of detail shown in Figure 12. This means that the enactment effort, i.e., the work of turning the model into a resource for coordination, is arduous. Several hundred decisions must be agreed upon by stakeholders involved in defining the model. In fact, it took more than a year of discussions, testing, modifying, and iterating the model and its implementation before it was put into use in a customer project at Ericsson. Even so, the evolution of the model and its implementation continued during and after the first project.

IMPLICATIONS

In Section 1 *The Practical Trail*, I described the development of the 3rd generation of mobile systems between approximately 1999 and 2003 at Ericsson. It was the experiences during this period that shaped the conceptualization of the anatomy-centric approach as an integrated way of coordinating complex development projects. The experiences are mainly collected from two activity domains: the S-domain in Stockholm, Sweden and the A-domain in Aachen, Germany. These domains developed different parts of the 3G system, and the same IS platform was used at both sites: the Matrix PDM system from MatrixOne (MatrixOne, 2008). As it turned out, the A- and S-domains were enacted in completely dif-

Figure 11. An example of an information model in the Ericsson practice (1999)

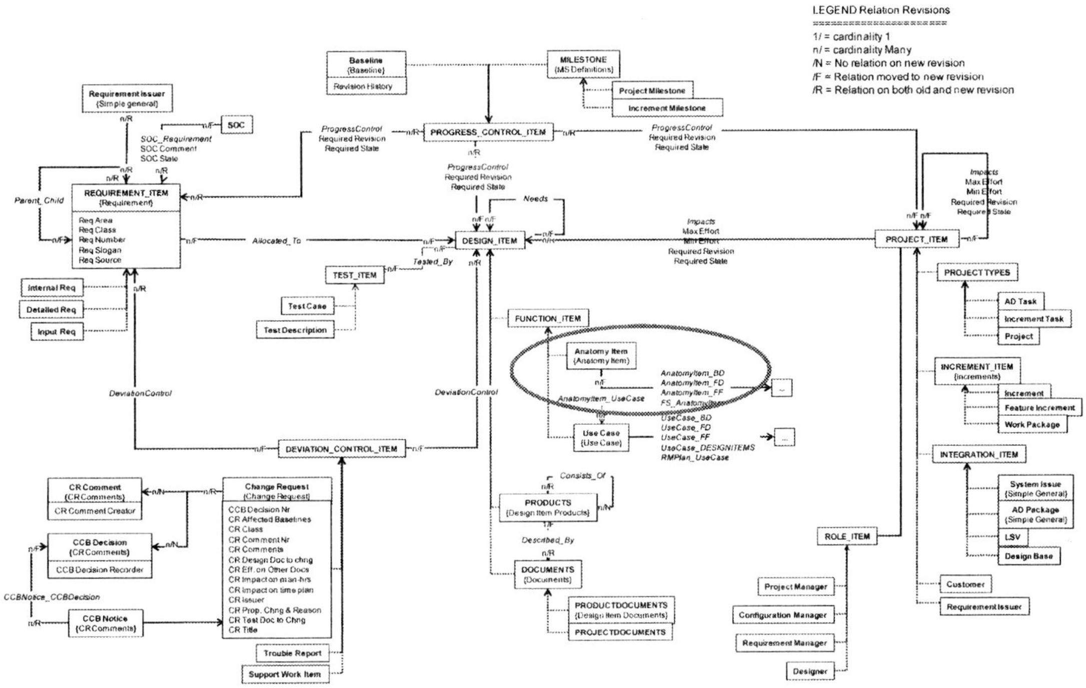

Figure 12. A detailed view of the requirement context in Figure 11

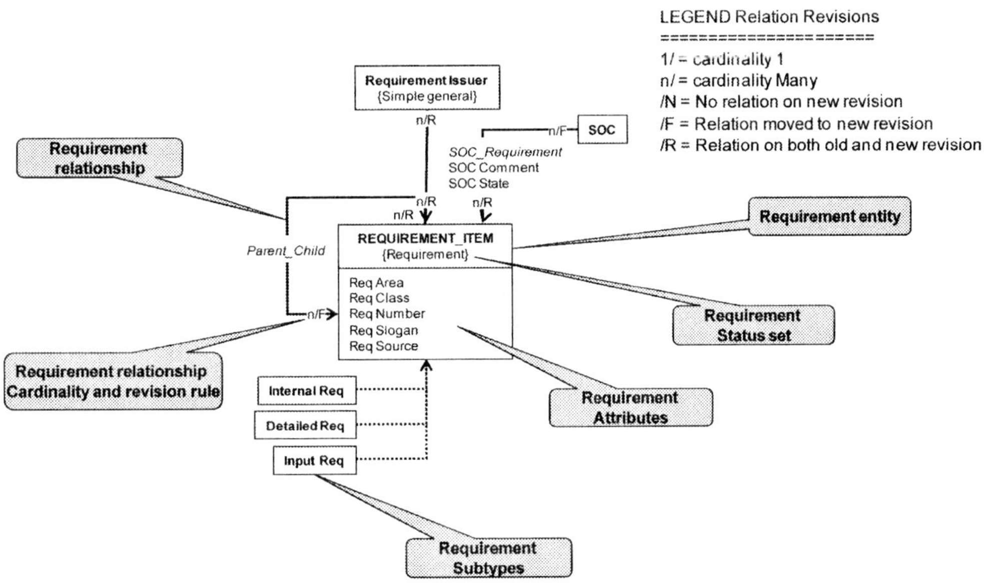

ferent ways, although there was a clear need for maintaining some commonality across both domains. This was however not achieved during the 3G development period that I followed.

In this section, I discuss what kind of implications can be expected from applying the anatomy-centric approach[2]. The experiences from Ericsson indicate that at least the following concerns need to be addressed in order to provide sufficient coordination support for complex development tasks:

- Managing the integration of increments.
- Managing dependencies between capabilities.
- Achieving common understanding about how coordination shall be conceived.
- Allocating work packages to various organizational units.
- Managing requirements and their dependencies.
- Planning and controlling the project.
- Coping with cultural diversity.
- Managing change.
- Providing proper IS/IT support.

In Figure 13 the framing of the discussion in this section is illustrated. The concerns are addressed through the various parts of the approach, which in turn is informed by the ADT. As the double directed arrows indicate, the influence between the challenges, the anatomy-centric approach and the ADT goes

Figure 13. Framing the discussion

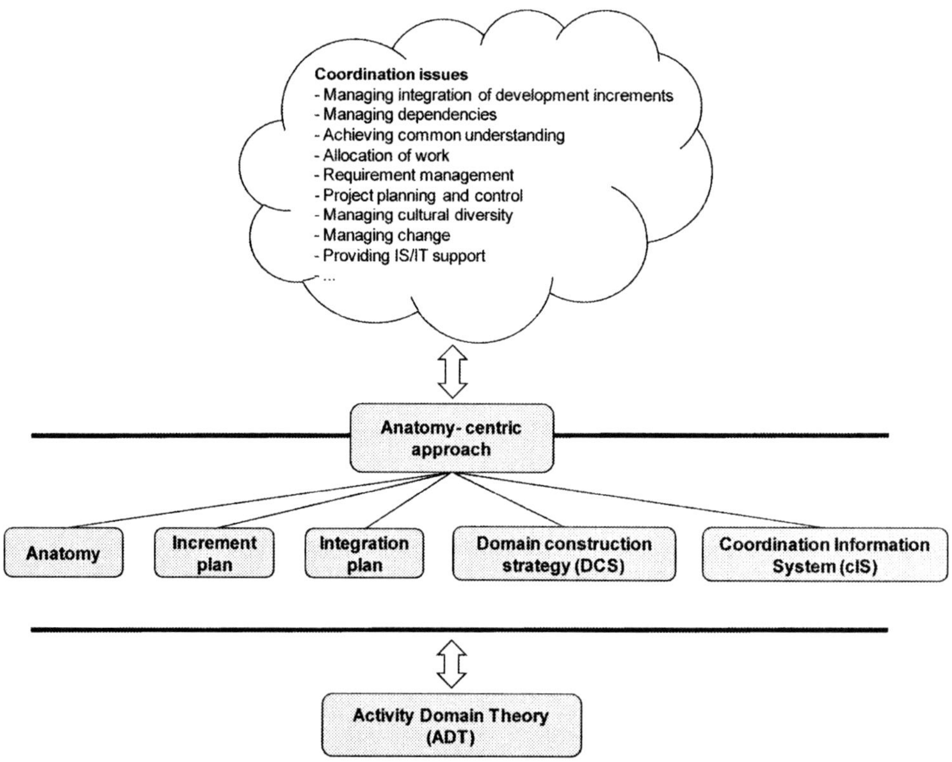

in both directions. New issues discovered in the practical application of the approach may result in a modification of the ADT and its operationalization.

Managing Integration

This issue has its origin in the transition from the waterfall method to an incremental way of working. Managing the integration of increments is a central task in the anatomy-centric approach.

Managing Dependencies

When coordinating complex system development tasks, it is crucial to know how things depend on each other (e.g. Malone & Crowston, 1994; Toffolon & Dakhli, 2000). The most crucial dependencies are expressed by the anatomy, the increment plan and the integration plan:

And also based on the anatomy chart [...] you see a lot of dependencies both time wise and product wise. [...]. Of course that is also a key issue. (Project manager, A-domain)

In addition to these dependencies, the implementation of the information model in Matrix makes it possible to manage other types of dependencies, for example, from requirements to system elements and test cases. Traceability is provided by browsing the information in Matrix or generating tailor cut reports with traceability information. The scope of traceability is defined in the information model, where the types and their relationships frame what can be traced. In principle, all kinds of dependencies between entities in the information model can be visualized:

I think for the MSC [Mobile Switching Centre node] there is a clear need for the tool and I have met now a couple of people who have said: without the tool we would not have survived the projects. [This is due to] the complexity of the node, the complexity of the dependencies, time pressure people have. If you visualize dependencies it is far easier to take decisions. (Project manager, A-domain)

Achieving Common Understanding

The need for common understanding has been a recurrent theme in many contributions (e.g. Battin, Crocker, Kreidler, & Subramanian, 2001; Komi-Sirviö &Tihinen, 2005; Hersleb & Moitra, 2001; Mc-Chesney & Gallagher, 2004; Ovaska, Rossi, & Marttiin, 2003; Prikladnicki, Audy, & Evaristo, 2003; Sakthivel, 2005). One element around which common understanding may grow is the architecture: "The coordination of multi-site work needs a common understanding of the architecture of the system to direct the development work toward a coherent, working system" (Ovaska, Rossi, & Marttiin, 2003, p. 245). The architecture should be transparent in the sense that it can be easily communicated across the entire project (Battin, Crocker, Kreidler, & Subramanian, 2001; Prikladnicki, Audy, & Evaristo, 2003).

The basic mechanism in the anatomy-centric approach for achieving common understanding is the domain construction strategy where reflection and action are continuously interwoven as described above. Similar approaches have been advocated by Schon (1983), Kolb (1984), and Mathiassen (1998) among others. The enactment in an activity domain provides a common view of the project for all participants, regardless of how they are distributed across geographical and organizational borders:

For we didn't have one object type just being on its own, it was always to show impacts of relationships to other, and [...] all the different roles together, they build the overall picture then. (Method & Tools coordinator, A-domain)

Allocating Work

When allocating work, *group cohesion* and *task coupling* should be considered (Sakthivel, 2005). Group cohesion refers to variations in language, culture, organization, etc. A group with large variety in these respects is defined as having low group cohesion. Task coupling refers to the need of co-presence, proximity and face-to-face communication.

If possible, subprojects in the integration plan should be based on high group cohesion and high task coupling. Increments borders should be placed at nodes in the anatomy where a large number of dependencies converge or diverge. Either such nodes indicate opportunities for parallel development (diverging dependencies) or potential critical tasks (converging dependencies) on which many other tasks depend.

The definition of the anatomy requires a lot of interaction, communication and face-to-face meetings among participants from various sites. Thus, anatomy definition has a high degree of task coupling and low group cohesion. By participating in this work, the participants will communicate the common understanding of the anatomy at their home sites. In other words, they will act as liaisons in the project, something that have been identified as quite important in distributed software development (Battin, Crocker, Kreidler, & Subramanian, 2001).

The anatomy and increment plan can be utilized interchangeably when allocating work. For example, the architecture of the system can be defined in such a way that it alleviates the definition of tasks with low coupling. This is of course only possible in situations where the architecture is not settled.

Managing Requirements

Insufficient requirement management is a main source of error in software development (e.g. Damian, 2003; Komi-Sirviö & Tihinen, 2005; Prikladnicki, Audy, & Evaristo, 2003). For example, "failure to fully understand the required system features, as well as reduced trust and an inability to effectively resolve conflicts result in budget and schedule overruns and, ultimately, result in damaged client–supplier relationships" (Damian, 2003, p. 180).

Several features of the anatomy-centric approach address requirement management issues:

- The approach was originally developed in order to mitigate the effect of requirement instability. For example, stable requirements can be allocated to early increments and unstable ones to later increments.
- Requirements traceability is provided by the cIS.
- Changes in requirements are subject to the engineering change order process, which also is supported by the cIS.

Figure 14. A screen dump from the Matrix cIS

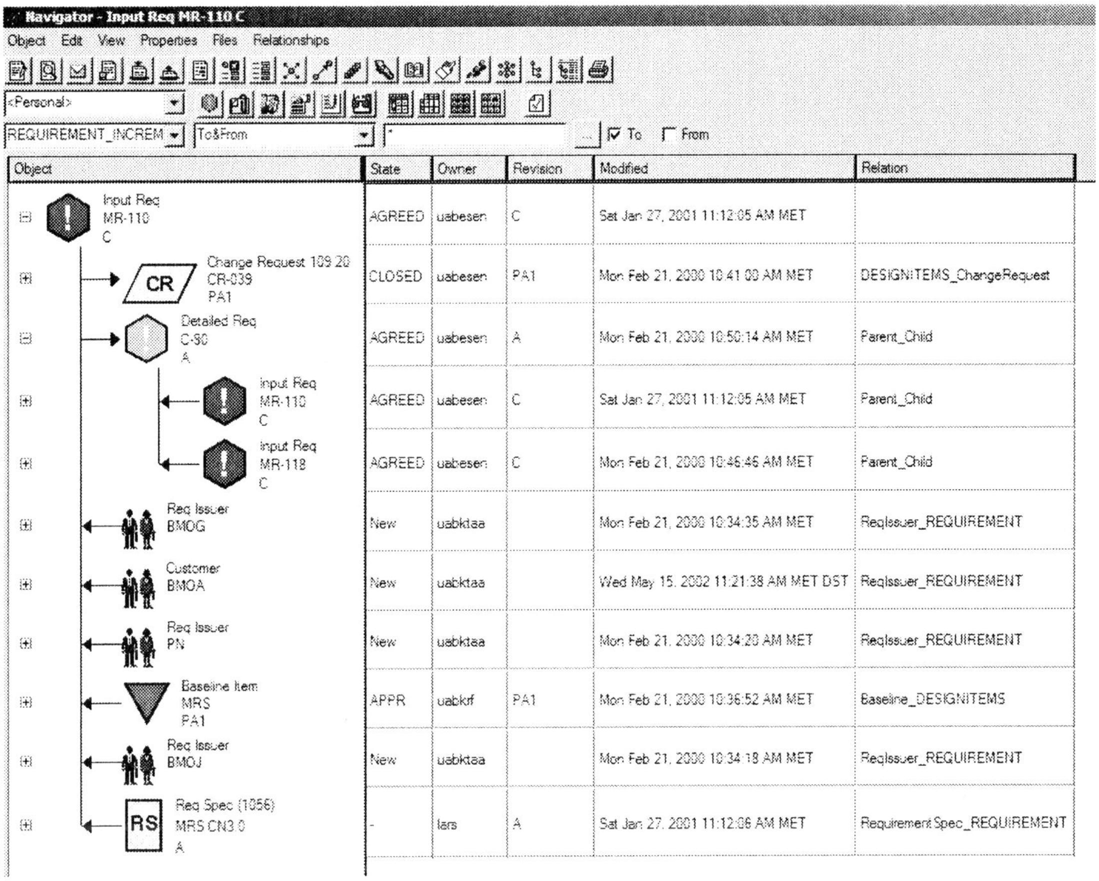

An example of the kind of information that can be recovered from the cIS with respect to requirement management is given in Figure 14.

This particular view shows the context of a customer requirement "MR-110", which is issued by a set of customers "BMOG", "BMOA", "PN", "BMOJ". The requirement "MR-110" has been subject to a change request "CR-039" (the Ericsson internal name for an engineering change order), which is "CLOSED", i.e. analyzed and implemented. The view also shows that the "MR-110" requirement is base-lined, i.e., subject to change restriction rules, and further allocated to a detailed requirement "C-90". From the table to the right, the Ericsson internal name of the person owning a particular item is shown, as well as the revision of the item, when it was created, and the names of the relations between items.

In a nutshell, the screen dump shows indications of most activity modalities and how these are related: contextualization – the "MR-110" in focus and relevant items around it; spatialization – the different relationships; temporalization – dates for creation of the items; stabilization – rules for base-lining an item. The transition modality would come into play if, for example, the Ericsson internal way of naming customer requirements ("MR-110", "MR-118", etc.) would have to be related to the customers ways of identifying the same requirements.

In short, the capabilities for requirement management were summarized in this way:

Figure 15. A web-based report showing information related to engineering change orders (CRs)

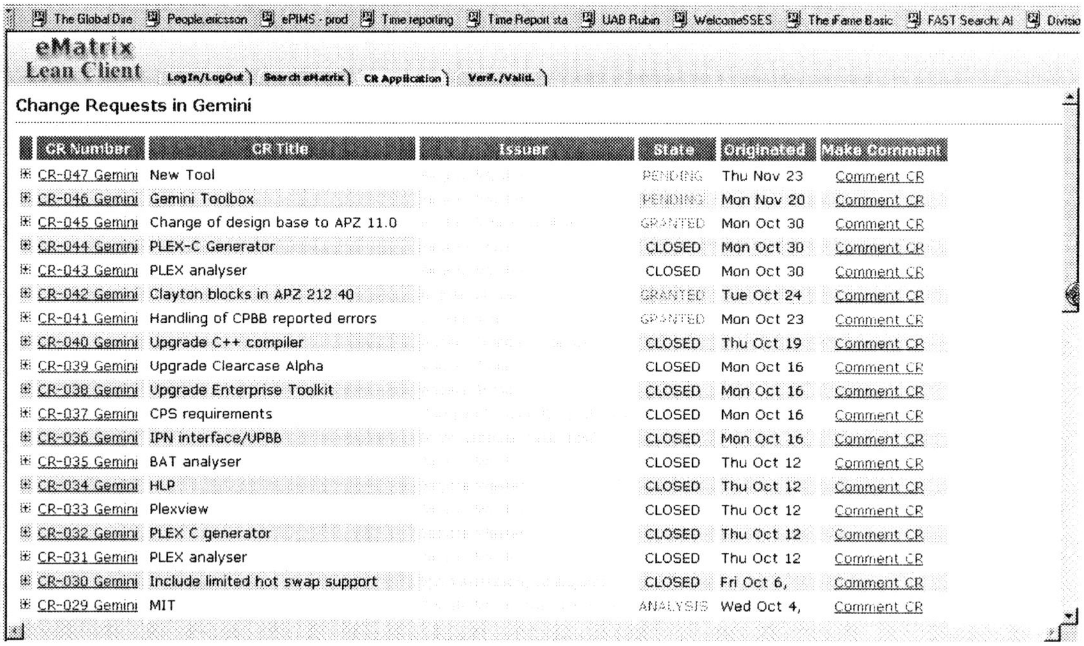
eMatrix Lean Client

LogIn/LogOut | Search eMatrix | CR Application | Verif./Valid.

Change Requests in Gemini

CR Number	CR Title	Issuer	State	Originated	Make Comment
CR-047 Gemini	New Tool	[illegible]	PENDING	Thu Nov 23	Comment CR
CR-046 Gemini	Gemini Toolbox	[illegible]	PENDING	Mon Nov 20	Comment CR
CR-045 Gemini	Change of design base to APZ 11.0	[illegible]	GRANTED	Mon Oct 30	Comment CR
CR-044 Gemini	PLEX-C Generator	[illegible]	CLOSED	Mon Oct 30	Comment CR
CR-043 Gemini	PLEX analyser	[illegible]	CLOSED	Mon Oct 30	Comment CR
CR-042 Gemini	Clayton blocks in APZ 212 40	[illegible]	GRANTED	Tue Oct 24	Comment CR
CR-041 Gemini	Handling of CPBB reported errors	[illegible]	GRANTED	Mon Oct 23	Comment CR
CR-040 Gemini	Upgrade C++ compiler	[illegible]	CLOSED	Thu Oct 19	Comment CR
CR-039 Gemini	Upgrade Clearcase Alpha	[illegible]	CLOSED	Mon Oct 16	Comment CR
CR-038 Gemini	Upgrade Enterprise Toolkit	[illegible]	CLOSED	Mon Oct 16	Comment CR
CR-037 Gemini	CPS requirements	[illegible]	CLOSED	Mon Oct 16	Comment CR
CR-036 Gemini	IPN interface/UPBB	[illegible]	CLOSED	Mon Oct 16	Comment CR
CR-035 Gemini	BAT analyser	[illegible]	CLOSED	Thu Oct 12	Comment CR
CR-034 Gemini	HLP	[illegible]	CLOSED	Thu Oct 12	Comment CR
CR-033 Gemini	Plexview	[illegible]	CLOSED	Thu Oct 12	Comment CR
CR-032 Gemini	PLEX C generator	[illegible]	CLOSED	Thu Oct 12	Comment CR
CR-031 Gemini	PLEX analyser	[illegible]	CLOSED	Thu Oct 12	Comment CR
CR-030 Gemini	Include limited hot swap support	[illegible]	CLOSED	Fri Oct 6,	Comment CR
CR-029 Gemini	MIT	[illegible]	ANALYSIS	Wed Oct 4,	Comment CR

I think that we have the possibilities to manage requirements in a good way and make them obvious and we can achieve a very clear traceability all the way from customer requirements one might say. (Project manager, S-domain)

Project Planning and Control

Globally distributed projects bring about more demanding requirements on project management (Komi-Sirviö & Tihinen, 2005). The need for coordination and communication increases when people cannot meet face to face. Geographical separation means that effort estimations, project planning and control become more difficult. In addition, the management of issues like trust, responsibilities, commitments, incitements, etc., is aggravated.

Communication

The increment and integration plans are the main project communication mechanisms in the anatomy-centric approach. Based on these plans, responsibilities, progress, delivery dates, release plans, etc. can be communicated in a uniform manner to the entire project. Another communication mechanism is the various report generators in Matrix (see Figure 15).

The report shows a list of Change Requests for a particular project, which has issued the CR, what state it is in, and when the CR was issued. There is also an option to make comments to the CR, comments that can be made by anyone in the project regardless of organizational unit and geographical site. Reports like this contribute to the construction of a common understanding of the project:

I presented this way of working to them, to the mobile people. And [they thought] this is nice and simple and easy to understand for everybody and very effective.... the project managers [got] the information they needed directly from the web with one click in these pre-defined tables. (Method and tools coordinator, A-domain)

Decentralized Coordination

Besides providing the overall coordination support, one important effect is that coordination is carried out at the most relevant level:

Yes, what is the great benefit is that you have one common place where all the project area stored the information. It means that a lot of the coordination, which previously went via the main project, now can go directly. A lot of coordination is now happening on the level it should be. (Project manager, A-domain)

This means that the need for the main project management to engage in coordinating subprojects is decreased.

Commitments and Responsibilities

The anatomy-centric approach enhances commitments and responsibilities. The increment and integration plans show which groups are responsible for what as well as dependencies between these groups. These plans together with their implementation in the Matrix cIS make it more difficult for participants to escape their responsibilities:

I will never forget I when was in a meeting where they were actually pointing in a circle. Nobody in this meeting had a problem. But at the end of the day the build process didn't work. And without a proof that it is their problem they would not do anything about it. Now this tool makes it so damn visible that they have a problem. (Project manager, A-domain)

Separation of Concerns

With the anatomy-centric approach the total project manager can concentrate on the status of each increment and the dependencies between them. This means that he/she can focus on the "whole" picture rather than controlling the details of each increment, which is the responsibility of subproject managers and team leaders. If needed, the focus between these levels can be changed:

Of course there is also something that is maybe also a benefit of the tool is that we do have different levels of projects. And the tool can really provide support for various levels, [...]. It is one common database with everything in it ... that is very valuable. (Project manager, A-domain)

Trust

A major difficulty in both the A- and S-domains was to convince the participants of entering data into Matrix. There were various reasons for this: lack of support resources, resistance to change established ways of working, poor performance of early versions of Matrix, awkward user interface, etc. However, a main reason was that the participants did not understand why they had to enter the data. From their point of view, this task was just an extra burden. This situation created a sense of distrust regarding the data in Matrix.

The mechanism to boost the trust is to create a chain of dependencies among the participants in such a way that if someone fails to enter the data, the others will suffer:

As soon as different roles take the reports as input for their activities you must trust them. That is the driver to motivate all the people to enter the information into the tool. (Method and tools coordinator, A-domain)

Managing Cultural Diversity

The issue of culture in relation to organizations has been extensively treated in the literature, e.g. (Smircich, 1983). From the ADT point of view, culture is regarded as the activity domain specific ideology that each domain evolves. In the anatomy-centric approach, cultural diversity emphasizes the balance between central and local ways of working. On the one hand, development tasks must be distributed to various sites that are specialized in a particular type of development. On the other hand, there is a need to integrate the contributions of the various sites into a working system. Basically, this leads to two different strategies for handling cultural issues. For example, Ebert & De Neve from Alcatel (2001) recommend that all sites working with the same type of products should use common processes, methods and terminology. On the other hand, Battin, Crocker, Kreidler, & Subramanian (2001) from Motorola suggest that as a key strategy in distributed development is *not* to impose a common process. Even though it is desirable to use the same tools, methods and processes throughout the project, this is seldom, if ever, an option. This is especially so if the project consists of a conglomerate of otherwise independent organizations.

A key insight from the empirical experiences is that the rapid development of the coordination support was achieved by *not* enforcing a common way of working in both domains. This was confirmed by statement such as:

I know there are some initiatives and try to come up with one object model for all projects within Ericsson. And I think that's a little bit trying to search for the Holy Graal. [...]. I mean you can even see how different the ways of working are within our subprojects at this moment [...]. To even keep one common way of working for that is really quite a challenge. (Project manager, A-domain)

The drawback of the autonomous ways of working was that the coordination *between* the A and S-domains was not achieved. This coordination had to be done without tool support. Thus, the balance between centrality and locality was too much biased towards the locality side.

Providing IS/IT Support

Distributed development puts demanding requirements on the IS/IT support. There is a need for a common project repository where requirements, engineering change orders, system builds, etc., are globally available (Gorton & Motwani, 1996). The infrastructure at distant sites may be inadequate, which may cause problems with IS performance and stability. In addition, problems related to tool support, upgrading and maintenance are aggravated (Komi-Sirviö & Tihinen, 2005).

Common Repository for Coordination Data

The common project repository is provided by the cIS. Having all coordination data in the same IS is advantageous in several ways:

- Often, support for management is provided by several systems, for example, one for requirement management, one for engineering change orders, one for bug tracking, etc. Consequently, several interfaces between those systems must be maintained if support for managing all different kinds of items shall be possible. Those interfaces are not needed if the same system is used for all items.
- The information kept in the cIS is consistent. There will be no two copies of the same information record, something which is quite common when several management systems are used.
- There is a homogeneous management world-view of the entire domain. Each separate system brings with it some in-built world-view of what is relevant to manage, and how these things should be characterized. This may cause problems of reconciling the views, both technically and cognitively.
- Changes, which are common to all items, are easier to implement. This concerns, for example, names of attributes shared by several item types, changes in forms for entering data, and the like. In addition the ability to quickly react to changes is improved.

There will still be a need to interface the cIS to other tools, for example, special purpose tools for configuration management of SW code files during the development, or ERP tools such as SAP. However, interfaces to these tools will be on the border of the activity domain instead than internal.

Performance and Stability

The preferred cIS architecture is a central server in each domain interacting with web-based clients at development sites. This was the architecture at both the A- and S-domains. However, such architecture may face severe problems. The performance of the early versions of Matrix turned out to be insufficient for global access. The reasons were poor quality of the IT infrastructure on a global basis, and inadequate technical solutions in the Matrix system. Thus, a major lesson is that performance and stability should be secured early in the construction of the activity domains.

Managing Changes

Two kinds of changes can be identified in the anatomy-centric approach. The first one concerns the mathetic phase in the domain construction strategy (DCS). In this phase, changes in the context model and its implementation in Matrix are made quite frequently. Between 1999 and 2002, around 200 and 500 changes were made in the A- and S-domains respectively. A necessary prerequisite for this type of change is the ease by which the implementation in the cIS can be modified. However, frequent changes have the drawback that the implementation may turn out to be a patch work, which has to undergo major reconstructions. This happened once in the A-domain. The experience was, nonetheless, that this problem could be mastered.

The other type of change concerns a long-term evolution in the pragmatic phase of the DCS. This evolution is related to changes in the environment of the domain, providing coordination support for new types of items, improving the way of working, providing maintenance, etc. An example of this was the inclusion of test-case management in the midst of one project in the S-domain. The results from Ericsson indicate that this type of evolution is indeed possible due to the iterative way of working and the flexibility of the cIS:

The positive effects that we have, an integrated project support system where we have tremendous possibilities to improve, continuously improve our operations. (Project manager, S-domain)

ANATOMY-CENTRIC AGILE DEVELOPMENT OF SOFTWARE

In this section, I will discuss possible implications from using the anatomy-centric approach in connection with agile development of software. Broadly speaking, there are two different approaches to software development: the plan-driven and the agile ones (e.g. Boehm & Turner, 2003b). The world-view of the plan-driven approach is characterized by predictability and stability. Requirements are supposed to be stable, complete and consistent. The architecture is laid out early and used for assigning work packages, preferably in a concurrent and parallel manner. Testing is requirement driven, progress is base-lined, and communication relies heavily on documentation.

In contrast, during the last decade, software development is increasingly done according to agile methods such as Scrum, XP (Extreme Programming), and others (see e.g. Meso & Jain, 2006, for a comprehensive overview). Agile development started off in comparatively small projects, where Microsoft was one of the forerunners. Recently, however, agile methods have gained more influence in large, globally distributed software development projects (e.g. Karlsson, Andersson, & Leion, 2000; Karlsson & Andersson, 2001; Larman & Vodde, 2008; Olsson, 1999).

The world-view of the agile approach emphasizes dynamism and change. It favors individuals and interactions over processes and tools; working software over comprehensive documentation; customer collaboration over contract negotiation; and responding to change over following plans (Agile Manifesto, 2006). Requirements are considered inherently unstable, and development is done iteratively. The focus is on frequent integrations and verifications, often on a daily basis: so called "daily build" or "continuous integration" (Fowler, 2006). Iterations comprise a full development cycle, including planning, requirements analysis, design, coding, unit testing, and acceptance testing when a working product is demonstrated to stakeholders (Wikipedia, 2009). In order to enhance quality, a process called *refactoring* is utilized, in which the internal consistency and clarity of the code is improved (Wikipedia, 2009b). In essence, the agile way can be seen as the anti-thesis of the plan-driven way.

Figure 16. A principle sketch of feature based development

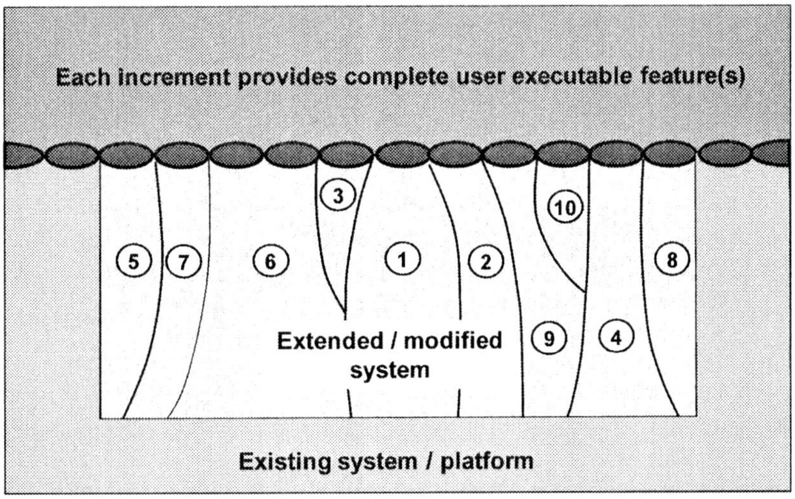

When combining incremental development with daily builds, it is near at hand to base the contents of the increments on features, i.e., customer executable capabilities of the system. A principle sketch of this kind of development is illustrated in Figure 16.

Each increment develops one or several features (represented by ovals in the figure). The order in which increments are developed can be based on characteristics such as (Olsson, 1999):

- cost/benefit ratio for customer
- stability of requirements
- basic - advanced (from a user point of view)
- simple - complex (from an implementation point of view)
- critical - non-critical (from a safety point of view)

The ordering of increments is constrained by the dependencies among increments, which in turn are due to the dependencies between capabilities in the system. For example, in Figure 16, increments 3 and 10 are dependent on increments 1 and 4 respectively.

No doubt, agile methods have proven viable in industrial development of large systems. However, with increasing scale of applications, issues begin to materialize:

In recent agile methods workshops with our large-company industry affiliates, the participants have unanimously agreed that agile methods have helped them become more flexible and adaptive to change. But they have also agreed that scalability and legacy practices have limited their range of adoption of agile methods. (Boehm, in Fraser, Boehm, Järkvik, Lundh, & Vilkki, 2006, p. 226)

Scalability issues are, among other things, "team-of-teams coordination and change management, independent–team product interoperability, multi-customer change coordination and unscalable COTS[3] or architectural suboptimization on early increments" (ibid).

In this section, I want to discuss the role that the architecture may have in agile development; a subject that so far seems to have been downplayed (e.g. Babar & Abrahamsson, 2008). If attended, architectures emerge as the development progresses through refactoring. This has cast doubts about the need for and quality of architectures in agile development. For example, concerning XP, Jensen, Møller, Sönder, & Tjørnehøj (2006, p. 135) ask: "Does XP have a weak side concerning establishing a viable architecture?" Other methods, for example, Scrum appear to be neutral on the subject of architecture: "[Teams] can do as much modeling or upfront 'architecture' work as they find useful" (Larman & Vodde, 2008, p. 135).

In any case, there is a growing insight that the role of architecture in agile development should be advanced. Software architectures are considered as means to reduce development cost and time, and to increase quality. There is a "vital need for devising a research agenda for identifying and dealing with architecture-centric challenges in agile software development" (Babar & Abrahamsson, 2008, p. 242). In the context of such an agenda, it is interesting to explore the role of the anatomy as the architecture for software systems.

Architectural Areas of Concern

From the literature, it is possible to conceive a number of areas where architectures in general and the anatomy in particular, is interesting in connection with agile development.

Balancing Agile and Plan-Driven Development

A consequence of scaled-up applications is a need for mixing agile and plan-driven methods in the same project. To this end, Boehm & Turner (2003) have proposed a risk-driven process where the project can be compared to basic characteristics of agile or plan-driven "home grounds". The process is based on five critical factors by which a classification of parts can be done: size (number of personnel), criticality (loss due to impacts of defects), culture (thriving on chaos versus order), personnel (skill of work force), and dynamism (frequency of requirement changes). If either agile or plan-driven risks dominate, pure agile and plan-driven methods can be safely used. However, if neither dominates, a balance has to be struck where agile methods are used for agile parts and plan-driven elsewhere. The architecture is a means for encapsulating the agile parts from the plan-driven ones. Thus, the importance of architectural considerations increases when projects cannot be easily classified as either agile or plan-driven.

With two different methods at work in the same project, the need for managing dependencies between the development tasks increases. This can be done using the anatomy. Moreover, increments can be defined over the anatomy, based on their classification as either agile or plan-driven.

Planning

A further role for the anatomy in agile development is in planning the project on a coarse level in terms of which capabilities shall be delivered when. In principle, there is no contradiction between rigorous planning and agile methods. However, it does not make sense to plan what should be delivered every day. A more appropriate procedure seems to be planning by the week and beyond. For example, Ericsson is working at some radio network units according to a steady-rhythm method called "One Track", where daily build of code, weekly builds of radio nodes, and complete build of the entire radio network every five week, are constantly iterated (Kristofersson & Lindeberg, 2006; see Figure 17).

Figure 17. The "One Track" agile process (adapted after Kristofersson & Lindeberg, 2006, p. 46)

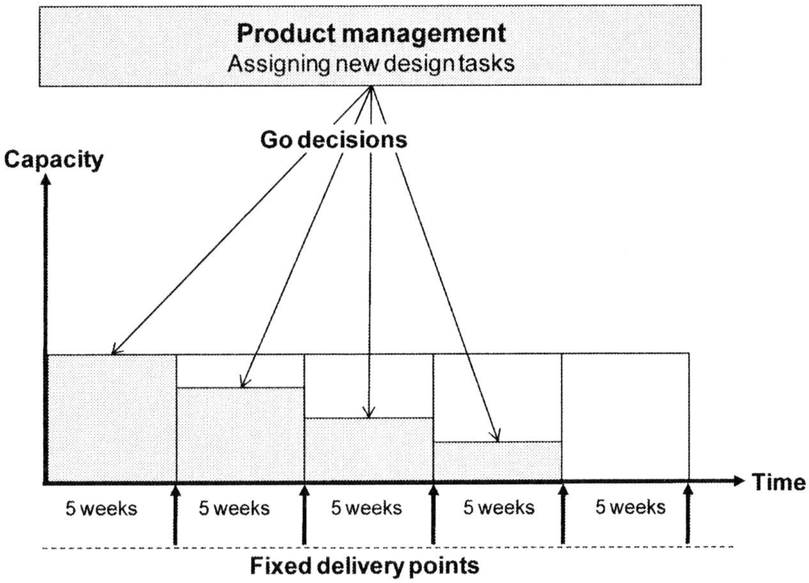

The "One Track" way of working puts high requirements on detailed and continuous planning (Ny teknik, 2006). Every five weeks, a fully functional network is ready to be shipped to a customer, regardless whether there is one or not. This gives freedom for product management to assign new tasks to the development units on a five-week regularity basis. With every delivery, capacity in the form of people and equipment may be released for other assignments. Started tasks are always completed, and the focus is on minimizing disruptions: "keeping hooligans from throwing things into the playground during the game". Although no details are available of how this planning is carried out, it is near at hand to consider the anatomy as a means for supporting the planning process.

An interesting observation is that unplanned daily build is strongly intertwined with strictly planned 5-week deliveries; a mixture of agility and stability. In this sense, the "One Track" method is an expression of the stabilization activity modality; when managing complex systems, there must be some stable element present. If properly defined, the anatomy may assist in striking a balance between agility and commonality; a task that is central to scaling up agile methods according to Vilkki[4]:

Being successful in large-scale product development requires finding ways to enable self-directing teams to work towards a common goal without compromising empowerment and feeling of ownership in the teams - a task easier said than done. Finding the balance between agility and commonality is a dynamic process. (Vilkki, in Fraser, Boehm, Järkvik, Lundh, & Vilkki, 2006, p. 228)

Finding Critical Points

Another role for the anatomy concerns the impact of several features on the same module. Suppose we have a situation like the one in Figure 18:

Figure 18. An example of several features impacting the same module

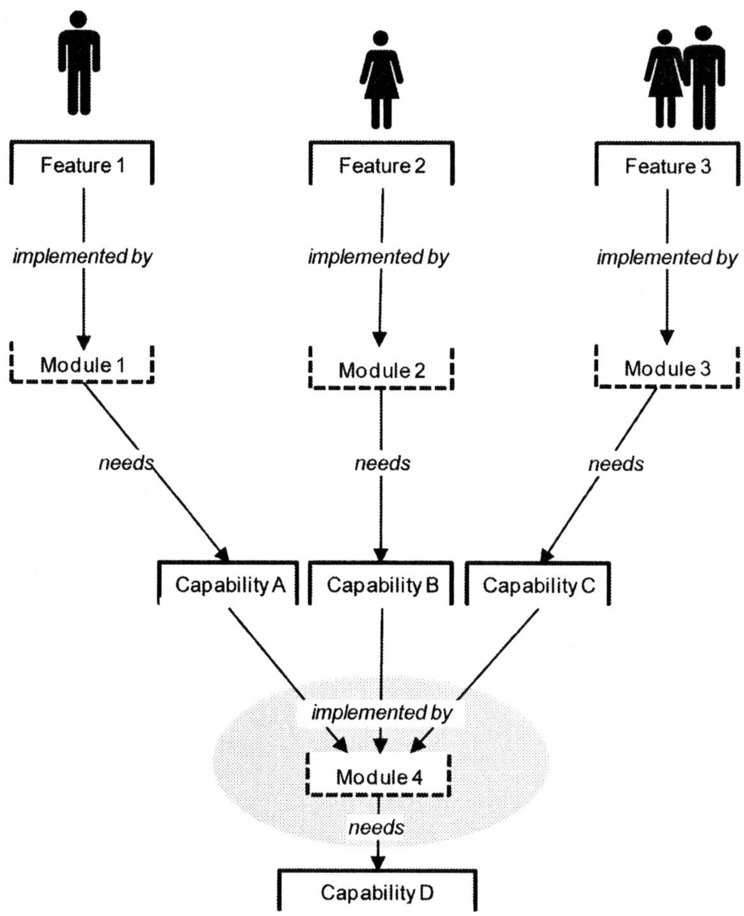

All features need capabilities A, B and C implemented by the same module 4, which in turn needs capability D. Thus, all teams code in the same module, which means that they need to coordinate their work. In agile development, this coordination is usually done by the teams themselves. However, if the number of features impacting the same module becomes large, the coordination may become very complex. Such impact points are called "critical points" by Olsson (1999), who suggest that the responsibility for maintaining the integrity of such modules should be assigned on a module basis rather than feature. Another kind of critical points occurs when many capabilities are dependent on the same capability (as in Figure 7, p. 168). In both cases, critical points are easily identified in the anatomy.

Communicating the Big Picture

According to Olsson (1999) feature based development puts specific requirements on the architecture. Feature teams must understand the system architecture since their feature may be impacting many modules. Therefore, the architecture has to be easy to understand. One reason for the "4+1" architecture proposed by Krutchen (1995) was that previous architectures tried to squeeze too much into a single diagram illustrating the architecture (Krutchen, 1995, p. 42). The "4+1" model approach to this problem is viewing the architecture from different vantage points, thus suppressing irrelevant aspects in each view.

The views provided by the "4+1" model are no doubt needed to manage all aspects of the development of a software system. Nevertheless, it goes without saying that it is hard to achieve common understanding among all stakeholders about the five different views of a complex system. This is not even the purpose of the views; these are meant to be aligned with the interests of different stakeholders.

The anatomy offers an alternative/complementary view of the architecture that is easily comprehensible, yet focused on a main issue: to understand the dependencies between capabilities in the system. The anatomy can be seen as a return to the single diagram; a "canonical" architecture in the sense that if these dependencies are not attended to, the system will not work. As such, the anatomy may play an important role as a common instrument for communicating the "big picture" to all participants; a picture that can be used for coordinating and planning the "big chunks" of the project such as deliveries:

Shifting the interest from activities (and subsystems) toward system-level dependences requires software architects and developers to have a common understanding of the software architecture [...]. On the basis of the findings, we claim that architecture could be used to coordinate distributed development. (Ovaska, Rossi, & Marttiin, 2003, p. 233)

Alleviating Re-Use

Usually, systems are not developed anew for each project. Most product development companies have ambitious product programs in place, where new releases are added to already existing design bases. Architectures expressed as anatomies or any other way, may be seen as a "carry-overs" from one release to another, making it possible to reuse already developed capabilities. Moreover, keeping in mind that the anatomy is a social construct, the anatomy for a new release will always be modified by the current project. However, with an easy to comprehend image of the system, the enactment of the modified anatomy, including the introduction of new project members, is greatly alleviated.

Eschew Refactoring

Refactoring refers to the improvement of the code without changing its overall results. The purpose is to improve the understandability of the code, change its internal structure and design, and remove dead code in order to make it easier to comprehend, more maintainable and amenable to change. However, in connection with large-scale development, frequent refactoring becomes less desirable:

Refactoring is a good solution that relies on a large test framework as a safety net. There is nothing wrong with this, refactoring is very efficient in general. There are, however, large refactorings which are difficult and expensive to perform – so we want to minimize these refactorings. (Elshamy & Elssamadisy, 2006, p. 165)

Elshamy & Elssamadisy (ibid) suggest that agile development should be seen as a "conquer phase" followed by a "divide phase". In the conquer phase, a stable working example of the architecture and system design is built iteratively with constant refactoring, ensuring that the design works for the current requirements. This provides a kind of scaffolding or skeleton system that is stable enough for subteams to follow and build upon in the divide phase. According to Elshamy & Elssamadisy, this approach reduces the risk of redesigning the system later on.

A similar approach is advocated by Boehm & Turner (2003b), who claim that XP should be adapted for developing complex, large-scale applications by introducing plan-driven method elements such as high-level architectural plans to provide essential big-picture information, and the use of design patterns and architectural solutions. Such an approach might delay refactoring.

Mitigating Risks

Mitigating risk appears to be a common concern when scaling up agile methods. As discussed above, Boehm & Turner (2003) have proposed a risk-driven process for matching project characteristics with agile or plan-driven methods. In this process, the architecture plays a central role. Another example is Henry (2007) who suggests that "taking an architecture first approach to managing large software projects can reduce a significant amount of the uncertainty present in project estimates [...]. Research suggests that there exists a significant drop off in risk when the architecture is developed" (ibid, p. ii).

Recognizing change

With increasing scale, the need for recognizing and managing change increases:

[Recognition] of change is greatly affected by the size of the team and the size of the artifacts (code, use cases, tests). As we have more people, whether or not we have multiple subteams, it is more difficult to determine if a change in one part of the system affects other parts of the system. (Elshamy & Elssamadisy, 2006, p. 165)

This is definitely an area where the anatomy may be useful. If a certain capability is changed, it is immediately clear by looking at the anatomy which other capabilities might be affected, since the anatomy shows their dependencies. By tracing capabilities to impacted modules, the consequences of changes and design decisions can be evaluated. As systems become more complex, the importance of how parts interrelate will accrue:

We need a way to understand how design decisions made during the creation of a complex system architecture will interact. XP doesn't address evaluating the design explicitly; code is tested continuously and offers feedback to the design. (Nord & Tomayko, 2006, p. 52)

Summary

It is likely that agile methods will gain influence in large, volatile software development projects. In order to meet increasing issues raised by this trend, architecture-centric agile methods will become more interesting. I have indicated some areas where the anatomy-centric approach may play a role in addressing these issues. The main motivation for introducing the anatomy in agile development would be the same as in other types of projects: to provide an easily understandable, concentrated, and relevant image of the architecture that can be complemented with other, more specific architectural aspects as in the "4+1" model. In a general sense, the anatomy may be a means to balance between agility and commonality, thus providing a synthesis between the two extremes of agile and plan-driven methods. However, the viability of this conjecture remains to be validated empirically.

THE GIST OF THE ANATOMY-CENTRIC APPROACH

In this section, I elaborate on what I believe is the gist or quintessence of the anatomy-centric approach from two perspectives: integration of multiple meanings of architecture, and alignment with activity modalities.

Integration of Architectural Meaning

The undisputable importance of the anatomy as a means for reconciling various stakeholder interests at Ericsson makes it interesting to position the anatomy-centric approach in a broader context of architectural discussions in academia and practice. The traditional objective for architectures has been to specify the system to be implemented. This meaning of architecture is, in a metaphorical sense, called "blueprint" by Smolander, Rossi, & Purao (2008), and is of primary interest to engineers developing the system. Blueprint architectures are high-level representations, usually expressed in UML (Unified Modeling Language) or other ADLs (Architecture Description Language) notations, which are progressively detailed and refined during the implementation of the system.

The blueprint meaning of architectures is prevalent in academia (ibid). Research has focused on finding representations emphasizing correctness, consistency, unambiguity, and coherence. This line of research has been developed into a mature academic discipline over the last decades (Shaw & Clements, 2006). However, the approaches "to representing, designing and communicating software architectures continue to be narrowly focused, with an emphasis on software architects and engineers as the key stakeholders, and continue to employ highly technical representations" (Smolander, Rossi, & Purao, 2008, p. 575). As a consequence, the adoption of research findings in practice has been slow, and "sustained benefits from academia to practice are still difficult to identify" (ibid, p. 578).

Motivated by this observation, Smolander, Rossi, & Purao (2008) formulated the somewhat obvious research question. "How is software architecture developed and used in an organization?" (ibid, p. 578). Their main finding is that the architecture in practical settings is a concern for several groups of stakeholders, not only the implementers of the system. Salesmen, production people, managers, customers, quality responsible persons, project managers, etc., all have an interest in, and may contribute to, the development of architecture. Thus, in contrast to the prevalent academic view of architectures as "factual" or objective, its social constructional aspects are emphasized.

The analysis carried out by Smolander, Rossi, & Purao indicates that there are basically four different meanings associated with architectures in practice; the already mentioned blue-print metaphor, the "language", the "decision", and the "literature" ones. The objectives associated with these metaphors are as follows (ibid, p. 581):

- Blueprint: Specification of the system to be implemented.
- Language: Medium of communication for achieving common understanding.
- Decision: Choices about the system to be implemented and rationale.

- Literature: Documentation for current and future generations of users and developers.

Besides these objectives, Smolander, Rossi, & Purao also identified six major aspects or dimensions associated with the metaphors: time-orientation, formality, detail, activity, customer focus, and business focus (ibid, p. 580):

- Time-orientation: Descriptions of past architectural solutions versus current design situation versus prescriptions about future implementations.
- Formality: Descriptions for enabling understanding versus those meant for generating executables.
- Detail: Descriptions of technical details or descriptions that purposefully constrain the level of detail.
- Activity: Nature of typical activities associated with the descriptions such as recording versus negotiating versus sense-making.
- Customer focus: Frequency and strength of interaction between the development organization and customers utilizing the software architecture.
- Business focus: Extent of reasoning the development group must make about the business area of the system.

The multiple meanings of the architecture and its characterization according to the dimensions are summarized in Table 1.

The framework in Table 1 is useful for analyzing and positioning the anatomy-centric approach. It can be noted that this approach, as well as the framework, are both derived from practice and not from academia. Thus, the research interest is rather the opposite of the traditional one; instead of transferring results from academia to practice, the research question is how the results from practice can be explained, elucidated, made transferable, and so on.

One way to approach this question is to evaluate the anatomy-centric approach according to the multiple meanings and dimensions in Table 1. A first observation is that the anatomy is primarily "language" oriented. The objective is above all to reach a common understanding about the capabilities in the system and how these depend on each other. It is future-oriented in the sense that it signifies a system to be developed. The formality and details of the anatomy are low, and it is subject to intensive negotia-

Table 1. Multiple meanings of software architecture (adapted after Smolander, Rossi, & Purao, 2008, p. 581)

	***Blueprint*: Specification of system, technical details**	***Language*: Communication, common understanding**	***Decision*: Resources, evaluating alternatives, work force**	***Literature*: Documentation, re-use, "carry-over"**
Time	Future	Present/future	Future	Past
Formality	High	Low	Usually low	Varies
Detail	High	Low	Usually low	Usually high
Activity	Implementation	Negotiation	Evaluating choices	Reading, analyzing
Customer focus	Low	Possibly high	High	Usually low
Business focus	Low	Possibly high	High	Usually low

Table 2. The anatomy-centric approach positioned in the Smolander, Rossi, & Purao (2008) framework

	Blueprint: **Specification of system, technical details**	*Language*: **Communication, common understanding**	*Decision*: **Resources, evaluating alternatives, work force**	*Literature*: **Documentation, re-use, "carry-over"**
Time	Future	Present/future	Future	Past
Formality	High	Low	Usually low	Varies
Detail	High	Low	Usually low	Usually high
Activity	Implementation	Negotiation	Evaluating choices	Reading, analyzing
Customer focus	Low	Possibly high	High	Usually low
Business focus	Low	Possibly high	High	Usually low

tion among various stakeholders during its definition. Since the anatomy also shows the capabilities the customer can expect from the system, it also has a high customer focus and business orientation.

Moreover, the anatomy is "blueprint" oriented in the sense that is specifies, on a coarse level, the capabilities to be developed while postponing implementation details[5]. During the implementation, the progress can be reported by, for example, coloring the capabilities in the anatomy according to states such as "not-started", "in progress", "ready", or any other relevant set. When the project is finished, the anatomy may be used as documentation of the system for re-use and "carry-over" to future generations of the system. Thus, the anatomy is also "literature" oriented.

There is no direct "decision" orientation of the anatomy. On the other hand, the increment plan and the integration plan are clearly "decision" oriented. In the increment plan, the increments are laid out over the anatomy, and the work effort of implementation is estimated. In the integration plan, the increments are assigned to subprojects, resources allocated, and the schedule for deliveries set. In Table 2, the position of the anatomy-centric approach in the Smolander, Rossi, & Purao (2008) framework is illustrated:

To sum up, the anatomy-centric approach covers, on an overarching level, the different meanings the architecture can take according to Smolander, Rossi, & Purao (2008). This is achieved by three images held together by the anatomy. In a sense, this is akin to the viewpoint approach, as exemplified by the "4+1" architecture related earlier. However, while the "4+1" architecture is primarily an elaboration of the blueprint meaning of architecture, the anatomy-centric approach provides an integrated view of the different meanings associated with the architecture. At the same time it "holds the center together" through a common understanding among stakeholders about the architecture; an understanding which can and should be complemented with other, more detailed viewpoint representations.

Activity Modalities Alignment

The viability of the anatomy-centric approach as a central way of managing projects in the Ericsson practice makes it interesting to investigate its nature more in detail. The basis of the approach: the three images the anatomy, the increment plan, and the integration plan, turned out to be useful means for planning, executing and monitoring extraordinary complex development projects in the telecom industry. An intriguing question is what qualities these images have that make them suitable for this task. Also, why exactly three images instead of two or four or just one?

Figure 19. A detail of the anatomy of the processor in Figure 8

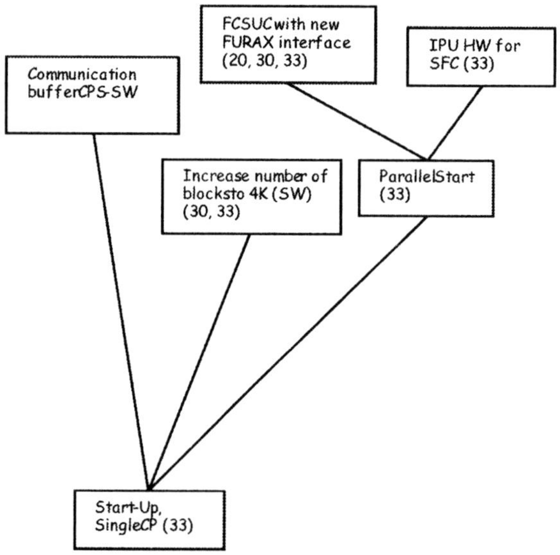

To begin with, these images highlight critical dependencies of different types: dependencies between capabilities, between increments, and between deliveries by subprojects. What is maybe more interesting is that each image appears to be aligned with a dominant activity modality. In Figure 19, a detail of an anatomy is shown:

The anatomy has spatial structure since only static dependencies between capabilities in the system are shown. The capability "Start-Up Single CP" is needed for the capabilities "Communication Buffer CPS", "Increase number of blocks to 4K" and "Parallel Start" to "come alive". The capability "Parallel Start" in turn is a prerequisite for "FCSUC with new FURAX Interface" and "IPU HW" to be realized, and so on. There is no indication of time in the anatomy, just an illustration of things and how these are related. Thus, the anatomy is clearly aligned with the spatialization modality.

In Figure 20, a detail of an increment plan for the same anatomy detail is shown.

Here, the capabilities are grouped into increments that develop capabilities that for some reasons are grouped together. With the increment, the element of action is introduced. Something is going to be acted upon by humans in order to produce an outcome. The increments signify a framing of these actions; in other words – the increments can be conceived as activity domains. Thus, the increment plan shows how different activity domains are dependent on each other and how they interact. For example, increment "Green" developing the 'Start-up Single CP' capability is needed in order for increments "Blue", "Brown" and "Orange" to complete their tasks. This means that there has to be common understanding about how these increments/activities shall interface each other. Thus, the increment plan can be conceived as an image of the activities – the contextualization modality – and how these interact – the transition modality. It is as if these two modalities are overlaid on the spatialization modality – the anatomy.

In Figure 21, finally, a detail of the integration plan is shown.

The position of each increment on the time axis indicates when capabilities being developed by a certain activity shall be ready. Moreover, the capabilities in each increment are subdued; only the dependencies between increments/activities are shown. In many cases, information about dates and

Figure 20. A detail of the increment plan in Figure 9

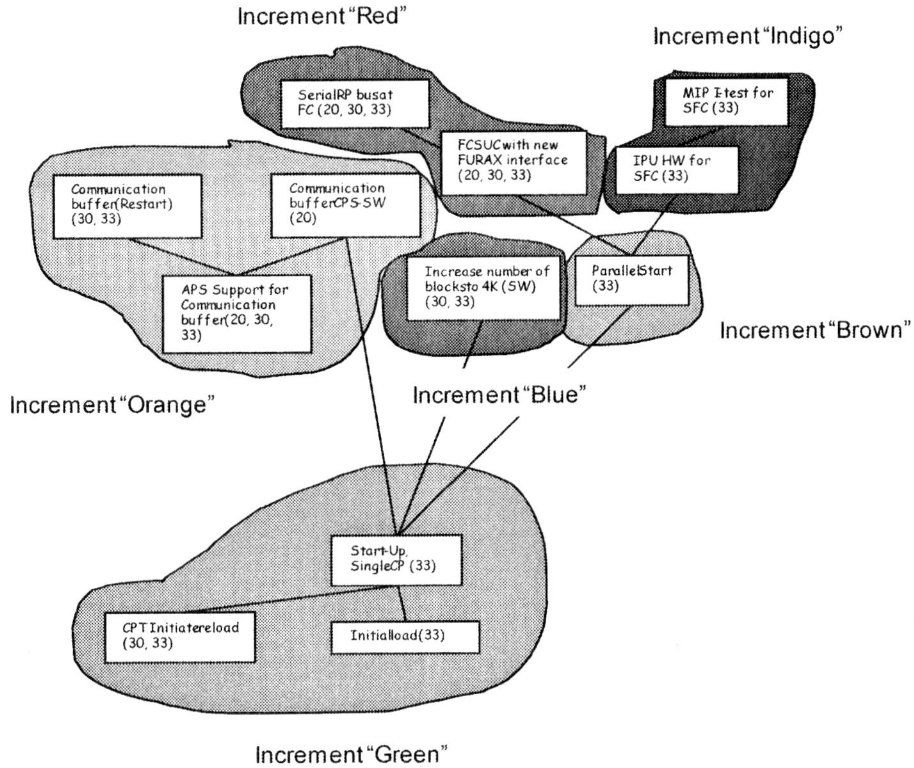

subprojects are shown in the integration plan. In this image, the element of time is emphasized – the temporalization modality.

Thus, each of the images signifies one or two dominant modalities: the anatomy: spatialization, the increment plan: contextualization and transition, and the integration plan: temporalization. In addition, the images are tightly interrelated through the anatomy that holds them together. A change in any image will impact the others. For example, an additional capability required from the customer will lead to changes in both the increment and integration plan. Another example is when the project discovers that some capability cannot be delivered on time. In such cases, a decision may be taken to remove that capability from the project or postpone it for delivery in a later release. This will impact the integration plan and lead to a subsequent adjustment of the increment plan and the anatomy.

The images used in the anatomy-centric approach have been compared to traditional project images such as Gantt diagrams, Work Breakdown Structures (WBS) and PERT/CPM (Taxén & Lilliesköld, 2008). Some findings are (ibid, p. 534):

- Traditional images are focused on control rather than on action and coordination.
- The anatomy-centric images are focused on dependencies and integrations, and emphasize comprehensibility and informality over formality and rigor.
- The basic planning unit in the Work Breakdown Structure is the "work package". In the anatomy-centric approach this corresponds to the increment. Both the work package and the increment are used to estimate the development effort it takes in terms of man-hours, and assigning the work to

Figure 21. A detail of the integration plan in Figure 10

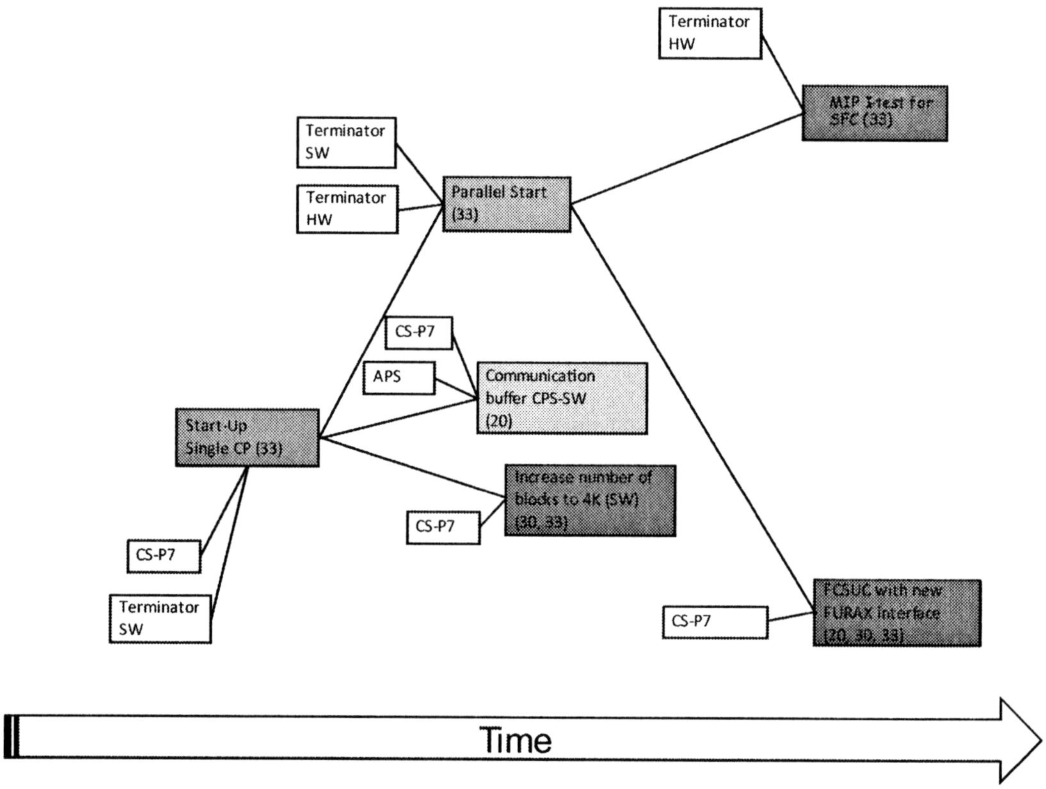

project teams/groups. However, the ways in which the work packages are defined are quite different from how increments are defined. The increment plan is based on the anatomy, which lays a firm ground for the project in terms of dependencies between capabilities in the system to be developed. This is lacking in traditional WBS diagrams.

In general, the anatomy-centric images are strongly oriented towards action. They are not used only for reporting purposes – the images are means for anticipating possible actions and foreseeing the consequences of these actions. The action aspect of traditional images is much less evident.

While the anatomy-centric images are aligned with dominating modalities – mainly one modality per image – traditional images seem to compress different modalities into a single image. In Figure 22, an example of a WBS image is shown.

Traces of contextualization, temporalization and spatialization can be seen in one and the same image. At the top level the purpose of the project is shown: "Make a new RNC" (Radio Network Controller, a node in the 3G telecom network controlling the base stations). The second level shows three organizational functions: Market, Design and Production. These can be apprehended as activity domains, each of which provides some outcome that the project needs. Thus, this level shows remnants of the contextualization modality. The following levels show various actions, like "Do preliminary market analysis"; "Create the advertisement"; "Plan a sales campaign"; and so on. Each of these actions has the form "do something with something"; that is a verb signifying action – temporalization – and a noun

Figure 22. A Work Breakdown Structure

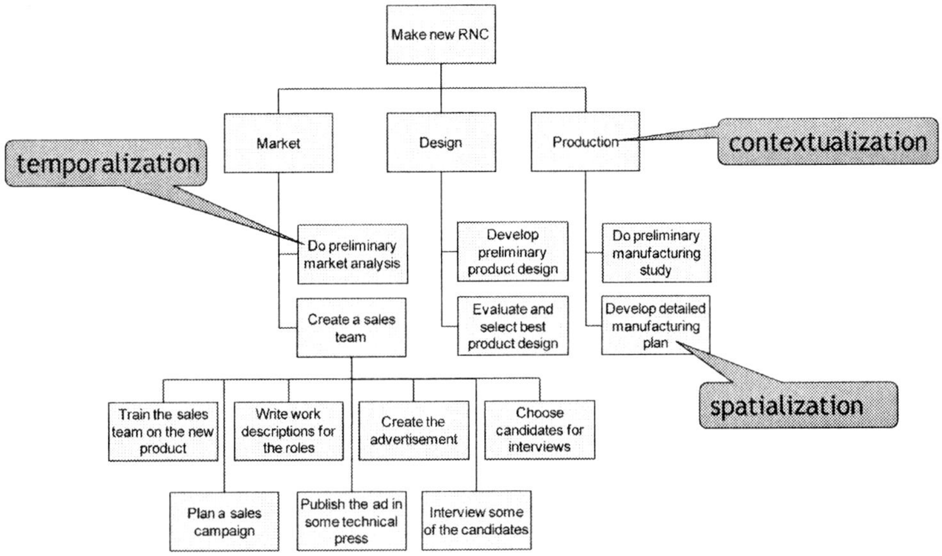

signifying a thing – which is a remnant of spatialization since things and their relations are elements in a spatial structure.

It is near at hand to comprehend the anatomy-centric images as more congruent with the modalities than the traditional ones. If the activity modalities indeed represent dominant coordinative dimensions developed during the phylogenetic evolution of mankind, this would indicate that the alternative images are more resonant with the way our cognitive system experiences and constructs the world. Thus, they are more apt to understand and control complex developments.

The construction of the images is a social accomplishment. The images turn out slightly different depending on the composition of the group of actors developing the images. Moreover, these are constantly revised during the project due to various reasons such as changed requirements, new insights, improved ways of working, errors and mistakes, etc. This indicates that the actors involved are indeed enacting the images; they are turning the images into means of action while simultaneously constructing communal meaning of the use of these means. Thus, the enactment process goes beyond mere learning where actors are passively appropriating the use of already existing artifacts.

THE ANATOMY-CENTRIC APPROACH: SUMMARY

As with all methods, the anatomy-centric approach faces some concerns and limitations that need to be considered:

- **The approach is considered too costly:** The fast construction of an activity domain relies on the premise that the scope of the domain is not too encompassing. Thus, in large organizations, several autonomous domains must be constructed. However, maintaining more than a few domains may be considered too costly. At a first glance, this seems obvious for several reasons. The cost of the cIS support becomes higher since more licenses have to be acquired, and the maintenance costs of the cIS applications increase. The capabilities of users trained in one domain cannot easily be transferred to other domains since these are constructed differently. Different ways of working may cause problems with misunderstandings and erroneous deliveries, and so on.

These aspects are all relevant and need to be carefully considered. After the crisis years, Ericsson is in fact doing quite well. This seems to indicate that the strategy of maintaining one coordination domain only is successful. However, costs associated with more intangible aspects, such as achieving consensus, is seldom, if ever, included in corporate analyses. These costs may be substantial.

- **An activity domain view of the organization is needed:** The approach presumes that the organization is apprehended in terms of activity domains having certain autonomy to construct their own ways of working. This may be problematic, especially during cost reduction efforts when values such as centralization, uniformity and commonality are honored. Thus, an alternative mind setting of what constitutes an organization is needed.
- **IS replacements may cause conflicts:** As a rule, management data in organizations are kept in separate ISs for requirement management, engineering change order management, etc. As stated above, a common repository for such data is preferable in several ways. Thus, it is natural to replace the many ISs by a single one, the cIS. However, actually carrying out such a replacement may be problematic due to efforts already invested in existing ISs and resistance from users and other stakeholders.
- **The anatomy-centric approach challenges the traditional view on sound engineering practice:** The success of the domain construction strategy is based on an agile approach of making frequent changes in the information models and the cIS. From established, plan-driven engineering practice, making many changes indicates a badly specified system. Thus, a new mind setting concerning what characterizes good engineering practice is required.
- **Determining an optimal set of activity domains:** Identifying an optimal set of activity domains may be problematic. On the one hand, the construction of common understanding is aggravated if the domain becomes too large and heterogeneous. On the other hand, many small domains bring about several cIS implementations, which is costly to set up and maintain.
- **Coordinating activity domains:** The coordination between the A- and S-domains was not implemented according to the intentions in the approach. During the construction of these domains, there was no organizational unit in place at Ericsson to enforce the necessary commonality. Thus, a lesson learnt is that the construction of new domains must go together with the construction of the coordination between them.
- **Forgetting quality attributes:** The inherent focus on capabilities in the anatomy might relapse into a focus on functionality, losing attention to quality attributes such as performance, maintenance, usability, and the like. However, the mere use of the concept of "capability" instead of "functionality" denotes a broader scope in which quality attributes can naturally be included when specifying a certain capability.
- **Transferability:** The anatomy-centric approach was conceived in the practice of Ericsson and has not been implemented elsewhere. This raises questions about the transferability of the results to other organizations. However, the development of telecom systems comprises most, if not all, of the challenges in distributed large-scale and global development, which would indicate that the approach is indeed transferable. This is a topic for further research.

- **One IS platform:** Another issue concerns the cIS used in the domain construction strategy. The experiences so far come from one particular system: Matrix. It is desirable to identify general criteria by which other cISs could be evaluated for use in the approach.
- **Software configuration management tools:** The coordination of software codes files is a major issue in distributed and agile software development (Battin, Crocker, Kreidler, & Subramanian, 2001; Ebert & De Neve, 2001; Komi-Sirviö & Tihinen, 2005; Herbslev & Moitra, 2001). Different groups may be concurrently working on variants of the same piece of code, which much be consolidated and synchronized for global integration. This aspect of is not treated here. However, there are several interaction points between the coordination domain and code writing activities. For example, when the code implementing the capability in an increment is tested and verified, this should be notified to the activity domain. This implies that the cIS and code configuration tools need to be interfaced (Crnkovic, Asklund, & Persson-Dahlqvist, 2003).

IN CONCLUSION

The anatomy-centric approach is an attempt to come to grips with the challenges of managing the development of exceptionally complex systems during volatile circumstances. The core construct of the approach is the anatomy, which is an image – preferably on one page – illustrating the dependencies between capabilities in the system. With the anatomy as a basis, the total scope of the development can be divided into a series of steps – increments – in such a way that each step can be integrated and tested on the way. The increments can be developed according to plan-driven or agile methods depending on the character of the increment.

A remarkable observation is that such projects may be coordinated and monitored by, in principle, very simple images that are laid out differently from traditional project management images. It is plausible that these images have advantages over traditional ones when managing complex systems since they have prevailed in the Ericsson practice for more than a decade now. I have suggested one reason for this is their conformance with dominant activity modalities and their interdependencies, thus sustaining the claim that the modalities play a central role in coordinating human activity.

In conclusion, the anatomy-centric approach addresses many of the challenges inherent in the development of complex projects. The domain construction strategy has demonstrated its capability to construct communal meaning of coordination and its cIS support in the development of the 3rd generation of mobile systems. These results indicate that the ADT is a viable guiding framework for the coordination of complex projects. However, the application of the approach is so far limited to Ericsson, and more experiences have to be collected in order to substantiate this claim.

REFERENCES

Adler, N. (1999). *Managing Complex Product Development – Three approaches*. Stockholm: EFI, Stockholm School of Economics.

Agile Manifesto (2009). Retrieved Jan 29, 2009, from http://agilemanifesto.org/

Anderstedt, J., Anderstedt, U., Karlsson, M., & Klasson, M. (2002). *Projekt och helhet – att leda projekt i praktiken.* [Project and whole – to manage projects in practice (in Swedish)]. Stockholm: Författares Bokmaskin.

Babar, M. A., & Abrahamsson, P. (2008). Architecture-Centric Methods and Agile Approaches. In P. Abrahamsson, R. Baskerville, K. Conboy, B. Fitzgerald, L. Morgan, & X. Wang (Eds.), *Proceeding of 9th International Conference, XP 2008* (pp. 242–243). Limerick, Ireland, June 10-14, 2008. Heidelberg: Springer.

Battin, R., Crocker, R., Kreidler, J., & Subramanian, K. (2001). Levering Resources in Global Software Development. *IEEE Software*, (March/April): 70–77. doi:10.1109/52.914750

Boehm, B., & Turner, R. (2003). Using Risk to Balance Agile and Plan-Driven Methods. *IEEE Computer*, *36*(6), 57–66.

Boehm, B. W., & Turner, R. (2003b). *Balancing Agility and Discipline: A Guide for the Perplexed.* Boston: Addison-Wesley.

Booch, G. (2006). On Architecture. *IEEE Software*, *23*(2), 16–18. doi:10.1109/MS.2006.52

Booch, G., Rumbaugh, J., & Jacobson, I. (1999). *The Unified Modeling Language User Guide.* Harlow, UK: Addison-Wesley.

Brooks, F. P. (1995). *The mythical man-month: essays on software engineering*. Reading, MA: Addison-Wesley.

Crnkovic, I., Asklund, U., & Persson-Dahlqvist, A. (2003). *Implementing and Integrating Product Data Management and Software Configuration Management.* London: Artech House.

Damian, D. (2003). Global Software Development: Growing Opportunities, Ongoing Challenges. *Software Process Improvement and Practice*, *8*, 179–182. doi:10.1002/spip.182

Ebert, C., & De Neve, P. (2001). Surviving Global Software Development. *IEEE Software*, *18*(2), 62–69. doi:10.1109/52.914748

Ekstedt, M., Johnson, P., Lilliesköld, J., & Jonsson, N. (2003). Making Project Complexity Understandable: the Elegance of Notations. In *Proceedings of the 12th International Conference on Management of Technology,* Nancy, France, 2003.

Elshamy, A., & Elssamadisy, A. (2006). Divide *After* You Conquer: An Agile Software Development Practice for Large Projects. In P. Abrahamsson, M. Marchesi, & G. Succi (Eds.), *XP 2006 LNCS 4044* (pp. 164–168). Berlin: Springer-Verlag.

Fowler, M. (2003). Who needs an architect? *IEEE Software*, *20*(5), 11–13. doi:10.1109/MS.2003.1231144

Fowler, M. (2004). *Is design dead?* Retrieved Jan 19, 2009, from http://martinfowler.com/articles/designDead.html#id58351

Fowler, M. (2006). *Continuous Integration.* Retrieved Jan 19, 2009, from http://martinfowler.com/articles/continuousIntegration.html

Fraser, S., Boehm, B., Järkvik, J., Lundh, E., & Vilkki, K. (2006). How Do Agile/XP Development Methods Affect Companies? In P. Abrahamsson, M. Marchesi, & G. Succi (Eds.), *XP 2006 LNCS 4044* (pp. 225–228). Berlin: Springer-Verlag.

Gandhi, M., & Robertson, E. L. (1992). A Specification-based Data Model. *Indiana University Computer Science Department Technical Report TR344*. Bloomington, IN: Indiana University. Retrieved Sept 30, 2008, from http://www.cs.indiana.edu/ftp/techreports/index.html

Gandhi, M., & Robertson, E. L. (1995). SDBM as a Model for Codesign Data. In J. Rozenblit & K. Buchenrieder (Eds.), *Computer Aided Software/Hardware Engineering* (pp. 338–355). Piscataway, NJ: IEEE Press.

Gorton, I., & Motwani, S. (1996). Issues in co-operative software engineering using globally distributed teams. *Information and Software Technology, 38*(10), 647–655. doi:10.1016/0950-5849(96)01099-3

Grant, R. M. (1996b). Prospering in Dynamically-competitive Environments: Organizational Capability as Knowledge Integration. *Organization Science, 7*(4), 375–387. doi:10.1287/orsc.7.4.375

Heeks, R., Krishna, S., Nicholson, B., & Sahay, S. (2001). Synching or Sinking: Global Software Outsourcing Relationships. *IEEE Software, 18*(2), 54–60. doi:10.1109/52.914744

Henry, T. S. (2007). *Architecture-Centric Project Estimation.* Master of Science thesis in Computer Science & Applications. Blacksburg, VA: Virginia Polytechnic Institute and State University. Retrieved Jan 26, 2009, from http://scholar.lib.vt.edu/theses/available/etd-05142007-214700/unrestricted/thesis.pdf

Herbsleb, J., & Moitra, D. (2001). Global Software Development . *IEEE Software, 18*(2), 16–20. doi:10.1109/52.914732

Järkvik, J., Berggren, C., & Söderlund, J. (2007). Innovation in project management: A neo-realistic approach to time-critical complex systems development. In T. Brady (Ed.), *IRNOP VIII Conference, Projects as Innovation, Innovation as projects*. Brighton, UK: University of Sussex. September 19–21, 2007.

Jensen, R. N., Møller, T., Sönder, P., & Tjørnehøj, G. (2006). Architecture and Design in eXtreme Programming; Introducing "Developer Stories". In P. Abrahamsson, M. Marchesi, & G. Succi (Eds.), *XP 2006 LNCS 4044* (pp. 133–142). Berlin: Springer-Verlag.

Jönsson, P. (2006). The Anatomy-An Instrument for Managing Software Evolution and Evolvability. In *Second International IEEE Workshop on Software Evolvability (SE'06)* (pp. 31–37), Philadelphia. September 24, 2006.

Karlsson, E. A. (2002). Incremental Development - Terminology and Guidelines. In *Handbook of Software Engineering and Knowledge Engineering,* (Vol. 1, pp. 381–401). Singapore: World Scientific.

Karlsson, E. A., & Andersson, L. G. (2001). XP and Large Distributed Software Projects. In G. Succi, M., & M. Marchesi, (Eds.), *Extreme Programming Examined* (pp. 119–134). Boston: Addison-Wesley.

Karlsson, E. A., Andersson, L. G., & Leion, P. (2000). Daily build and feature development in large distributed projects. In [Los Alamos, CA: IEEE CS Press.]. *Proceedings of the International Conference on Software Engineering ICSE, 2000*, 649–658.

Kolb, D. A. (1984). *Experiential Learning: Experience as the Source of Learning and Development.* Englewood Cliffs, NJ: Prentice Hall.

Komi-Sirviö, S., & Tihinen, M. (2005). Lessons Learned by Participants of Distributed Software Development. *Knowledge and Process Management, 12*(2), 108–122. doi:10.1002/kpm.225

Kraut, R., & Streeter, L. (1995). Coordination in Software Development. *Communications of the ACM, 38*(3), 69–81. doi:10.1145/203330.203345

Kristofersson, A., & Lindeberg, C. (2006). *Lean Product Development in Swedish Industry - An Exploratory Study.* Master Thesis in Operations Management, Stockholm School of Economics, Stockholm. Retrieved Jan 20, 2009, from http://arc.hhs.se/download.aspx?MediumId=161

Kruchten, P. (1995). Architectural Blueprints – The "4+1" View Model of Software Architecture. *IEEE Software, 12*(6), 42–50. doi:10.1109/52.469759

Larman, C., & Vodde, B. (2008). *Scaling lean & agile development: thinking and organizational tools for large- scale Scrum.* Upper Saddle River, NJ: Addison-Wesley.

Lilliesköld, J., Taxén, L., Karlsson, M., & Klasson, M. (2005). Managing complex development projects – using the system anatomy. In *Portland International Conference on Management of Technology and Engineering, PICMET '05,* July 31st – Aug 4th, 2005.

Maier, M. W., Emery, D., & Hilliard, R. (2001). Software architecture: introducing IEEE Standard 1471. *Computer, 34*(4), 107–109. doi:10.1109/2.917550

Malone, T., & Crowston, K. (1994). The Interdisciplinary Study of Coordination. *ACM Computing Surveys, 26*(1), 87–119. doi:10.1145/174666.174668

Mathiassen, L. (1998). *Reflective Systems Development.* Dr. Techn. Thesis, Aalborg University.

MatrixOne. (2008). Retrieved Feb 2, 2005, from http://www.matrixone.com/index.html

McChesney, I. R., & Gallagher, S. (2004). Communication and co-ordination practices in software engineering projects. *Information and Software Technology, 46*(7), 473–489. doi:10.1016/j.infsof.2003.10.001

Meso, P., & Jain, R. (2006). Agile Software Development: Adaptive Systems Principles and Best Practices. *Information Systems Management, 23*(3), 19–30. doi:10.1201/1078.10580530/46108.23.3.2006 0601/93704.3

Nord, R. L., & Tomayko, J. E. (2006). Software architecture-centric methods and agile development. *IEEE Software, 23*(2), 47–53. doi:10.1109/MS.2006.54

Ny Teknik. (2006). Ericsson utvecklar på löpande band [Ericsson develops on a conveyor Belt]. *Ny Teknik,* Jan 25, 2006 (In Swedish). Retrieved Jan 20, 2009, from http://www.nyteknik.se/nyheter/it_telekom/allmant/article38813.ece

Olsson, K. (1999). *Daily build – the best of both worlds: rapid development and control.* Stockholm: Sveriges verkstadsindustrier. Retrieved Jan 14, 2009, from http://www.spin-sweden.org/dailybuildreport.pdf

Ovaska, P., Rossi, M., & Marttiin, P. (2003). Architecture as a Coordination Tool in Multi-site Software Development. *Software Process Improvement and Practice*, *8*(4), 233–247. doi:10.1002/spip.186

Prikladnicki, R., Audy, J. L. N., & Evaristo, R. (2003). Global Software Development in Practice Lessons Learned. *Software Process Improvement and Practice*, *8*(4), 267–281. doi:10.1002/spip.188

Sakthivel, S. (2005). Virtual workgroups in offshore system development. *Information and Software Technology*, *47*(5), 305–318. doi:10.1016/j.infsof.2004.09.001

Schon, D. (1983). *The Reflective Practitioner. How Professionals Think in Action*. New York: Basic Books.

Shaw, M., & Clements, P. (2006). The golden age of software architecture. *IEEE Software*, *23*(2), 31–39. doi:10.1109/MS.2006.58

Smircich, L. (1983). Concepts of Culture and Organizational Analysis. *Administrative Science Quarterly*, *28*(3), 339–358. doi:10.2307/2392246

Smolander, K., Rossi, M., & Purao, S. (2008). Software architectures: Blueprint, Literature, Language or Decision? *European Journal of Information Systems*, *17*(6), 575–588. doi:10.1057/ejis.2008.48

Taxén. L., Lilliesköld, J. (2005). Manifesting Shared Affordances in System Development – the System Anatomy. *ALOIS*2005, The 3rd International Conference on Action in Language, Organisations and Information Systems* (pp. 28–47), 15–16 March 2005, Limerick, Ireland. Retrieved Feb 6, 2008, from http://www.alois2005.ul.ie/

Taxén, L. (2006). An Integration Centric Approach for the Coordination of Distributed Software Development Projects. *Information and Software Technology*, *48*(9), 767–780. doi:10.1016/j.infsof.2006.01.007

Taxén, L., & Lilliesköld, J. (2008). Images as action instruments in complex projects. *International Journal of Project Management*, *26*(5), 527–536. doi:10.1016/j.ijproman.2008.05.009

Toffolon, C., & Dakhli, S. (2000). A framework for studying the coordination process in software engineering. In *Proceedings of the 2000 ACM symposium on Applied computing - Volume 2* (pp. 851–857). Como, Italy.

Wikipedia (2009). Retrieved Jan 19, 2009, from http://en.wikipedia.org/wiki/Agile_software_development

Wikipedia (2009b). Retrieved Jan 30, 2009, from http://en.wikipedia.org/wiki/Code_refactoring

ENDNOTES

1 This type of development has also been named "global software development" (Damian, 2003) and "offshore development" (Sakthivel, 2005).

2 This section is a revised and extended version of Taxén, L. (2006). An Integration Centric Approach for the Coordination of Distributed Software Development Projects. *Information and Software*

Technology, 48(9), 767-780, published by Elsevier: http://www.elsevier.com/wps/find/homepage.cws_home.

3 Commercial, off-the-shelf (COTS) is a term for software or hardware products that are ready-made and available for sale, lease, or license to the general public.

4 It is interesting to note that Vilkki holds a position as Manager for Agile Coaching at Nokia Siemens Networks, one of the fiercest competitors to Ericsson. Both companies stress, in their own ways, the importance of attending both agility and commonality.

5 Of course, a preliminary understanding of the implementation must exist, otherwise the anatomy cannot be defined (you need to know what capabilities a certain implementing module needs).

Chapter 11
Enterprise Architectures:
An Alternative View

With the increased pressure on organizations to be agile with respect to changes in markets, products, re-organizations etc., the interest in so called "enterprise architectures" (EA) has gained momentum. The definition of enterprise architecture usually starts with a definition of "architecture". Basically, architecture is that which is essential or unifying in a system. An EA would then simply be the architecture for the enterprise seen as a system. Thus, the architecture concept is in principle transferred from "technical" systems (buildings, software programs, telecom systems, and the like), and applied to systems where humans are included in the system; a complication to which I will return later on.

The most renowned and influential EA is the Zachman Framework for Enterprise Architecture (Sowa & Zachman, 1992; Zachman, 1987). I will discuss this framework more in detail later on from the ADT point of view. Other prominent EAs are The Open Group Architecture Framework (TOGAF: Open Group, 2008), The Department of Defense Architecture Framework (DoDAF, 2004), and The Federal Enterprise Architecture (FEA, 2008). All these frameworks propose that abstract models of the enterprise and its IT systems should be the basis for enterprise analysis, design, understanding and communication (Lindström, Johnson, Johansson, Ekstedt, & Simonsson, 2006, p. 82). In addition, several other architectural frameworks exist (Greefhorst, Koning, & Vliet, 2006).

In this chapter, I reconceptualize enterprise architectures by using the activity domain as the basic architectural entity, thus emphasizing the elements of communal meaning and transition between domains. I compare this view of the enterprise with influential EA frameworks such as the one proposed by Zachman. I discuss implications of the ADT approach and suggest how to operationalize the construction of enterprise architectures from the ADT perspective.

DOI: 10.4018/978-1-60566-192-6.ch011

WHY ENTERPRISE ARCHITECTURES?

A number of reasons have been put forward to motivate investments in defining and maintaining an EA for the organization.

Integrated Approach Indispensable

The most down-to-earth reason is simply that an organization cannot do without it. An integrated approach to business and IT is indispensable (Jonkers et al., 2006, p. 63). Integration has been conceived as "the act or process of making something whole and entire" (Kirsilä, Hellström, &Wikström, 2007, p. 715). "By referring to integration, we thus mean bringing or joining together a number of distinct things so that they move, operate and function as a harmonious, optimal unit" (ibid). Integration can only be achieved if an enterprise-wide point of view is taken (Goethals, Snoeck, Lemahieu, & Vandenbulcke, 2006, p. 70).The EA is the means by which an integrated view of the organization can be achieved. It provides "a coherent whole of principles, methods and models that are used in the design and realization of the enterprise's organizational structure, business processes, information systems, and infrastructure. EA captures the essentials of the business, IT and its evolution" (Jonkers et al., 2006, p. 64). The importance of an integrated view is reinforced by the tendency of organizations to cooperate in networks. An EA can provide a means to understand the dependencies between the organization and its customers, suppliers, partners and competitors.

Managing Change

Organizations are ever changing entities. An EA is vital when it comes to managing these changes. It helps stakeholders to evaluate new circumstances and respond appropriately. Without an EA it is, for example, difficult to estimate the consequences of replacing legacy IT systems with an enterprise resource planning (ERP) system. Such systems (as offered by, for example, SAP) come with in-built EAs for specific types of businesses or industries. By comparing the architectures for the organization and the ERP-package, the effort of adapting either the organization or the ERP-package can be estimated.

A decision to introduce an ERP-system is one example of a wider issue, that of aligning the business strategy of the organization and its ICT-resources. New technological inventions such as the Internet, global restructuring of development and production, outdated and incompatible legacy systems, company down-sizing, changes in customer demands, are but some circumstances that put an ever increasing pressure on alignment. The "availability of architecture descriptions enhances agility. After all, it is easier to change something you know well, and architecture descriptions are of major help in getting to know and understand existing system" (Goethals, Snoeck, Lemahieu, & Vandenbulcke, 2006, p. 70). An EA provides a mechanism that enables the organization to relate business strategies to technological means. This however, requires that the EA is seen as a living entity that needs to be constantly updated and aligned to new circumstances.

Trade-Off Between Stability and Flexibility

Even if the EA must change, an important aspect of the EA is that it should provide stability to the organization. Without stability, an ordered change process is impossible. Thus, contrary to what is sometimes claimed (e.g. Appleton, 2004), flexibility and agility presume stability. The issue is rather the extent and detail to which the architecture prescribes the operations of the organization. An architectural description should focus on those entities of the organizations that are stable and can be expected to change at a slow pace. Examples of such entities are basic business rules for naming and identifying parts in products or services. Usually, these rules have been elaborated over a long time and are, so to say, part of the taken-for-granted in the organization.

Communication Among Stakeholders

If the EA shall be a sharp instrument in managing the organization, it must become a common asset among stakeholders. Only then can the EA provide a common foundation for discussions, evaluations and action. Each stakeholder brings his or her world-view to the arena of the EA. They might have different backgrounds, experiences and incentives for taking part in a joint definition of the EA. Moreover, each stakeholder is presumably interested in the particular view of the enterprise that concerns him or her the most.

These aspects put strict requirements on models used for conceptualizing the EA. The level of abstraction has to be carefully selected. If the model statements are too general, such as "applications executes on platforms, manipulates information objects, and supports business processes", everyone agrees. However, this will severely hamper the impact potential of the EA. It will become useless as a management instrument. On the other hand, if the models are too numerous, incoherent and detailed, it will be impossible to reach consensus about them. Thus, the communicative qualities of the EA descriptions are crucial.

Compliance Requirements

Some drivers for EAs are regulative acts and laws. For example, in order to increase the transparency of firms and protect investors from frauds, the Sarbanes-Oxley Act of 2002 was enforced as a result of the Enron scandal. This act, formally known as the Public Company Accounting Reform and Investor Protection Act, compels firms to make company executives personally accountable. One means to do so is the EA, which clarifies and makes transparent the responsibilities of employees in general and executives in particular.

Another example is the Clinger-Cohen Act of 1996. This act demands that every government agency has an information technology (IT) architecture, which is defined as: "an integrated framework for evolving or maintaining existing information technology and acquiring new information technology to achieve the agency's strategic goals and information resource management goals" (Jonkers et al., 2006, p. 64).

In some cases, suppliers of products and services are mandated to use a particular EA. The Department of Defense in the US mandates the use of DoDAF, which also has been widely influential within NATO and Great Britain's Ministry of Defense. Another example is FEA which is targeted to the United States federal agencies in order to harmonize investments in information technology.

ANALYSIS OF THE ZACHMAN FRAMEWORK FROM THE ACTIVITY DOMAIN THEORY PERSPECTIVE

One of the pioneers within the EA area is Zachman, who suggested an architecture framework in 1987 (Sowa & Zachman, 1992; Zachman, 1987). About 20% of EA-related activities are derived from the Zachman framework (ZFK), which is the highest rate amongst all the other frameworks (Fatolahi & Shams, 2006). It is considered as the most referenced framework and a basis for evaluating other enterprise architecture frameworks, methods and tools (ibid). Thus, it is appropriate to use the ZWK as a touchstone against which the ideas of ADT can be contested.

The original framework consists of 6 columns and 6 rows (see Figure 1).

Rows represent "perspectives" of different roles in the enterprise: planner, owner, designer, builder, programmer and user. These roles are inspired by roles or stakeholders in building architectures, from where Zachman collected his inspiration for the framework. In the context of EA the corresponding perspectives of the rows are: Scope (contextual), Business Model (conceptual), System Model (Logical), Technology Model (physical), Detailed Representation (out of context) and Functioning Enterprise (the actual instantiation of the enterprise).

The columns of the framework represent "different abstractions from or different ways to describe the real world. The reason for isolating one variable (abstraction) while suppressing all others is to contain the complexity of the design problem" (Sowa & Zachman, 1992, p. 592). The abstractions are: what the product is made of (material), how it works (process), where the components are relative to each other (geometry), who does the work, when things do happen, and why various choices are made (Zachman, 2008b). In the context of EA these abstractions are associated with DATA (what), FUNCTION (how), NETWORK (where), PEOPLE (who), TIME (when), and MOTIVATION (why) respectively.

Each cell of the framework has its particular representation, sometimes as text but in most cases as a unique model. For example, the row 2, column B cell (Business Process Model) is described as: "a

Figure 1. The original Zachman Framework (after Zachman, 2008)

		WHAT A	HOW B	WHERE C	WHO D	WHEN E	WHY F	
SCOPE	1	List of things important to business	List of processes the business performs	List of locations which the business operates	List of organisations / agents that are important	List of significant events	List of business goals / strategies	Planner
BUSINESS MODEL	2	Semantic Model	Business Process Model	Business Logistics Syst.	Work Flow Model	Master Schedule	Business Plan	Owner
SYSTEM MODEL	3	Logical Data Model	Application Architecture	Distributed System Architecture	Human Interface Architecture	Processing Structure	Business Rule Model	Designer
TECHNOLOGY MODEL	4	Physical Data Model	System Design	Technology Architecture	Presentation Architecture	Control Structure	Rule Design	Builder
DETAILED REPR.	5	Data Definition	Program	Network Architecture	Security Architecture	Timing Definition	Rule Specification	Programmer
FUNCTIONING ENTERPRISE	6	Usable Data	Working Function	Usable Network	Functioning Organization	Implemented Schedule	Working Strategy	User
		DATA	FUNCTION	NETWORK	PEOPLE	TIME	MOTIVATION	

Figure 2. Business and technological architectures

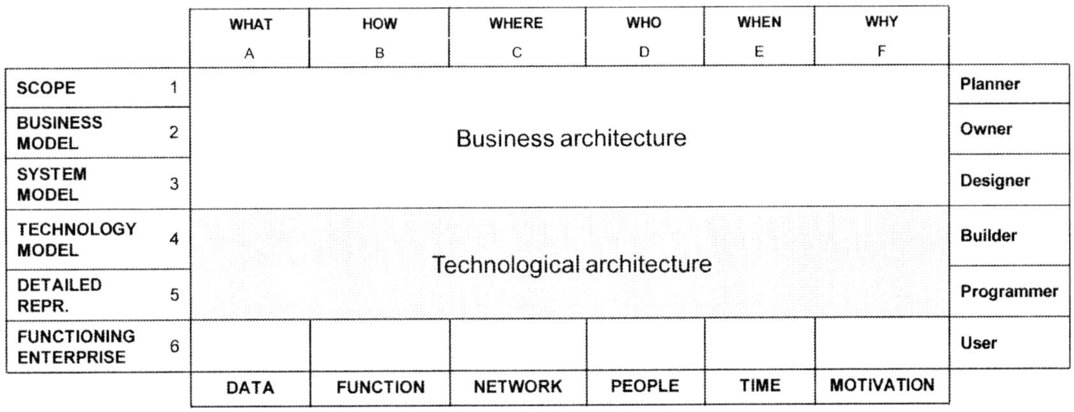

model of the actual Business Processes that the Enterprise performs, quite independent of any 'system' or implementation considerations and organizational constraints" (Zachman, 2008b, p. 2).

The ZFK lacks a theoretical basis such as the ADT: "The main disadvantage of Zachman's approach is the lack of a scientific foundation on top of which the framework is constructed" (Beznosov, 2000, p. 8). Thus, a comparison of theoretical foundations cannot be done. Since ZFK prescribes a unique model for each cell in the framework, it is against the operationalized ADT that a comparison can take place, i.e., against the spatial, temporal and transitional information models; the information systems (IS) and the domain construction strategy (DCS) (see Section 3).

The ZFK is all-inclusive in the sense that it comprises both a "business" architecture that focuses on business models and a "technological" architecture that is geared towards the technology implementing these models (Smith, O'Brien, Barbacci, & Coallier, 2002). The separation between these architectures goes roughly between rows 3 and 4 (see Figure 2). For example, cell (3, A) "Logical Data Model" is described as "a model of the logical (implementation – technology neutral) representation of the things of the Enterprise about which it records information" (Zachman, 2008b, p. 1). This would correspond to a spatialization model detailed enough for direct implementation in an IS like Matrix. The row 6 "Functioning Enterprise" is a particular case since this represents the "instantiated" enterprise defined in the two architectures. This means that the comparison between ADT and ZFK will be restricted to the "business" architecture; roughly the area indicated in Figure 2.

The Work Object

The work object entity in ADT is, together with the motive, of fundamental importance for the structure of the activity domain. In ZFK the work object is to a large extent subdued. It can be found in the column D "WHO/PEOPLE", cells (1, D), (2, D), and (3, D) under the concept of "work product", which is defined as "a kind of user view of materials or process inputs/outputs" (Sowa & Zachman, 1992, p. 597). However, the actual work object motivating the existence of the enterprise is remarkably elusive. It is as if the most important entity of the enterprise – what product or service is being worked at – is so self-evident that it does not deserve attention.

The motive of the activity domain can be mapped to the first two rows in the "WHY/MOTIVATION" column in ZFK. Cell (2, F) "Business Plan" is described as "a model of the business objectives and strategies (the "ends" and "means") of the enterprise that constitute the motivation behind enterprise operations and decisions" (Zachman, 2008b, p. 6). However, the next cells in column F ("Business Rule Model", "Rule Design", and "Rule Specification") would be traced to the stabilization modality in ADT and not considered part of the motive of the activity domain.

The Actors

Actors in the activity domain would most appropriately be associated with the "WHO/PEOPLE" column in ZFK. However, as discussed above, this column includes a number of other things such as organizational charts, work products, work flows, and roles (cell 3, D). This signifies that the action perspective of organizations is to a large extent missing in ZFK. People are seen as one element among others that make up the enterprise system, which is a reflection of the basic system engineering perspective of the ZFK. This in turn means that cognitive and sense-making issues are not emphasized in this framework, which, according to ADT is a severe neglect.

Activity Modalities

Vestiges of the activity modalities can be found in various places in the ZFK as follows.

Contextualization

The contextualization modality is at play in the formation of activity domains. The target of ZFK, the enterprise, would be considered an activity domain in ADT. In addition, activity domains can be traced in the cells (1, C) "List of locations where the business operates" and more specifically in cell (2, C) "Business Logistics System" that models these locations and their connections. This model includes "identification of the type of facilities at the nodes like branches, headquarters, warehouses, etc" (Zachman, 2008b, p. 2). "Locations" can be understood as sites (activity domains) with particular motives and work objects in the enterprise.

Other places where the activity domain can be traced are in column D "WHO/PEOPLE", cells (1, D), (2, D), and (3, D). Cell (2, D) "Work Flow Model" is described as "the model of the actual enterprise allocation of responsibilities and specification of work products. Typically, an organization chart expresses the allocation of responsibilities but other supporting documents describe the work products" (Zachman, 2008b, p. 4).

In ADT, an organizational chart may or may not signify an activity domain. It depends on how well the chart can be mapped to the actual work context framed by the domain, which in turn is determined by the motive and the work object. The focus on the organizational chart in the ZFK is a consequence of the lack of theoretical basis, which means that the "surface" phenomena (the organizational chart) is emphasized rather than the underlying theoretical construct of the activity domain. This might have severe consequences when there is a misfit between the structure signified by the chart and the structure of activity domains; something that might occur in times of organizational changes like out-sourcing, acquisitions and the like. If strategists take actions based on the organizational chart, serious misalignment between conceived allocations of work and the actual work allocation may result.

From the ADT point of view, the naming of cell (2, D) "Work Flow Model" in the WHO column is a misnomer. A work flow model indicates a temporal dimension and would thus be seen in ADT as a manifestation of the modality temporalization.

Spatialization

The "WHAT/DATA" column in ZFK corresponds to the activity modality spatialization in the sense that all models in this column are "things" and "relations" of concern for the enterprise. For example, the cell (2, A) "Semantic Model" is described as follows:

This is a model of the actual Enterprise things (objects, assets) that are significant to the Enterprise. It typically would be represented as an "E/R"-type model and would be at a level of definition that it would express concepts (terms and facts) used in the significant business objectives/strategies that would later be implemented as "Business Rules". (Zachman, 2008b, p. 1)

The "WHAT/DATA" column "contains data throughout the entire enterprise" (Fatolahi & Shams, 2006, p. 136), which implies that all the data used in the entire enterprise should be included in this model. This model becomes more detailed as we move down the rows in column A. When the "Logical Data Model" in cell (4, A) is reached, we have a complete, enterprise wide model that can be implemented in an IS.

Temporalization

The ZFK separates between processes (the "HOW" column) and time (the "WHEN" column). In ADT manifestations, both processes and time would be considered as derived from temporalization since they both have a temporal dimension. The concept of "function" is not associated with processes in ADT; rather, a function is considered as a capability of some means in a context. Moreover, "HOW" indicates a way of doing things and would be associated with the stabilization modality in ADT.

Stabilization

According to Ross (2003) "...a business rule is a directive, intended to influence or guide business behaviour, in support of business policy that has been formulated in response to an opportunity, threat, strength, or weakness." Thus, business rules are statements that describe, constrain and control the structure, operations and strategies of businesses. These rules are found in cells (3, F) "Business Rule Model", (4, F) "Rule Design", and (5, F) "Rule Specification", which is where the stabilization modality is most salient.

Transition

The transition modality is inherently associated with coordination between activity domains in ADT. No traces of transition can be found in ZFK, presumably because the ZFK is focused on The Enterprise – the single organization. In spite of claims that the framework can be applied to any size firm, there are no constructs that enable the modeling of the interactions between firms.

Interdependencies between Models

In the ZFK different models are made for the "abstractions" data, function, network, people, timing, and motivation. A complete model for the enterprise needs to include how these models are interrelated (Goethals, Snoeck, Lemahieu, & Vandenbulcke, 2006). For example, the cell (2, B) "Business Process Model" expresses "the business transformations (processes) and their inputs and outputs" (Zachman, 2008b, p. 2). However, these inputs and outputs are found as entities in the "Semantic Model" in cell (2, 1). Thus, these cells are interrelated and cannot be modeled independently of each other.

Another example is found in the "Rule Design" cell (4, F), which is defined as follows:

[Rule Design] is a physical specification of the Business Rules. The rules are not presently factored out from their implementations and therefore are found as cardinality and optionally in the data models (Column A), as procedural code (Column B) or a policy specification (Column D). (Zachman, 2008b, p. 6).

Thus, there exist interdependencies between the columns that are not explicitly emphasized by Zachman.

Summary

As can be seen from the analysis in the preceding section, the ZFK and the ADT are structured very differently. In spite of this, there are some striking similarities if we restrict the analysis to the "business" architecture part of the ZFK (rows 1-3). The WHAT (data) column can be associated more or less directly to the spatialization modality. Likewise, the HOW (function) column can be associated with the temporalization modality since the cells in this column are all focused on processes.

Stabilization is present in some cells. However, a straightforward mapping to the interrogatives of the ZFK would assign stabilization to the HOW, since HOW expresses valid and meaningful actions in an activity domain. Due to the dialectical interdependencies stabilizing elements will be found in other modalities and manifested in, for example, processes – the 2nd column of ZFK (and, for that matter, in cardinality rules manifested in the E/R models in column A (WHAT).

The column WHERE is clearly related to contextualization. However, the actual geographical location of sites should not be included in the EA, no matter how important they might be. It seems more reasonable to associate WHERE with the instantiated enterprise, i.e., the row 6 (FUNCTIONING ENTERPRISE) in ZFK. It is only after the definition of activity domains with their work objects and motives that the location of these domains can be worked out. A single domain may be geographically co-located or distributed across many sites. A similar reasoning can be applied to the WHEN column, which is obviously a manifestation of temporalization. However, the actual instant a certain event take place, belongs more properly to the instantiated enterprise. Concerning transition, there is no indication of this modality in the ZFK.

Thus, the mapping between modalities and the columns of the ZFK is far from straightforward. It is however interesting to note that Zachman has recently proposed a version 2 of the framework (Zachman, 2008c):

As can be seen, the framework is now more cleanly structured along two classifications. In the columns we find the same interrogatives as earlier: What, How, When, Who, Where, and Why. In the rows we find

Figure 3. The modified Zachman Framework (after Zachman, 2008c)

		WHAT Entity Relation A	**HOW** Transform Input B	**WHERE** Location Connection C	**WHO** Role Work D	**WHEN** Cycle Moment E	**WHY** Ends Means F	
SCOPE Identification	1	INVENTORY Types	PROCESS Types	NETWORK Types	ORGANIZATION Types	TIMING Types	MOTIVATION Types	**Strategists**
BUSINESS Definition	2	BUSINESS Entity Relation	BUSINESS Transform Input	BUSINESS Location Connection	BUSINESS Role Work	BUSINESS Cycle Moment	BUSINESS Ends Means	**Executive Leaders**
SYSTEM Representation	3	SYSTEM Entity Relation	SYSTEM Transform Input	SYSTEM Location Connection	SYSTEM Role Work	SYSTEM Cycle Moment	SYSTEM Ends Means	**Architects**
TECHNOLOGY Specification	4	TECHNOLOGY Entity Relation	TECHNOLOGY Transform Input	TECHNOLOGY Location Connection	TECHNOLOGY Role Work	TECHNOLOGY Cycle Moment	TECHNOLOGY Ends Means	**Engineers**
COMPONENT Configuration	5	COMPONENT Entity Relation	COMPONENT Transform Input	COMPONENT Location Connection	COMPONENT Role Work	COMPONENT Cycle Moment	COMPONENT Ends Means	**Technicians**
OPERATIONS Instantiation	6	OPERATIONS Entity Relation	OPERATIONS Transform Input	OPERATIONS Location Connection	OPERATIONS Role Work	OPERATIONS Cycle Moment	OPERATIONS Ends Means	**Workers**
		INVENTORY	**PROCESS**	**NETWORK**	**ORGANIZATION**	**TIMING**	**MOTIVATION**	

the transformation of an abstract idea into an instantiation: Identification, Definition, Representation, Specification, Configuration and Instantiation (Zachman 2008c). The second column can now be directly associated with the temporalization modality since it is geared towards various aspects of processes.

It is my guess that the new version of the framework is a result of experiences collected during the 20 years that the framework has been in use, although I have not been able to find any account of this in the literature. It is tempting to see the re-orientation as an indication of strength in the activity modality construct; a move towards a more modality aligned framework.

IMPLICATIONS: ENTERPRISE ARCHITECTURES

A striking observation is that still today there is an obvious need for means by which EAs can be operationalized:

The instruments needed for creating and using enterprise architectures are still in their infancy. (Jonkers et al., 2006, p. 65)

[A] complete enterprise model based on the Zachman framework is still beyond reach. In other words, after more than a decade of development, the predominant approach to enterprise architecture still does not provide a pragmatic solution to the problem of developing an enterprise wide model! (Khoury & Simoff, 2004, p. 65)

What are the reasons for these rather amazing statements? After all, the Zachman framework has been around for more than 20 years, numerous other frameworks have been proposed, and still, there seem to be no effective instruments available for creating and using EAs.

I propose that the reason for this state of affairs is insufficient attention to a number of "bearing elements" in established EA approaches:

- Theoretical foundation
- Integrative construct
- EA attractor
- Construction of communal meaning
- Dependencies between resources
- Coordinated flexibility
- Scalability
- Dimension decompression

Most of these bearing elements can be traced in the EA literature as important aspects of EA design. However, they have not been systematically treated, presumably because the theoretical instruments have been missing. I argue that the ADT provides such an instrument by which these elements can be analyzed and included in strategies and methods for developing EAs. All bearing elements are mutually dependent and there is no particular preference between them except for the primacy of the theoretical foundation. In the following, I will discuss these elements in turn.

The Activity Domain Theory as Theoretical Foundation for Enterprise Architectures

As has been stated by several scholars (e.g. Beznosov, 2000) the ZFK lacks a firm theoretical foundation. The inspiration for the ZFK comes from the traditional system engineering field, and the use of architectures in building processes, which Zachman transferred to the EA area. However, there is a fundamental difference between buildings and organizations: humans are elements in organizations but not in buildings. It is a main error of inclusion to consider humans as system elements on par with other, non-human elements. Humans cannot be treated as bricks or buttresses in a building. Once humans enter the picture a completely new point of departure has to be taken.

The neglect of acknowledging the specific role of human actors in the framework has decisive consequences. The enterprise system cannot be treated as a deterministic one (Appleton, 2004). When people are involved, things change as the EA is being defined. The EA system behavior "cannot be architected in the traditional sense of systems engineering. And, attempts to do so at the scale of the enterprise are doomed to failure" (ibid, p. 4). The ADT, however, is from the outset based on human action and, consequently, may provide a foundation for EAs where humans are related to, yet distinct from other elements in the enterprise.

Integrative Construct: The Activity Domain

In many organizations an "integrated view of the entire enterprise is still far off" (Jonkers et al., 2006, p. 63). Obviously the ZFK and other frameworks do not supply such integrated views; in any case, they do not come easily. The huge number of frameworks being proposed is an indication of the lack of a common basis. This has spurred inquiries for regular patterns or other fundamental structures that may be hidden behind the various frameworks (e.g. Greefhorst, Koning, & Vliet, 2006).

An integrated view is important for several reasons. Changes in goals and strategies will affect all "domains of the enterprise, such as the organisation structure, business processes, software systems, data management and technical infrastructure" (Jonkers et al., 2006, p. 64). An integrated view, where the relationships and dependencies between various elements of the enterprise are clearly visible, is necessary to assess the consequences of actions taken or contemplated. Jonkers et al. (ibid) point out that integrated views might be present in restricted parts of the enterprise. However, since these are often expressed in non-compatible ways, the integration over the entire enterprise is aggravated.

A central rationale for elaborating the notion of the activity domain is that it provides an integrating construct which can be applied to all organizational units. The integrands, the parts that are involved in the integration, are those that make up the activity domain: the work object and the motive, the actors, the outcome, and the enactment of capabilities along the activity modalities. All these parts and the whole, the activity domain, are dialectically related, meaning that parts constitute the whole and the whole constitutes the parts. The nested character of the activity domain provides a stable core that enables the integration of organizational units of any kind, be that a group, team, organizational function, business unit, the enterprise, an extended enterprise, or whatever. All these work settings can all be considered as activity domains as long as they have a motive and a work object.

A common methodological technique in the EA community is to construct EAs from views or viewpoints (Balabko & Wegmann, 2006) such as the five interrogatives of What, How, Where, When, and Why in the Zachman framework. However, one might ask: "A view of what?" A viewpoint presumes that there is something to view. This something, the essence of the enterprise, is strangely absent in most contributions in the EA area. It is as if the enterprise is defined by its very viewpoints. For example, Sowa & Zachman claim that "each column is related to every other" (Sowa & Zachman, 1992, p. 611). This means that goal, entity, process, location, role, and time are all related. However, the nature of these relationships is not clear (see Figure 4).

I claim that working with views of "some-thing" elusive is not enough if we want to understand how to work out useful instruments to handle this something. We must understand the essence of the phenomena we are investigating. It is here that the activity domain, or any other similar construct, becomes indispensible. The viewpoints do still make sense, but now we know what we are looking at (see Figure 5).

Just by comparing Figure 4 and Figure 5, it is intuitively clear that Figure 5 is easier to comprehend, especially if we take the recursive nature of the activity domain into consideration, whereas the structure in Figure 4 may quickly become incomprehensible when scaled up.

In EA discussions, the organizational unit is often taken for granted and unproblematic:

As companies are operating in complex environments, enterprises need to be segmented in units, each of which has as its major task the problem of dealing with a part of the environment [...]. Each of the departments within a company does part of the total work that needs to be performed by the company in order to achieve the goals of the enterprise. Of course, all of the departments are expected to work together. (Goethals, Snoeck, Lemahieu, & Vandenbulcke, 2006, p. 68)

Figure 4. The six viewpoints in the Zachman framework (adapted after Sowa & Zachman, 1992, p. 611, figure 18)

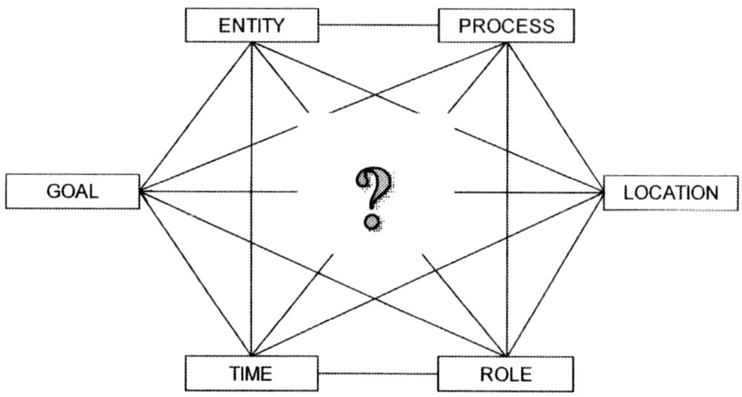

Figure 5. Viewpoints of the activity domain

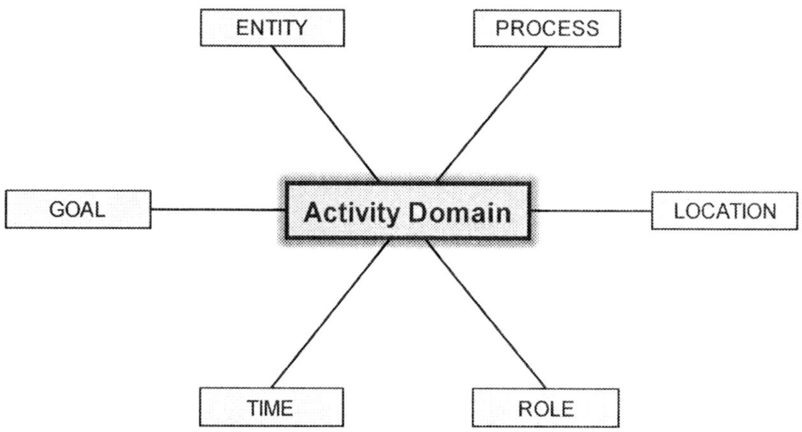

Terms like units, departments and companies are used indifferently since these terms are assumed to be more or less common knowledge. The introduction of the activity domain is an attempt to problemize the organizational unit and provide a common foundation for these units.

An interesting observation is that contributions discussing integration make use of various terms resembling the activity domain construct. For example, Jonkers et al. (2006) talk about "restricted domains", "different domains", "individual domains"; however without articulating these further. Examples of such domains, where mature architectures exist, are information, process, product, application and technical architectures (Jonkers et al., 2006, p. 64).

An interesting contribution in line with this thinking is given by Versteeg & Bouwman (2006). They take as their starting point the concept of "business architecture". Business architectures contribute to clarify the complexity within organizations, and groups "business functions and related business objects into clusters (business domains) over which meaningful accountability can be taken as depicted in the high level description of the related business processes" (Versteeg & Bouwman, 2006, p. 93). Such business domains can be production, distribution, marketing, manufacturing, assembly, transport, wholesale, and so on. These domains are assigned to owners which are responsible for the further development of the domains. The Business Architecture is derived from the business strategy, and forms a useful starting point from which subsequent functional, information, process and application architectures can be developed.

This view has many similarities with the ADT view. The concept of "business domain" has more than nominal resemblance with the concept of activity domain. Like the activity domain, business domains are placed at the center of the framework, not processes, information, technology or organization. Moreover, business domains are decoupled from the organizational chart. This indicates a turn towards the "practice" as the basic Unit of Analysis in the organizational discourse. "Business domains" such as production, distribution, marketing, etc., would most likely be considered activity domains in ADT. Within these, manifestations of spatialization (information), temporalization (process) and mediational means (applications) would be identified. Moreover, the activity domain is certainly an "area of responsibility" in the sense that the actors within a domain are responsible for their joint outcome.

Versteeg & Bouwman distinguishes between Enterprise Architectures and Business Architectures. "Enterprise" is used to indicate an enterprise-wide scope of the architecture, while Business Architecture is generic in the sense that it may stretch over multiple organizations (supply chain level), one organization (enterprise level) or part of an organization (business unit level) (Versteeg & Bouwman, 2006, p. 92). Thus, the ambition of the Business Architecture is similar to that of ADT.

From the ADT point of view the approach suggested by Versteeg & Bouwman is on the right track. However, there are fundamental differences between these approaches. The most striking one is that the drive behind business domains is accountability. Business domains are areas of responsibility that can be assigned to a person. Business domains reflect an individual centric perspective of organizations, where the personal responsibility over a certain area is in focus. Other examples of such roles are "concept owners", "process owners", and the like.

In contrast, the activity domain emphasizes the social nature of work. In this perspective, the development of the domain is due to social forces and not to single individuals who happens to have a formal role in the organization. The drives behind the activity domain are the motive and the work object. The responsibility as basis for the formation of domains instead of the work object is, in my view, a blind alley.

Another difference concerns the way business domains are developed. According to Versteeg & Bouwman the starting point is a formulated business strategy, which is subsequently embedded in the business architecture and functional, information, process and application architectures. This indicates a linear top-down approach where the opinions of CEO and his advisors are the point of departure. This view is alien to ADT, which takes a distinctly dialectical approach to the development of the activity domains. The business strategies and the ensemble of capabilities framed by activity domains are mutually dependent on each other.

The business domain does not seem to play a fundamental role of the EA like the activity domain does in ADT. Although not entirely clear, it appears that business processes are more basic than business domains. "Within the business architecture a high level description is provided of how the business processes are dealt with by these domains and which domain is responsible for specified business functions or objects" (Versteeg & Bouwman, 2006, pp. 92-93). Thus, the business process comes first. In ADT, the activity domain is basic. A business process would be seen as a manifestation of temporalization in the domain where the process is defined; usually at the enterprise level.

Another difference seems to be that the business domain construct is not recursively applied; business domains do not use other business domains as resources. Moreover, processes, functions, and objects are seen as separate "pillars" that can be decomposed independently in each pillar into application, process and information architectures. These architectures are subsequently linked in a sharable IT-unit linkage model. In contrast, in ADT, the activity domain always precedes and integrates the application process, and information architectures.

In spite of these concerns, the approach of Versteeg & Bouwman is, as far as I can tell from surveying the literature, the one that resembles ADT the most. I interpret this as a move towards taking a more insightful position concerning the often unreflecting use of ordinary domain-like constructs.

EA Attractors: The Work Object and the Motive

This bearing element concerns the drive of the structure of the EA; what ultimate impetus lies behind the formation of the architecture – its attractor, to stretch the metaphor slightly:

An attractor is a set to which a dynamical system evolves after a long enough time. That is, points that get close enough to the attractor remain close even if slightly disturbed. Geometrically, an attractor can be a point, a curve, a manifold, or even a complicated set with a fractal structure known as a strange attractor. Describing the attractors of chaotic dynamical systems has been one of the achievements of chaos theory. (Wikipedia, 2008c)

A general purpose of enterprise architectures is business/IT alignment and hence, the architecture should be driven by business strategy and market dynamics (Keltikangas, 2006). However, the dynamics of change makes this drive too volatile to be operational. Other, more stable attractors have to be found. For Versteeg & Bouwman (2006) the attractor is "responsibility". Their business domains are centered on the possibility to allocate a personal responsibility to the domain. For Sowa & Zachman (1992) the attractor is the five interrogatives What, How, Where, When and Who? For Ross (2005) the attractor is one of the "core" operating models: unification, coordination, diversification, and replication. Each of these operating models represents various levels of business process standardization and integration.

In the ADT, the attractors are the work object and the motive of the activity domain. The nature of the products or services that the domain provides to its clients will determine the nature of the capabilities needed of actors and mediational means. If the product includes modules implemented in software, the capabilities of software engineers, programming language editors, compliers, etc., are needed. If the service consists of training pilots, the capabilities of instructors, manuals and school planes are needed. In addition, capabilities for coordinating actions must be enacted.

The motive of the domain determines what emphasis is put on various aspects of the work object. Public and private hospitals have the same work object: humans in need of care. Thus, the same capabilities are needed in both types of hospitals. However, in private hospitals, the motive is to make a profit, while in public hospitals the motive is to treat people. This difference leads to dissimilar ways of doing things. For example, after stating a diagnosis the first priority of the doctor in private hospitals is to make sure that the patient can pay for the treatment, while in public hospitals the first priority of the doctor is to find a proper treatment.

For EAs, the focus of the work object and motive means that the structure of the EA will be mainly determined by the capabilities needed to produce the output of the domain. In particular, the EA will emphasize the dependencies between resources.

Communal Meaning: Taking the Issue Seriously

A major role of EA models is to serve as a means for communication among stakeholders. With the model as a common asset, various stakeholders can discuss alternatives and foresee consequences of taking/not taking actions. Does the model reflect all relevant things we want to have control over? Are there important aspects missing? If we out-source a certain unit, do we need to modify the model? And so on.

In order for the EA to become a resource, it must be enacted in the organizations. This means that stakeholders must achieve communal meaning about the EA artifact, i.e., they must go through the mathetic phase of meaning construction before they can enter the pragmatic phase where the EA model has become a useful instrument in the organization. The mathetic phase must be made as effortless as possible. However, most EA models suggested are hard to comprehend due to their complexity:

Given the wide scope of enterprise architecture, the size and scalability of EA models quickly becomes a major problem. (Jonkers et al., 2006, p. 65)

A quick survey of EA illustrations corroborates this impression (see e.g. Goethals, Snoeck, Lemahieu, & Vandenbulcke, 2006, p. 72; Sowa & Zachman, 1992, pp. 594-595). In cases where EA models are comprehensible, they are so simple that it is inconceivable how they can be scaled up to the complexity of an entire enterprise (e.g. Fatolahi & Shams, 2006, p. 136).

An example of a detailed architecture is given in Versteeg & Bouwman (2006, p. 96). It is hard to see how this architecture can become a common instrument for stakeholders since it is so complicated. Later on in their contribution, Versteeg & Bouwman gives an example of a business architecture that is difficult to trace back to the proposed nomenclature of business domains and application, process, and information architectures.

Similar concerns can be raised regarding the ZFK. The system engineering background of the ZFK brings with it an engineering approach to deep human issues. This is perhaps mostly noted in the devaluation of the effort to construct common understanding about the framework. Another way of expressing the same thing is that the enactment of the framework into an instrument for managing the enterprise is not an issue. This is most noticeable in the absence of a serious discussion of how stakeholder shall ever come to a consensus about the 36 models and their interdependencies in the ZFK. A similar critique has been launched by several scholars. For example, Khoury & Simoff state the following:

Because each view is modeled using disparate techniques and methods (developed independently of the Zachman framework) each segment interface presents a discontinuity. This creates a barrier to the understanding of how structures flow from one part of the enterprise to another. Thus, by dividing the organisation into distinct views, the Zachman framework defeats its goal, which is to provide a unified model of the organisation. (Khoury & Simoff, 2004, p. 65)

In order to grasp the complexity involved in the construction of communal meaning, let's consider an example from the telecom industry. Figure 6 shows an information model in UML-notation from Ericsson. This model would be found in the cell (2, A) "Logical Data Model" in the ZFK.

On the surface, this model looks fine. There are a number of "important things" represented by the boxes (classes) and attributes (text lines within classes). Sharp arrows indicate relationships and the numbers at the ends of the arrows cardinalities, i.e., how many exemplars (instances) of "things" can be related by the arrow. In addition, the diamond arrows indicate aggregation, i.e., a "consists-of" type of relation. All in one, the model in Figure 6 contains about 25 classes, 90 attributes, 45 relations and 90 cardinality values, implying that about 250 decisions have to be taken in order to define the model.

The experience from Ericsson and other organizations as well (see e.g. Bititci & Muir, 1997) is that it is exceedingly difficult to agree on a model of this size and detail. Enactment implies that the model must be proven useful in the domain where it is used. This is a process that may take years of testing, assessing and modifying the model. However, the enactment endeavor doesn't leave any traces on the model; in a sense it is just an illustration that can be drawn in a couple of days with a good tool at hand such as Rational Rose (Rational™, 2008). In fact, the model in Figure 6 was drawn by a couple of consultants after doing some reconnaissance work at Ericsson, thus completely ignoring the enactment process. The result was a lame duck; the model was in a literal sense – useless. In contrast, the model in Figure 7 is a high-level, enacted model that has been implemented in Matrix and proven useful in development projects.

Figure 6. Example of non-enacted ER-model in UML from Ericsson

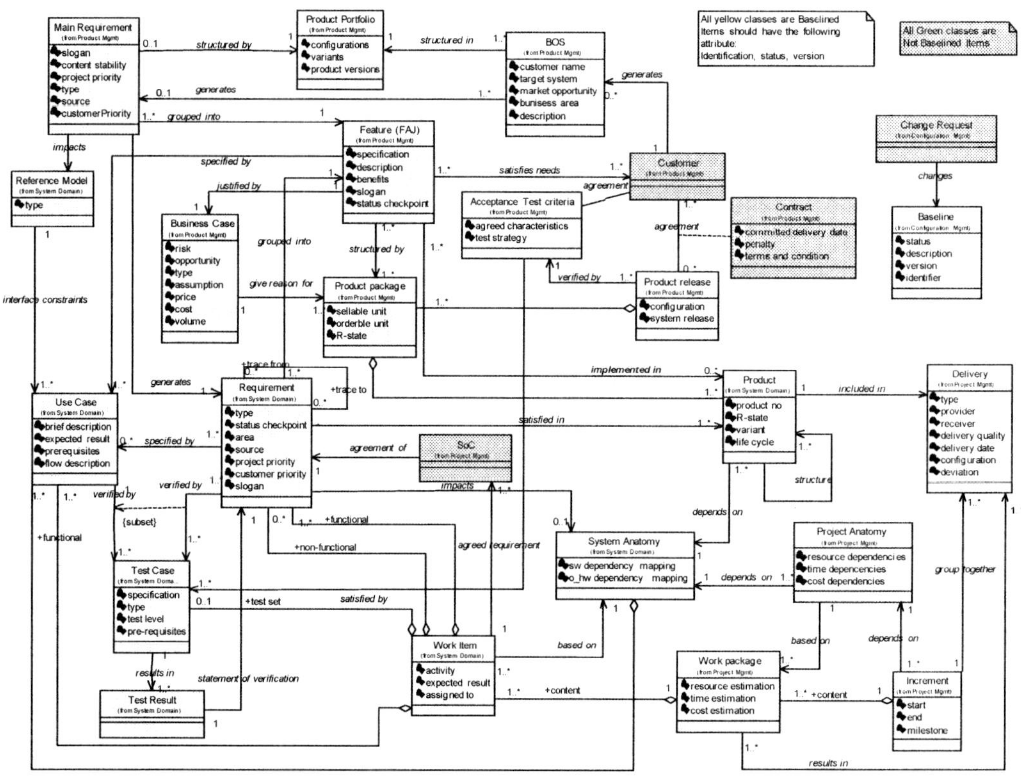

Figure 7. An enacted E/R-model from Ericsson (Remington & Pollack, 2007. ©Ashgate Publishing Ltd. Used with permission)

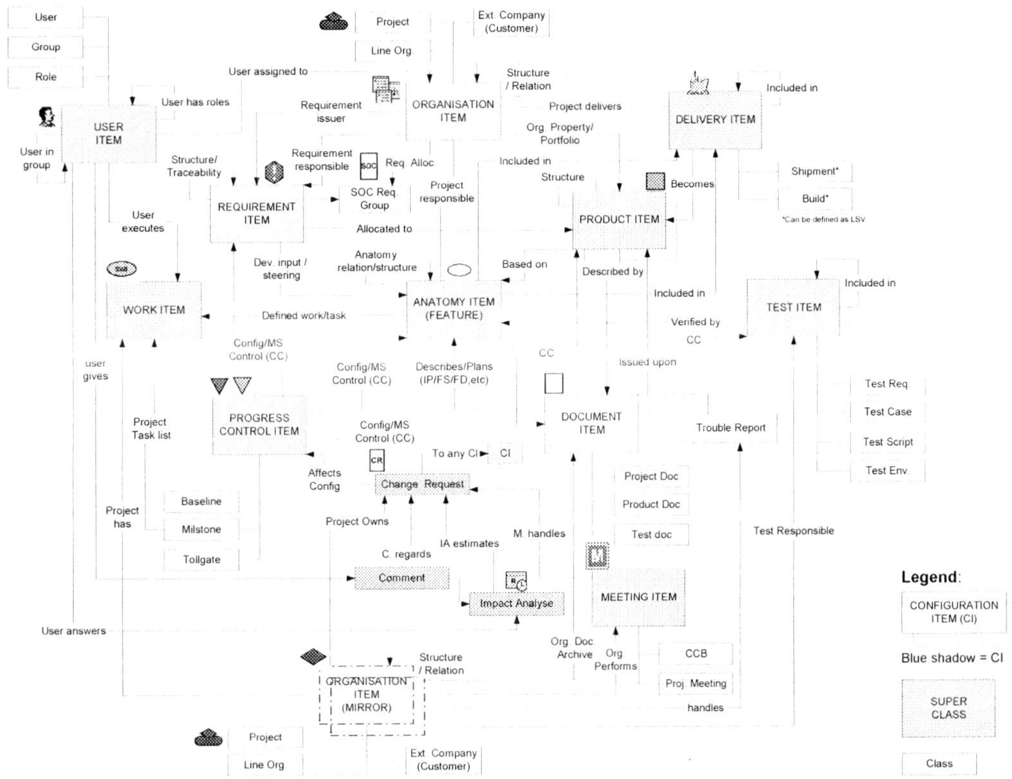

Still, the models in Figure 6 and Figure 7 represent only some development units of the entire Ericsson enterprise. For example, marketing, human relations, logistics and order, production and customer service units as well as other development units are not included. The task of defining one single, enacted model for the entire enterprise at the level prescribed by the "Logical Data Model" in ZWK is downright impossible; it simply cannot be done since there is no way stakeholders can agree on the myriads of details at that level. As a consequence, common enterprise-wide understanding can only be reached on a general level, where detailed decisions about implementation can be avoided.

From the ADT point of view, there is no such thing as *one* model of the entire enterprise along the columns in the ZFK. The A column (DATA) in ZFK cannot be captured in one E/R model (albeit at various levels of granularity); the same applies to the other columns. For each activity domain in the enterprise, there is a unique information model that is enacted according to the work object and motive of that domain. This means that communal meaning about the model is ultimately restricted to the activity domain. Thus, the complexity of enactment, or defining a useful model, is reduced.

However, this approach brings about another problem. In order to coordinate domains, some commonality of meaning must exist between domains. This can be done either through elements that are common to all domains, such as business rules stating how to identify items in the enterprise, or by translating and mapping activity domain specific concepts and terms between domains (the transition modality). The question of which elements should be involved in the commonalization process is a matter of maintaining an optimal balance between centralization and decentralization.

Several contributions in the EA literature have recognized the issue of achieving communal meaning about the EA models. For example, Balabko & Wegmann suggest a construct called "concern" in order to address this problem:

EA models have to model enterprises facets that span from marketing to IT. As a result, EA models tend to become large. Large EA models create a problem for model management. Concern-based design methods (CBDMs) aim to solve this problem by considering EA models as a composition of smaller, manageable parts—concerns. (Balabko & Wegmann, 2006, p. 115)

Another approach is suggested by Khoury & Simoff (2004) who introduces the concept of "elastic metaphors". They claim that the segmentation of the EA into disjointed, cell-specific models can be avoided by modeling organizations as elastic metaphors derived from ordinary, everyday occurring enterprise structures. However, no practical exhibits of the feasibility of this approach for complex enterprises are presented.

Coordinated Flexibility: Maintaining the Balance

Without a solid foundation such as the activity domain it is easy to focus on the wrong issues. Several contributions to the EA literature take the so called "stovepipe" syndrome as a problem that a good EA can resolve. The stovepipe syndrome or "applications silos" refer to the emergence of isolated IT solutions in the organizations:

[Integration] across departments is achieved by sending information between departments, a task for which ICT is highly appropriate. Unfortunately, in the past, computer applications have often been developed to function as standalone applications. These applications were paid by and built on behalf of departments and were tuned to their requirements. Each system had its own presentation layer, business processing logic, and database and formed as such a 'stovepipe' [...]. These stovepipes, thus, do not share data or logic. That is, they are disintegrated. (Goethals, Snoeck, Lemahieu, & Vandenbulcke, 2006, p. 68)

In general, the emergence of application silos can be traced to the need for an individual department to optimize its local information processing needs without interference from other departments (e.g. Ross, 2003). The drawback, of course, is that cross-enterprise information processing needs become hard to support since application silos are not designed to interact with other systems. Usually, this leads to the development of interface bridges between the systems, something which eventually turns into an overwhelming maintenance task.

It is close at hand to see the rise of application silos as a strategic organizational failure. This is also the general undertone in contributions discussing the stovepipe syndrome. Actually, one reason for the huge uptake of ERP-systems in contemporary organizations is that they promise to replace a number of "silo" systems with one, single system. The message from the EA community is that with powerful EAs in place, proper measures can be taken to avoid the chaotic situation many organizations face today (e.g. Ross, Weill, & Robertson, 2006; Zachman, 1997).

However, a word of caution is in place here. The mere fact that consolidation efforts always go from established silos towards integration (never the other way around) should raise a warning flag. It is as if silos have a growing power of their own; like weeds that defies every attempt to coerce them and make them more cultivated. From an activity domain point of view, this situation is quite natural. The rise of application silos has two sides. On the one hand, it is inevitable that different domains enact different means, simply because their work objects and motives are different. Enactment brings combined capabilities of actors and means into resources in the domain. Thus, domain specific systems are assets, not liabilities. In fact, the very existence of organizations is dependent on the existence of "silos" and "stovepipes":

[If] knowledge resides in specialized form among individual organizational members, then the essence of organizational capability is the integration of individuals' specialized knowledge. (Grant, 1996b, p. 375)

On the other hand, activity domains have to be coordinated across the enterprise and possibly between enterprises. There must be some commonality across domains. This commonality must, however, be limited to the absolute minimum necessary for coordination.

A hasty conclusion of this state of affairs would be to centralize application development. Enforcing a top-down, central solution, however, completely ignores the enactment process. Such processes must be confined to the very activity domain where it makes sense. The essence of the stovepipe syndrome is that the integrity of activity domains must be acknowledged if an optimal integration shall be possible. Thus, one task of an EA must be to support the establishment and maintenance of a balance between completely isolated application silos and the obliteration of these silos. This balance is what I refer to as coordinated flexibility, a term that I used already back in 1998:

In a world which is constantly changing, a centralized responsibility will simply be too slow, inflexible and insufficient. On the other hand, a completely decentralized responsibility will lead to other disadvantages such as lack of coordination, duplicated work and higher overall costs for maintenance. There must be a balance between these two extremes. To solve this dilemma we propose to make use of a principle called "Coordinated Flexibility". The idea of "coordinated flexibility" is to split the responsibilities for the tool development between central and decentralized organizations. [...]. The central organization develops a standard tool, which is adapted to local needs by the local development centers and possibly the projects. (LTX-1998-10-12, pp. 3-4)

Coordinated flexibility implies that certain elements in and between domains are stable. It is near at hand to see stabilizing elements as something that contradicts flexibility. However, the issue is not stabilization versus flexibility, but how encompassing the stabilizing elements are. Instructions, regulations, directives, etc., that prescribes in detail how actions should be carried out lowers flexibility because they are too rigid. On the other hand, an activity domain without any restrictions at all how to perform and coordinate actions is unable to produce anything at all, and thus cannot change in a controlled way. Both "end-points" on a stability "scale" are equally unproductive. An EA that provides optimum coordinated flexibility is a key instrument for change management:

A well-defined architecture is an important asset in positioning new developments within the context of the existing processes, IT systems, and other assets of an organisation, and it helps in identifying necessary changes. Thus, a good architectural practice helps a company innovate and change by providing both stability and flexibility. (Jonkers et al., 2006, p. 64)

Scalability: Recursive Core

Since most contributions to the EA literature are concerned with the enterprise level of organizations, it is quite natural that the question of recursiveness[1] is less emphasized. In Sowa & Zachman (1992) there is a lengthy discussion about the recursive character of the ZFK. The argumentation is arcane, but it appears that by recursion, Sowa & Zachman refers to three things: "related frameworks, framework versions, and nesting frameworks" (ibid, p. 605). Whatever meaning Sowa & Zachman assign to "recursive", it is clear that it differs from the ADT in the sense that there is no "core" construct in line with the activity domain, which can be applied to any constellation of organizational units.

The main consequence of the lack of a core construct is that the cognitive load of comprehending the framework and its associated models will increase. A stable core construct with a small number of entities that can be applied in various contexts, is much easier to understand for the human mind. With recursion, the task of keeping track of various organizational levels, both horizontally and vertically, is facilitated. Another consequence is that the need for considering how organizations interact may remain unnoticed, i.e., the transition modality is not attended to. This is evident from the absence of traces from this modality in the ZFK.

Dimension Decompression: Keeping Activity Modalities Distinct

The notion of "dimension decompression" refers to the fact that activity modalities are often "compressed" into entangled expressions or illustrative artifacts. For example, Jonkers et al. state that "Architecture is a process as well as a product" (Jonkers et al., 2006, p. 64). From an activity modality point of view a statement like this is an oxymoron. Process is an expression of temporalization, and product (in the sense of parts related to each other) an expression of spatialization, which implies that they are related, yet of quite different nature. Consequently, the architecture cannot be both a process and a product. However, it may have a static aspect – its constitution at a particular moment in time, and a dynamic aspect – its evolution over time.

Another example of dimension compression can be found in the ZFK. For example, the cell (2, D) "Work Flow Model" in the original framework bear traces of the following elements: temporalization modality (Work Flow), work object (work product), actors (people) and activity domain (organizational chart).

The compression of several modalities into one manifestation aggravates the separate recognition and management of these modalities. In particular, the "process" appears to be the modality onto which other modalities are projected:

Many people are vague about what is required to define a process. They often finish up requiring an identification of all the other supporting elements in the architecture. This tendency can result in the term "process" having very little value. (Veasey, 2001, p. 424)

Thus, even if the modalities are dialectically related in the sense that, for example, stabilization can influences all other modalities, it is important to analytically see these as distinct but tightly interrelated dimensions.

Enterprise Architectures as Dependencies Between Capabilities

Architectures of complex entities can be expressed in many ways. What things should be emphasized and what kind of relations should be included in the architecture? What particular perspective should be taken as the basic, ontological point of departure? What kind of notation should be used? And so on. The chosen expression of the architecture is decisive for the impact it will have as a means for mediating actions. This is valid also for EAs.

The particular perspective proposed from the ADT perspective is to base the EA on *dependencies between resources*. In order for an enterprise to function, it must have certain resources at its disposal. Resources in ADT are capabilities of humans and means in context. Resources presume enactment; only by actively engaging in technology at hand, by the inclusion of means in the social fabric of the activity domain resources come into being. Thus, resources must always be seen in relation to the activity in which they make sense.

Why dependencies? An insightful and experienced project manager at Ericsson (actually the one "inventing" the anatomy concept) once said:

The most important issue when working with complex things is to work from how things depend on each other (Järkvik)

If one resource needs another resource, and that resource is not at hand, things will not work. If the power generator providing electricity to your home is down, no electrical equipment in your home will work. Defining the architecture of technical systems from this simple principle has turned out to be a powerful instrument for developing complex systems (e.g. Järkvik, Berggren, Söderlund, 2007). The same principle would be valid for organizations as well; however with the crucial addition that the capabilities of humans are an indispensable part of the architecture.

In order to illustrate an EA based on the principle of dependencies between resources, consider an "enterprise" providing the service of giving care to elderly persons[2]. This might be conceived as follows:

- Elderly persons living in homes for retired may need to call for medical assistance.
- The elderly has an alarm button on a wristband that may be pressed in emergency situations.
- The alarm is relayed to a receptionist at a help-desk and to the nurse in charge. The alarm is registered in a data base.
- The receptionist tries to contact the elderly to determine the urgency of the alarm. Sometimes the elderly just feels lonely and want to chat with someone, or the medical condition may not be severe enough for a nurse to turn out and visit the elderly.
- If the alarm is severe, the nurse gets the spare keys to the home of the elderly and registers the access to the keys in a key access data base.
- The nurse gets her medical supplies, drives to the home of the elderly and gives the care.
- After care-giving, the nurse returns the keys and makes a journal note in the journal data base.

The EA is illustrated in the dependency map in Figure 8. The dependency map is similar to the anatomy, with the addition that capabilities of humans may be included in the map; not only capabilities of means as in the anatomy.

Figure 8. An enterprise architecture of the activity of giving elderly care

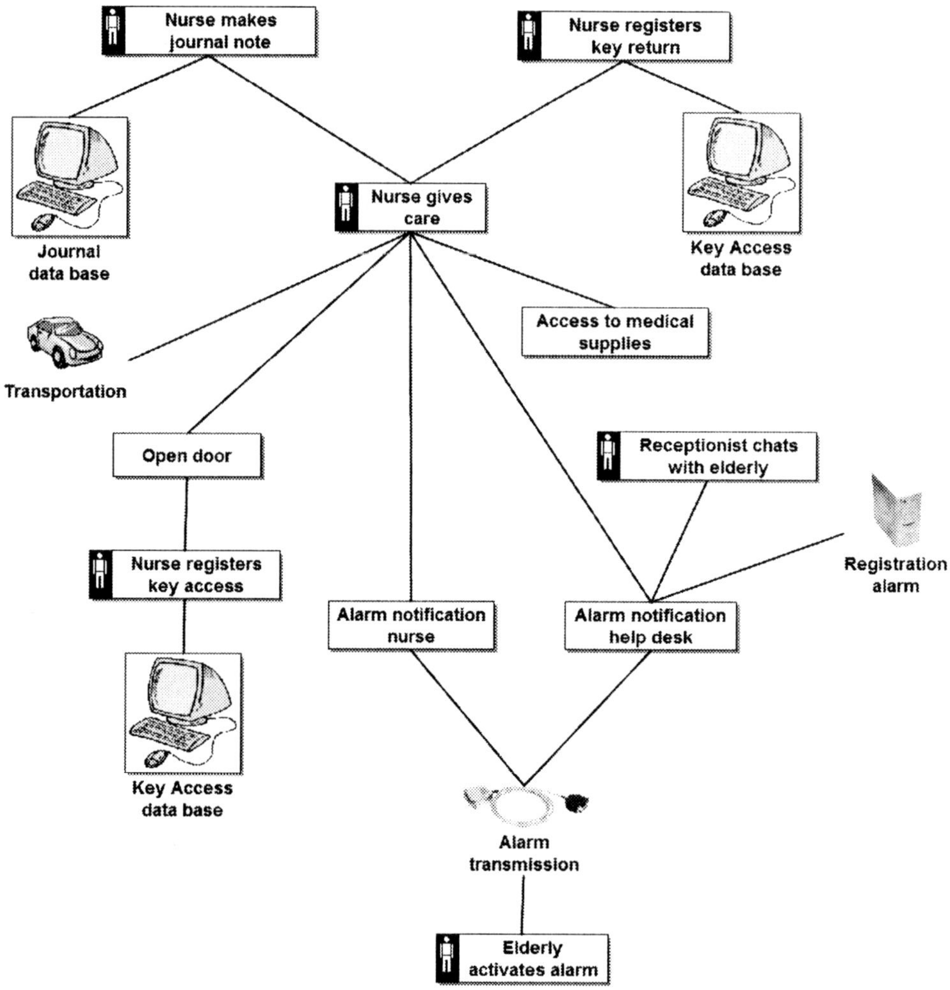

As can be seen the EA shows dependencies between capabilities of both means and humans. For example, the initial activation of the alarm requires that the elderly has acquired the capability of understanding and activating the alarm capability supplied by the equipment on the wrist. For example, the elderly must not take the wristlet off since she might not reach it in case of a fall.

The dependency map is read from the bottom up. The first resource that must be in place is the combined capabilities of the alarm and the elderly to set off the alarm. If one of these capabilities is not invoked, for example, if the alarm button is malfunctioning, nothing will happen. Similarly, if the alarm transmission fails, the capability of the nurse to treat the elderly will not be invoked. And so on.

The EA in Figure 8 might have been devised differently in terms of which capabilities to include and to what level of granularity. The final outcome of the EA artifact (the illustration) is dependent on the people developing the architecture and to what extent it is useful in, for example, contemplating various changes and modifications of the enterprise structure. For example, the activation of the alarm may be complemented by sensory equipment indicating that the toilet has not been flushed within a certain time limit.

A key point in modeling the EA as dependencies between resources is that the illustration of these dependencies is fairly easy to understand and discuss. Thus, the instrumentality of the architecture is strengthened. This, however, presumes that the illustration is confined to a single page. If there is a need to zoom in on some part of the architecture, this can be done in another, more detailed dependency map.

OPERATIONALIZATION

Much of the effort put down in defining various EA frameworks appears to be focused on the EA models themselves, rather than on how to define these models and how to use them in order to influence the workings of the organization (Goethals, Snoeck, Lemahieu, & Vandenbulcke, 2006). Existing strategies for operationalization may exist for disparate domains. However, the problem is to put together these into a coherent view that all stakeholders can understand. This requires "an integrated set of methods and techniques for the specification, analysis and communication of enterprise architectures" (Jonkers et al., 2006, p. 65). Guidelines for how to achieve this are lacking to a large extent:

[Enterprise] architecture frameworks, such as the Zachman framework [...] offer high-level guidance in identifying which areas of business and technology should be considered when creating an enterprise architecture, but they provide little assistance in creating the architectural artifacts themselves. (Jonkers et al., 2006, p. 65)

I claim that the ADT is capable of providing an integrated set of methods and techniques for intervening in organizations. The guidelines for operationalization are based on the view of EAs as dependencies between capabilities. The following general strategy is suggested:

- The starting point is to identify the activity domains from the characteristics of the work objects included in the product/service that the enterprise supplies.
- The dependencies between the activity domains are illustrated in a dependency map. In doing so, the activity domains are considered as resources that are needed by the client domains.
- Since the activity domain is a recursive construct, the pattern "dependencies between capabilities" can be applied for every domain until the granularity of the models is such that operational decisions can be taken. For example, a spatialization model should be defined to the level where a specification for its implementation in a particular IS can be defined.
- In each domain, manifestations of all the activity modalities need to be considered: spatial information models (e.g. expressed as E/R models), process models, business rules, and transition mechanisms between activity domains.
- Common capabilities used in several domains are identified. This might, for example, be enterprise wide standards, infrastructures, administration capabilities, etc.

In order to clarify this strategy, let's consider the following example. Suppose a municipal service office needs to take a decision in some matter. In order to do so a document with the relevant information is prepared. The process may look like the one in Figure 9.

In order to carry out the actions in this process, the actors of course must have the capabilities to do so. These capabilities are inexorably related to the capabilities of the means mediating these actions.

Figure 9. The process of taking a decision in a municipal office

- Write the document
- Register the document number in the document data base
- Distribute the document for comments
- Prepare the document for decision by the board
- Decide
- Distribute the document to whom it may affect
- Archive the document in the document data base

In Figure 10, a possible map over the dependencies between capabilities for the document activity is shown. At the top of Figure 10 the work object, the document, is depicted[3].The line represents the various states that the document can take during its life cycle (REGISTERED, DECIDED, and RELEASED). The arrows signify what actions change the state.

The dependency map should be read from the bottom up. At the bottom, the most basic capabilities are found. The dependency pattern between actions carried out (writing, registration, distribution, etc.) is a consequence of the fact that the capabilities each action realizes depend on each other. For example, the registration capability needs the writing capability in order to be realized; if nothing has been written, nothing can be registered.

Concerning the modalities, Figure 9 shows the temporal ordering of actions, the sequence in which these actions are carried out.

Figure 10. A possible capability dependency map for the decision making activity

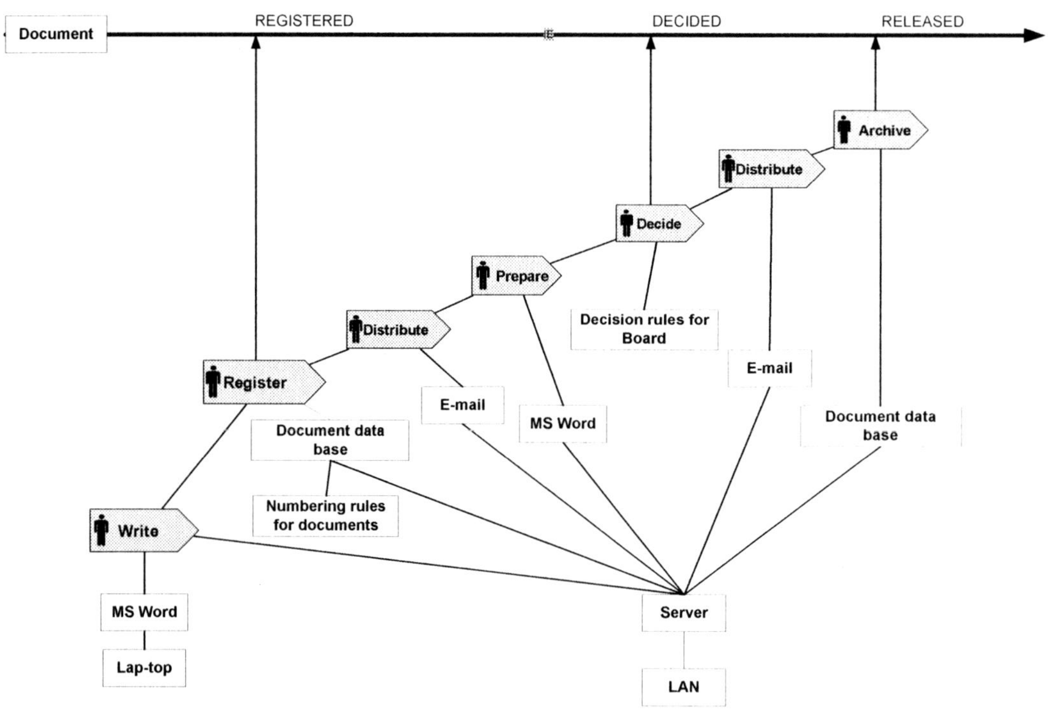

Figure 11. A manifestation of temporalization

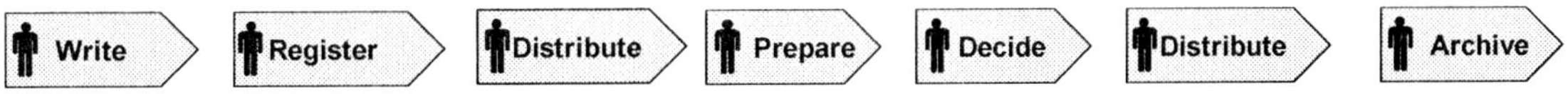

This sequence can be laid out only if the dependencies between the capabilities are understood. Thus, the dependency map is a more basic illustration of an activity than a process illustration. This becomes more evident if the process is not a simple, sequential one as in Figure 9.

The spatialization modality is manifested as the line representing the document entity:

In this case, there is only one entity involved. In more complicated situations, several entities might be involved, each with its own set of states. However, the same pattern as in Figure 12 is still valid. The entities (spatialization) are impacted by actions changing the state of the entity in a particular order (temporalization). Thus there is a dialectical relation between the two modalities; one cannot be conceived without the other.

Stabilization is manifested as the rules for numbering documents and taking decisions:

Again, these can only be defined in relation to the other modalities and in the context of the decision making activity where they make sense.

The different means used in the activity are illustrated in Figure 14.

Some of these means needs to be enacted in the activity. For example, actors writing the document and preparing it for decision must have engaged with the capabilities of MS Word in order for MS Word to become a resource in the activity. Usually, this enactment is restricted to a subset of all possible capabilities in MS Word; the capabilities that make sense in the activity where these are enacted.

The other means provide capabilities needed by other means. The registration of the document requires several capabilities of the document data base application. First, a document number has to be generated. This in turn requires the capability provided by the rules for numbering documents in the organization. In addition, the document data base application must have the capability of storing the new document in the archive. The computer based means (MS Word, e-mail, the document data base application) need the capability of the server, which in turn is dependent on the LAN.

Figure 12. A manifestation of the spatialization modality

Figure 13. Manifestations of stabilization

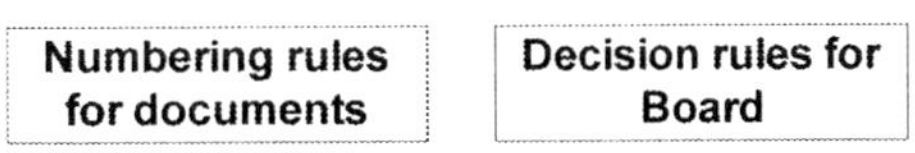

Figure 14. Means used in the decision making activity

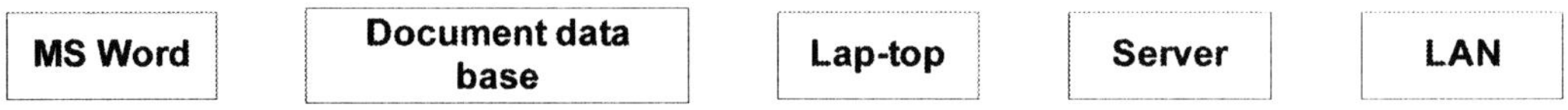

In summary, the operationalization of the EA takes as its point of departure the dependencies between activity domains, and proceeds within each domain until operational means can be specified and developed. In doing so, all activity modalities need to be considered; they form a complete whole which together constitutes the ideology; the capabilities needed to fulfill the goal of the domain.

ENTERPRISE ARCHITECTURES: SUMMARY

In this chapter, the subject of enterprise architectures is treated from the ADT perspective. As a representative for the state of art in this area, the Zachman framework is used. In spite of many similarities, there are a number of distinct differences as well. The main difference is that the Zachman framework does not acknowledge the agency of human beings other than as systemic elements; as "cogs in the wheel". Consequently, issues of interpretation and achieving common understanding about the enterprise architecture is not emphasized in the Zachman framework.

From the ADT point of view, there are a number of concerns that have to be addressed in order to advance the area of enterprise architectures: establishing a theoretical foundation, defining an integrated construct, deciding on the main attractor of enterprise architectures, attending the construction of communal meaning, emphasizing dependencies between resources, finding a proper balance between centralization and decentralization, the scalability of the architecture, and avoiding "dimension decompression" of activity modalities. I argue that the ADT is capable of addressing these concerns.

The chapter is concluded with a small example, indicating how the development of an enterprise architecture based on ADT would be operationalized. The architectural view suggested is to focus on the dependencies between resources in the enterprise.

REFERENCES

Appleton, D. (2004). Why Enterprise Architecture is an Oxymoron. *Business Rules Journal, 5*(4). Retrieved July 25, 2008, from http://www.BRCommunity.com/a2004/b191.html

Balabko, P., & Wegmann, A. (2006). Systemic classification of concern-based design methods in the context of enterprise architecture. *Information Systems Frontiers*, *8*(2), 115–131. doi:10.1007/s10796-006-7976-9

Beznosov, K. (2000). *Information Enterprise Architectures: Problems and Perspectives.* Technical Report 2000-06. Miami, Florida: Florida International University.

Bititci, U., & Muir, D. (1997). Business process definition: a bottom-up approach. *International Journal of Operations & Production Management, 17*(4), 365–374. doi:10.1108/01443579710159950

DoDAF (2004). *The DoD Architecture Framework Version 1.0.* Department of Defense, February 9, 2004.

Fatolahi, A., & Shams, F. (2006). An investigation into applying UML to the Zachman framework. *Information Systems Frontiers, 8*(2), 133–143. doi:10.1007/s10796-006-7977-8

FEA. (2008). Federal Enterprise Architecture Program Management Office website. Retrieved July 20, 2008, from http://www.whitehouse.gov/omb/egov/a-1-fea.html

Goethals, F. G., Snoeck, M., Lemahieu, W., & Vandenbulcke, J. (2006). Management and enterprise architecture click: The FAD(E)E framework. *Information Systems Frontiers, 8*(2), 67–79. doi:10.1007/s10796-006-7971-1

Grant, R. M. (1996b). Prospering in Dynamically-competitive Environments: Organizational Capability as Knowledge Integration. *Organization Science, 7*(4), 375–387. doi:10.1287/orsc.7.4.375

Greefhorst, D., Koning, H., & Vliet, H. (2006). The many faces of architectural descriptions. *Information Systems Frontiers, 8*(2), 103–113. doi:10.1007/s10796-006-7975-x

Järkvik, J., Berggren, C., & Söderlund, J. (2007). Innovation in project management: A neo-realistic approach to time-critical complex systems development. In T. Brady (Ed.), *IRNOP VIII Conference, Projects as Innovation, Innovation as projects.* Brighton, UK: University of Sussex. September 19–21, 2007.

Jonkers, H., Lankhorst, M., ter Doest, H., Arbab, F., Bosma, H., & Wieringa, R. (2006). Enterprise architecture: Management tool and blueprint for the organisation. *Information Systems Frontiers, 8*(2), 63–66. doi:10.1007/s10796-006-7970-2

Keltikangas, E. (2006). *Enterprise Architecture Documentation and Representation-A Pragmatic Documentation Framework.* Thesis, Degree of Master of Science, Helsinki University of Technology.

Khoury, G. R., & Simhoff, S. J. (2004). Enterprise Architecture Modelling Using Elastic Metaphors. In S. Hartmann, & J. F. Roddick (Eds.), *Conceptual Modelling 2004, First Asia-Pacific Conference on Conceptual Modelling (APCCM2004)* (pp. 65–69). Dunedin, New Zealand, January 18–22, 2004. CRPIT 31 Australian Computer Society.

Kirsilä, J., Hellström, M., & Wikström, K. (2007). Integration as a project management concept: A study of the commissioning process in industrial deliveries. *International Journal of Project Management, 25*(7), 714–721. doi:10.1016/j.ijproman.2007.02.005

Lindström, Å., Johnson, P., Johansson, E., Ekstedt, M., & Simonsson, M. (2006). A survey on CIO concerns-do enterprise architecture frameworks support them? *Information Systems Frontiers, 8*(2), 81–90. doi:10.1007/s10796-006-7972-0

Merriam-Webster On Line. (2008). Retrieved October 5, 2008, from http://www.merriam-webster.com/

Open Group. (2008). *The Open Group Architectural Framework Version 8*. Retrieved, July 21, 2008, from http://www.opengroup.org/togaf/

Rational™ (2008). Retrieved Oct 1, 2008, from http://www-01.ibm.com/software/se/rational/

Remington, K., & Pollack, J. (2007). *Tools for Complex Projects*. Burlington, MA: Gower Publishing Company.

Ross, J. W. (2005) Forget Strategy: Focus IT on your Operating Model. *MIT Sloan CISR Research Briefing, V*(3C). Cambridge, MA: Sloan School of Management, MIT.

Ross, J. W., Weill, P., & Robertson, D. C. (2006). *Enterprise Architecture As Strategy*. Boston: Harvard Business School Press.

Ross, R. (2003). Business Rules Manifesto - The Principals of Rule Independence. *Business Rules Group*, November 2003.

Smith, D., O'Brien, L., Barbacci, M., & Coallier, F. (2002). A roadmap for enterprise integration. *Software Technology and Engineering Practice, STEP 2002. Proceedings. 10th International Workshop on* (pp. 94–102). 6–8 Oct. 2002.

Sowa, J., & Zachman, J. A. (1992). Extending and formalising the framework for IS architecture. *IBM Systems Journal, 31*(3), 590–616.

Veasey, P. W. (2001). Use of Enterprise Architectures in Managing Strategic Change. *Business Process Management Journal*, *7*(5), 420–436. doi:10.1108/14637150110406803

Versteeg, G., & Bouwman, H. (2006). Business architecture: A new paradigm to relate business strategy to ICT. *Information Systems Frontiers*, *8*(2), 91–102. doi:10.1007/s10796-006-7973-z

Wikipedia (2008c). Retrieved Aug 10, 2008, from http://en.wikipedia.org/wiki/Attractor

Zachman, J. A. (1987). A framework for information systems architecture. *IBM Systems Journal, 26*(3), 276–292.

Zachman, J. A. (1997) Enterprise Architecture: The Issue of the Century. *Database Programming and Design, March, 1997*. San Francisco: Miller Freeman.

Zachman, J. A. (2008). *The Framework for Enterprise Architecture - Background, Description, and Utility*. Retrieved July 21, 2008, from http://www.zifa.com/

Zachman, J. A. (2008b). *The Framework for Enterprise Architecture - Cell Definitions*. Retrieved July 21, 2008, from http://www.zifa.com/

Zachman, J. A. (2008c). *The Zachman Enterprise Framework*[2]. Retrieved July 21, 2008, from http://www.zachmaninternational.com/index.php/home-article/34-zachman-framework/13-the-zachman-framework

ENDNOTES

[1] Recursive: "of, relating to, or constituting a procedure that can repeat itself indefinitely" (Merriam-Webster, 2008)

[2] This is of course a very simplified example; however, the general principles apply to EAs in general.

[3] This is usually not part of the dependency map. It is included here for illuminating the discussion.

Chapter 12
Product Lifecycle Management Revisited

The scope of this chapter is PLM (Product Lifecycle Management). I point out some concerns with the traditional way of understanding PLM, and discuss potential benefits from tackling these concerns from the Activity Domain Theory (ADT) perspective.

PLM is a wide-ranging information system (IS) that contains product data and product related data. Sometimes PLM is referred to as the product's "digital backbone" that defines all aspects of the product as it progresses through various phases of its lifecycle (see Figure 1).

PLM evolved from the need to manage large and complex data structures related to CAD (computer-aided design), CAM (computer-aided manufacturing), and CAE (computer-aided engineering) systems, usually referred to as CAx "authoring tools". Before 1999, CAx-systems were managed with Product Data Management (PDM) systems. Around 1999, vendors of PDM systems started to include capabilities for document management, configuration/Bill-of-material (BOM) management, revision control, and workflow management. The term PLM was quickly adopted by vendors, who wanted to differentiate PLM from the prevalent CAx data centric focus of PDM (Lindenthal, 2008).

Today, most PLM-systems offer the following capabilities:

DOI: 10.4018/978-1-60566-192-6.ch012

Figure 1. The scope of a PLM system (courtesy: Siemens PLM Software)

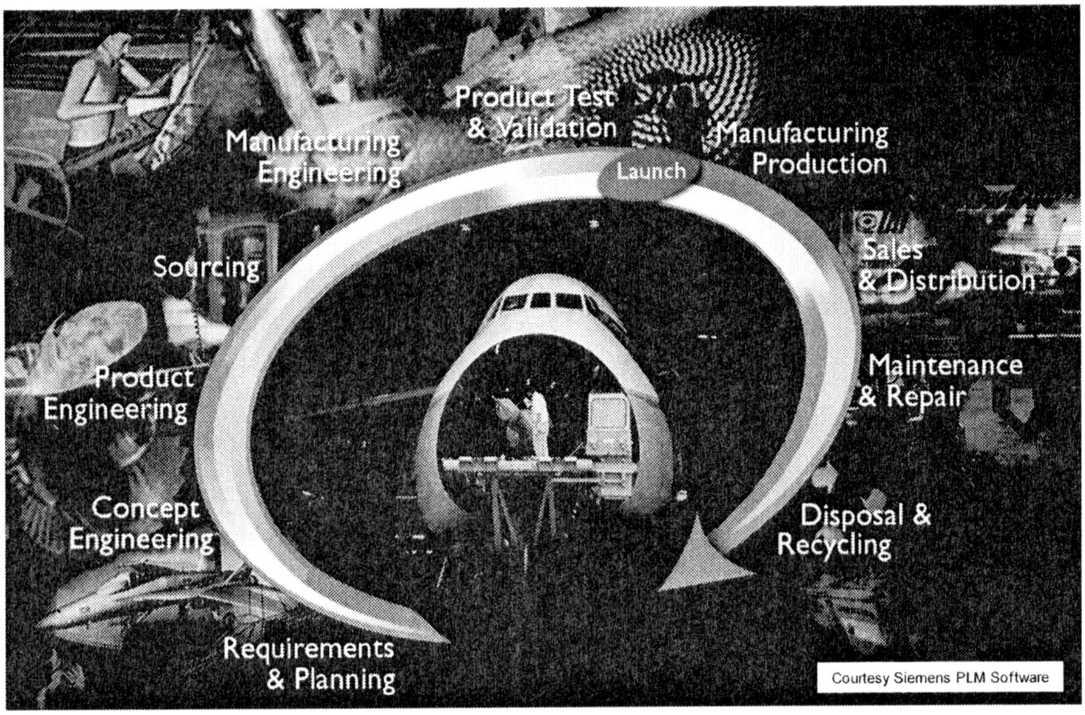

- *Product structure management*: structuring the product in various phases of its lifecycle: as-required, as-designed, as-planned, as-simulated, as-built, as-manufactured, as-delivered, and as-maintained.
- *Configuration control:* configuring a specific product from general product structures containing many possible options. This is often done by specifying a set of parameters and rules that pinpoints a precise configuration of the product.
- *Revision control:* managing various revisions of the same item.
- *Variant control:* managing variants of the same product, for example, a car built to customer specific features.
- *Document management:* managing various types of documents with revision control, definition of attributes, relations to other items, etc.
- *Classification:* structuring a particular type of item into various classes. For example, a car manufacturer might want to structure "cars" into the classes "sports cars", "load carrying cars", or "low emission cars".
- *CAx interfaces:* tight interfaces to authoring tools.
- *Viewing:* visualization of the product as 2-dimensional drawings or 3-dimensional solid geometries.
- *Work flows:* defining work-flows for repetitive tasks such as change management, release management, and registration of products.
- *Authorization control:* the allocation of rights to various roles concerning operations on the data items such as reading, writing, and deleting information.

- *Multi-Site management:* enabling globally distributed sites to access and manipulate the same data simultaneously.
- *Integration:* capabilities to interface the PLM system to other ISs, for example, legacy systems.

A DEFINITION OF PLM

Although the definition of the term "PLM" varies, the essence of PLM seems to be a widening perspective, from managing authoring data to managing all data that are pertinent to the definition of the product during its life cycle. CIMdata (2008), a leading consulting firm, states that:

- PLM is a strategic business approach that applies a consistent set of business solutions that support the collaborative creation, management, dissemination, and use of product definition information.
- PLM supports the extended enterprise (customers, design and supply partners, etc.).
- PLM spans from concept to end of life of a product or plant.
- PLM integrates people, processes, business systems, and information.

CIMdata emphasizes that PLM pertains to the entire business, not just the technology of an IS. PLM is as concerned with "how a business works" as with "what is being created" (ibid). According to CIMdata there are three fundamental capabilities of PLM:

- Provides universal, secure, managed access and use of product definition information.
- Maintains the integrity of product definition and related information throughout the life of the product or plant.
- Manages and maintains business processes used to create, manage, disseminate, share, and use the information.

My interpretation of PLM is summarized in Figure 2.

The central category is "Integration". PLM is about integration of a number of aspects: business, people, process, information and system. The key question is of course what integration means and how this integration shall be operationalized in practice.

The Evolution of PLM at Ericsson

The deregulation of the telecommunication market and the emergence of new technologies during the 1990s have created entirely new challenges for companies developing telecommunication systems. A long period of comparatively stable conditions has been succeeded by a turbulent period characterized by instability, uncertainty and an ever-increasing rate of change.

During the stable period, the customers of telecommunication systems were mainly government owned operators with long-term, nationwide infrastructure investment plans. These plans were typically updated every fifth year in tight collaboration with systems providers. Complete solutions were developed and manufactured in-house. The systems consisted of comparatively few technologies and evolved

Figure 2. The scope of PLM

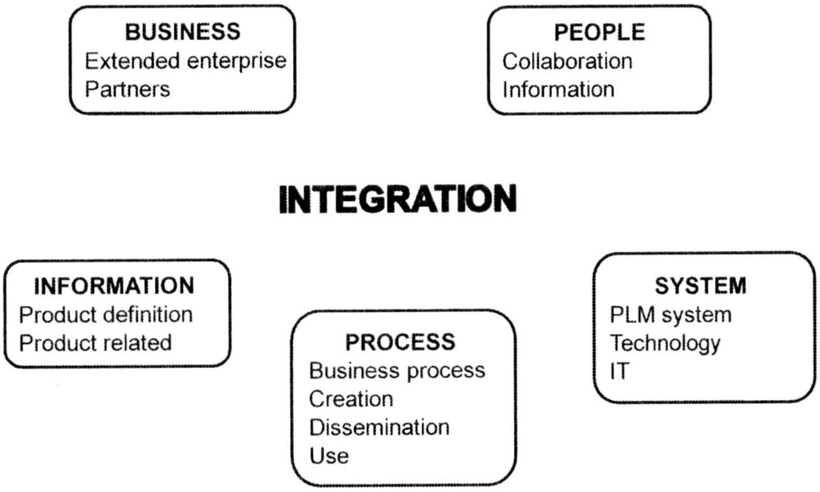

at a slow and steady pace. The management foundation during this period focused on the development and production functions and less on functions such as marketing and service. Since the entire lifecycle of the system was under the control of the same company, a comparatively homogeneous management foundation was established over time.

The long-term cooperation between system vendors and state-controlled customers on a regulated market gave way to a deregulation of the telecom area in the 1990s. After the deregulation, many new operators have been established in the telecommunication area. During the years around the millennium shift, operators invested heavily in licenses and networks for the 3rd generation of mobile systems. However, the incomes were less than expected, which caused the market to more or less collapse. This had tremendous impacts on development plans and organizational structures. A number of cost-reduction initiatives were launched such as outsourcing of production, product supply, and IT management. The number of employees at Ericsson was reduced from 107,000 employees to less than 50,000 in a few years.

Besides the immediate implications of the market collapse, several long-term effects could be seen for telecom system providers. More customers and stronger competition led to an increased focus on company functions such as marketing and service. The market turbulence resulted in frequently changed requirements on systems to be delivered. This period coincided with an ever-increasing complexity of the telecommunication systems themselves. The same trend can be seen in other industries as well such as the automobile, computer, mobile phones, and airplane ones. More products today also contain software, firmware, and electronic components. Some products, especially in the military sector, must be maintained for many years. In some sectors such as the pharmaceutical one, strict legislative compliance rules exist that must be upheld.

All these circumstances present an enormous challenge to PLM: how to secure the access to and quality of the terabytes of data produced and used by people around the world, 24 hours a day, 7 days a week in an ever increasingly turbulent world?

Concerns

The transition from the stable to the turbulent situation has brought to the surface a number of contradictions or tensions between the traditional management foundation and current needs. The contradictions are mainly due to the inadequacy of the traditional foundation to cope with diversification, change, and complexity. In addition, social issues such as the contextual dependencies of shared and commonly accepted conceptions, will have an increased influence (Martinsons & Davidson, 2003). More, specifically the following concerns need to be addressed:

- **Conceptual confusion:** The stable period made it possible to construct a consistent set of concepts pertinent to the management foundation. This set of concepts is part of each organization's specific language and cannot easily be translated to other organizations. The diversification of organizational capabilities means that the traditional language is inadequate for making sense of the turbulent situation, both within the company and across to cooperating partners. The result is confusion about the meaning of terms and concepts.
- **Surging effects of dependencies:** During the stable period, the components of the management foundation such as product structures, process models, and ISs could evolve in relative isolation from each other since the rate of change was small. In the turbulent situation however, the effects of the interdependencies between these components become more pronounced. A change in one component will impact other components more quickly. At Ericsson, failure to attend to the interdependencies has caused severe misalignment between product structures and business processes that operate on these structures. Moreover, this effect is worsened by the separation of responsibilities for the components into different organizational units.
- **Detached product structures:** In the traditional foundation, the product structure was developed for the one-in-all company situation; focusing on development and production. This structure cannot easily be adapted to the needs of other functions such as marketing and service. Moreover, outsourcing of production and supply functions may result in detached product structures, something which profoundly aggravates the information exchange between these functions.
- **Ad-hoc IS architectures:** At Ericsson the IS architecture in the traditional foundation consists of a central product IS with local client applications, a so-called Information Resource Management architecture (e.g. Trauth 1989). In addition to the central IS, a large number of other ISs have been introduced to meet various needs. Some examples of these are Product Data Management (PDM) systems, Software Configuration Management (SCM) systems (e.g. Estublier, Favre, & Morat, 1998; Persson-Dahlqvist et al., 2001; Svensson & Crnkovic, 2002), Requirements Management (RM) tools and Enterprise Resource Management (ERP) systems. Many of these systems contain product related information. These systems were introduced largely in an ad-hoc manner, which have resulted in a considerable overlap in functionality, managed information and an almost impenetrable IS architecture. With the arrival of the turbulent period, the fragmentation accelerated. Projects developing ISs are often launched without coordination of concepts, product structures, and processes. This state of affairs makes it difficult or even impossible to modify the IS architecture or even individual ISs.
- **Disjointed centralized control:** During the stable period, the control of the management foundation could be kept in-house. Central corporate units could decide which standards were to be used and how these should be interpreted. These standards were mainly about the rules for identification

and revision control of products and documents, and rules for structuring the product. The emergence of more independent in-house functions and cooperation with external partners erodes the authority of the central control function. Since the traditional standards basically did reflect the needs of the development and production functions, it is hard to convolute the standards to meet the needs of the marketing and service units. The cooperation with partners also brings completely new standards into focus. Thus, the whole issue of control needs to be reconsidered.

When these and other issues are surfacing, the natural strategy is to continue "business as usual". For some time, contradictions can be met by patches in the traditional foundation. However, since this foundation was established under different circumstances, only marginal improvements can be expected. In fact, many modifications will most likely impair an already confused situation since the traditional foundation lacks theoretical guidelines for coping with the issues of the turbulent period.

Current Approaches to PLM

Most approaches for implementing PLM so far have singled out one particular aspect of PLM. For example, during the 1990s, Business Process Reengineering (BPR) was the fad of the day. Although the term PLM was not invented at that time, BPR can be seen as an attempt to focus on the process aspect of PLM. The approach was to concentrate on implementing new, reengineered processes and supporting these with IT-systems. However, about 70% of these initiatives are believed to have failed (Cao, Clarke, & Lehaney, 2001). There are indications that these failures in part are due to an over-focus on the process aspect (ibid, p. 335). From the ADT perspective, it is near at hand to attribute the fairly high failure rate of BPR initiatives, at least partly to the inability to pay attention to the interdependencies between the activity modalities. By focusing on the process (temporalization) aspect, the others were left in oblivion.

Since the PLM-area is of major importance in many industrial sectors, it is not surprising that there is an abundance of initiatives around offering advice and software for implementing PLM. Often, these offerings come from vendors of PLM-systems like Dassault, PTC, Siemens (formerly UGS), Oracle/Agile, and SAP. The main interest of these vendors is, quite naturally, to sell their software products and offer consulting services for adaptation/configuration of the software to customer specific requirements. Thus, vendors often have a technology centric view of PLM. In addition, every PLM system brings with it a certain world-view concerning what things are worth attending to, and a preferred way of implementation. As a consequence, the selection of a particular system will over time be enacted into a resource that is geared to the particular world-view of the system. Thus, the choice of which PLM-system to implement has consequences far beyond the mere technology; something which is easily forgotten when the PLM-system is in focus.

In order to avoid the "one aspect as a time" approach to PLM, some frameworks have appeared that take a more holistic approach to PLM. Examples of these are ARIS (ARIS, 2008), the UML approach suggested by Eriksson & Penker (2000) and the framework for modeling and analyzing engineering information management systems proposed by Svensson, Malmström, Pikosz, & Malmqvist (1999), further elaborated by Hallin, Zimmerman, & Malmqvist (2004). All these frameworks consist of models of processes, information architectures, organizations, and information systems. The main purpose of the frameworks is to support a thorough analysis of the existing state of an organization in order to create a solid ground for future developments. Svensson, Malmström, Pikosz, & Malmqvist (1999) acknowledge the inherent complexity and difficulties in achieving common understanding. They also state that to master the complexity, several views are necessary.

In general, these frameworks lack a unifying theoretical perspective of human activity which may guide the selection of elements to be included in the framework. Consequently, human aspects such as achieving communal meaning and the enactment of capabilities are by and large ignored.

Moreover, there is no integrating category comparable to the activity domain. In the language of dialectics, this means that the relation between the parts making up the framework is considered as *external*; i.e., a change in one of them will have little effects on the others. Although the need for integration and a holistic view is emphasized, the dynamical interdependencies between them are less attended.

In summary the prevalent state of PLM initiatives can be characterized as follows. There is a commonly accepted view that PLM is an overarching business solution that includes people, information, process and system aspects. There is also an insight that the business solution is about integrating these aspects; to somehow make them jointly contribute to the goals of the organization. However, the prevalent driver of PLM initiatives after the BPR era is the PLM-system. This is the center of gravity around which everything else revolves in PLM today (see Figure 3).

Although business and people aspects are of course present in all PLM initiatives, these aspects have not achieved the same, operational attendance as the PLM-system. There is an apparent unawareness of the difficulties in constructing communal meaning in a PLM project. Clear indications of this are complex and incoherent models, impenetrable text-based specifications documents, and a focus on technical details.

Thus, the implementation of PLM systems must, in my opinion, be approached from entirely new directions where meaning, interpretations, and human actions are brought to the fore. Above all, this means a concentration on integration. PLM must be seen as a totality where the different aspects of PLM are *internally* related: it is in principle not possible to change one of them without affecting all the others and the whole (see Figure 4).

Figure 3. The prevalent focus of PLM initiatives today

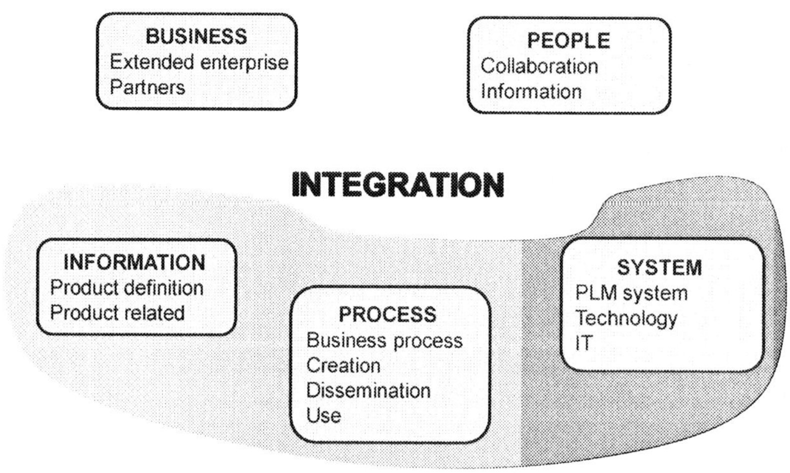

Figure 4. The integration centric view of PLM

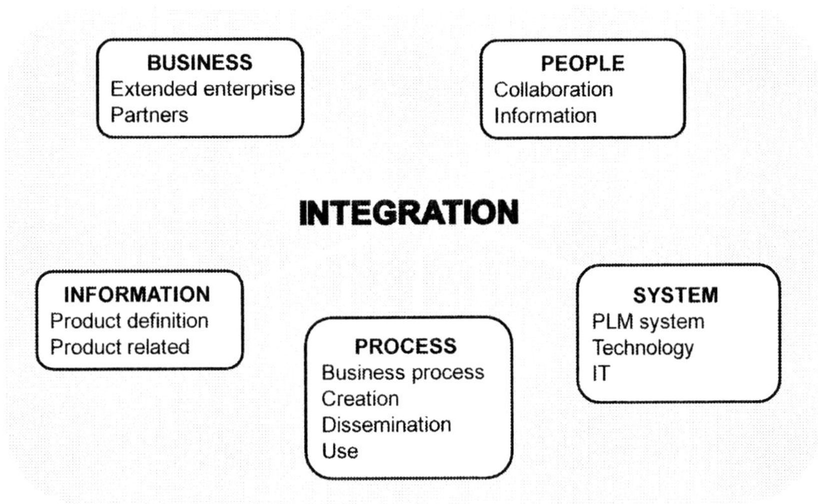

Figure 5. ADT elements in PLM

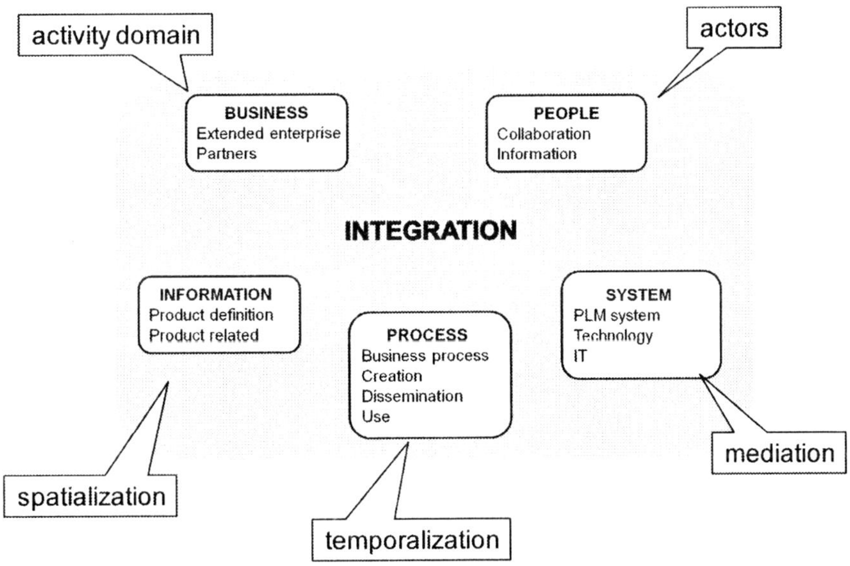

OPERATIONALIZING PLM: THE ACTIVITY DOMAIN THEORY PERSPECTIVE

From an ADT point of view, the presence of several ADT elements can be noticed from the CIMdata definition, (see Figure 5).

In addition to the more obvious elements, others can be read "behind the lines": "maintaining the integrity" needs stabilization in the form of, for example, business rules, "extended enterprise" needs transition in order to cooperate between units. The presence of ADT elements indicates that the activity domain may be a viable, integrative foundation for PLM. The contours of this approach are outlined in the following sections.

The PDM Core Project

In the years around the millennium shift it became increasingly evident that the traditional PLM platform was inadequate for coping with the situation Ericsson found itself in. Most of the concerns discussed earlier had been experienced. In order to address these issues a project called PDM Core[1] was launched in 2001. The purpose of this project, which I participated in, was to investigate an alternative foundation for PLM. This coincided with the period when the main elements of ADT were instigated. Thus, the PDM Core project provided a "once-in-a-life-time" opportunity to put the ideas of ADT to test in an industrial PLM improvement initiative.

In the following I will use the PDM Core project as an illustrative example of how to operationalize PLM with the ADT as a guiding framework. The experiences gained during this project were later reflected back into the further crystallization of the theory. Thus, when discussing the PDM Core project it is a view in hindsight, reflecting the elaboration of the ADT that has taken place since then. For example, the construct of activity modalities was not introduced in ADT until 2004 (Taxén, 2004).

No decisive conclusions can be drawn about the feasibility of the approach since the PDM Core project, to the best of my knowledge, did not leave any persistent traces in the organization. As with the anatomy-centric approach, the experiences gained never became "sedimented" in the aftermath of the turbulence and down-sizing of the company in the years that followed. However, it is my firm belief that the ADT-based approach towards implementing PLM is a step in the right direction for coping with the massive complexity inherent in such endeavors.

Establishing the Activity Domains

The drive for the PDM Core project was the need to decentralize the responsibility for different areas, while at the same time maintaining central control of vital elements. Early in the project it was clear that there was a need for a general concept denoting these "areas". Just using the organizational denomination did not seem satisfactory. I suggested using the then rather newly envisioned term "activity domain" for this purpose. The outcome of the discussions was to conceive Ericson as one main activity domain which coordinates four "supplementary" domains (see Figure 6).

The supplementary domains were described as follows:

- *Marketing:* The motive of this domain is to market and sell total telecommunication solutions to customers. The work objects are customers and tenders. The main drive is to become a total solution provider. The activities are carried out at market units worldwide. There are many markets, and each market unit works with many products and many supply and service units. The Marketing domain utilizes two other domains: Account management and Product Management.
- *Research & Development (R&D)*: The motive of this domain is to develop products. The work objects are different kinds of products. Drivers are to allow projects to adapt to new ways of working, to use incremental and model based development, and in-source components or total systems. R&D is performed at product development units worldwide. Each product is sold on many markets, produced by many supply units, and serviced by many service units. The R&D domain utilizes two other domains: System Management and Design & Test Management.
- *Sales & Supply:* The motive of this domain is to produce and install total solutions at customer sites. The work objects are orders and installed solutions. Drivers are global and out-sourced

Figure 6. The organization of Ericsson into main activity domains

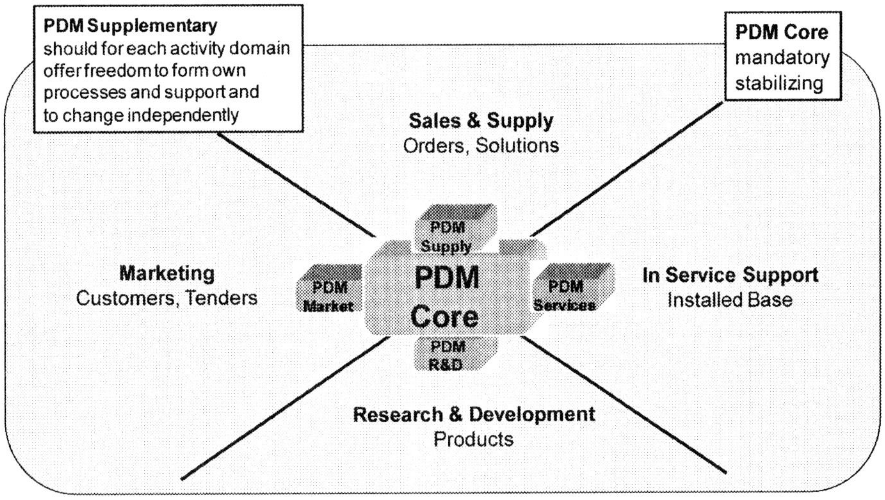

production. Production is done by supply units, so called Flow Control Centers, worldwide. Each center services many products and markets. The Sales & Supply domain utilizes two other domains: Supply Management and Global Services Management.

- *In Service Support*: The motive of this activity domain is to service and upgrade installed solutions at customer sites. The work objects are installed solutions. A main driver is to inform Sales & Supply about the needs of the installed bases. Service and support are provided by service units worldwide. Each service unit supports many products and is active on many markets. The In Service Support domain utilizes one other domain: Global Services Management.

As can be seen, there is an ambiguity about what shall be considered as an activity domain or just a convenient grouping of domains. Although no thorough analysis was made, the ambiguity is most likely due to an inability to discern a clear work object for the domains.

Clarifying Dependencies Between Activity Domains

In ADT, the capabilities of domains are regarded as resources. In line with the "anatomy" way of thinking, the most crucial aspect of managing complex situations is to have control of the dependencies between capabilities. Thus, after establishing which the main activity domains are, the next step towards implementing the PLM support is to clarify the dependencies between these domains. A good starting point for this analysis is provided by the business processes in the organization (see Figure 7).

As can be seen, this model highlights activities (the "fishes"). In addition some states are indicated (the circles), as well as some entity information (the black squares between the activities). The process model also shows that there are two major tracks: one called TTC (Time-To-Customer), which concerns the configuration of existing equipments into fast deliveries of solutions "off the shelf" to customers. The other track, TTM (Time-To-Market) concerns the development of new products, something which quite naturally takes a longer time and involves more activities.

Figure 7. The Ericsson Business Process around 2000

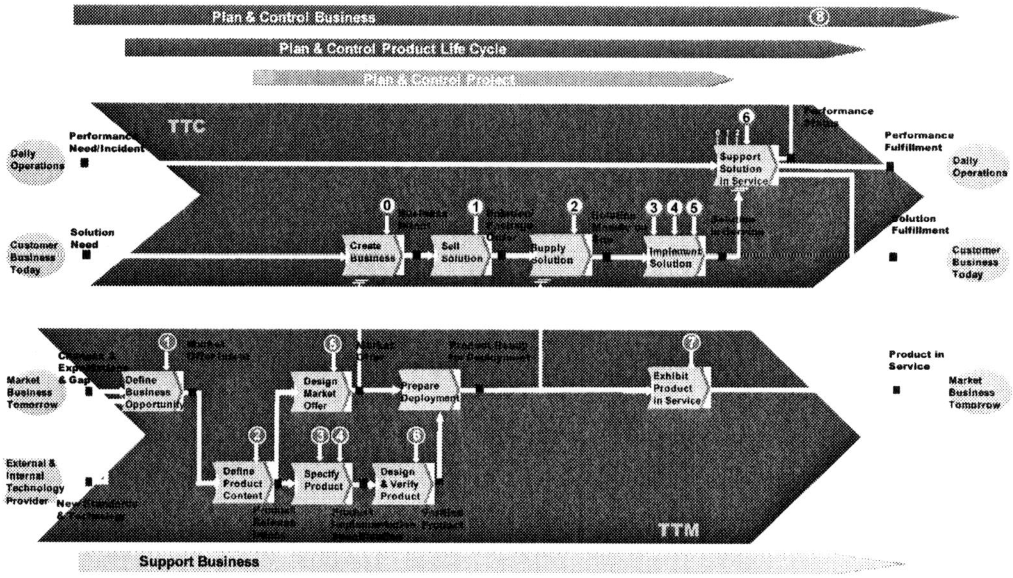

An elaborated (and slightly modified) illustration of the same business process is shown in Figure 8. An additional element is a grouping of activities (at the bottom) into areas of responsibility or inclusion:

From the ADT point of view, the illustration in Figure 8 is overloaded since it tries to express too much. A first task then is to "de-construct" the model in Figure 8, i.e., separate it into elements that are interrelated and more aligned with the activity modalities. From the business process model in Figure 8, it is possible to derive a dependency map between the activity domains (see Figure 9).

The dependency map should be read from the bottom up. At the bottom, the prerequisites are shown: "Performance Need or Incident" for the domain Global Services Management; "Solution Need" for Account Management; "Changes & Expectations & Gap" and "New Standards & Technology" for Product Management. At the top, the desired capabilities are found: "Performance Fulfillment", "Solution Fulfillment", and "Product in Service". In between the activity domains, the activities within each domain are shown, as well as the dependencies between them.

It is immediately clear from a visual inspection that the dependency map in Figure 9 is easier to comprehend than the process model in Figure 8. It is hard to imagine how to proceed from the model in Figure 8 towards the PLM implementation, since the model tries to squeeze too many things into the same image, resulting in an incomprehensible mess, void of action potential. With the dependency map as the basis, however, the next step is quite logical: to accentuate the information management perspective of the map.

Towards Implementing Information Management Capabilities

Almost all of the activities in the dependency map are phrased in the form "do something with something", for example, "Specify Product", "Design Market Offer", etc. This indicates the presence of two modalities: spatialization (the noun) and temporalization (the verb). From an information management point of view, it is natural to concentrate on the nouns, which will naturally map to entities in the infor-

Figure 8. An elaborated illustration of the business process (Taxén & Svensson, 2005. ©Inderscience Enterprises Limited. Used with permission)

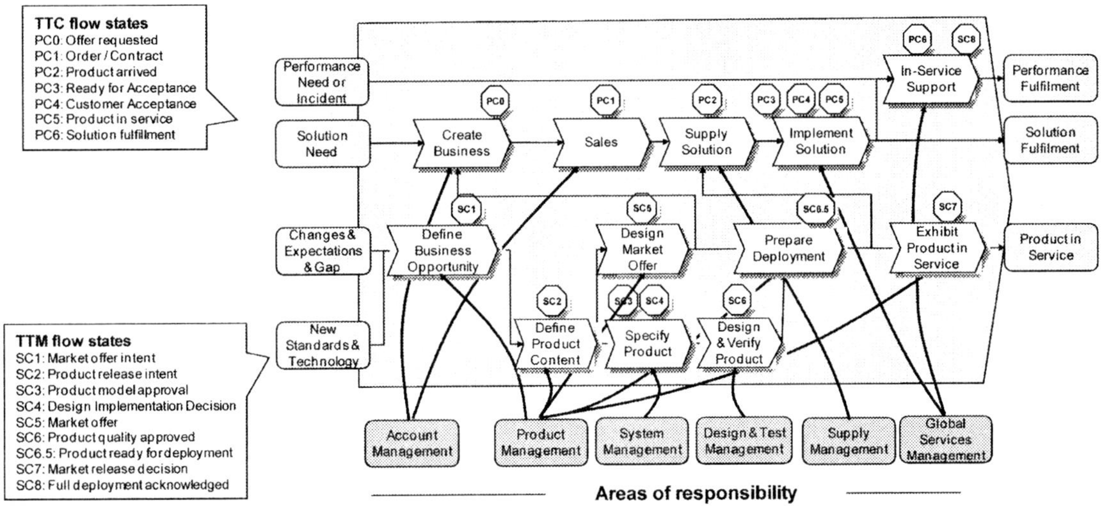

Figure 9. A dependency map for activity domains

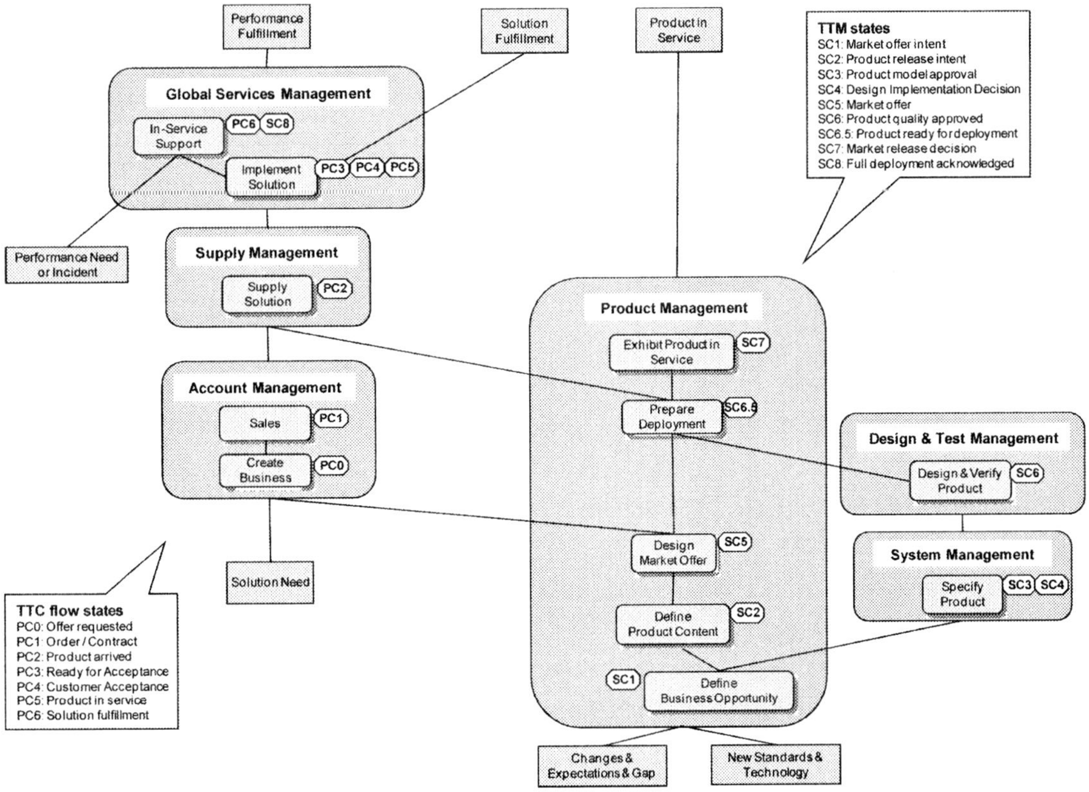

Figure 10. An Information Interaction Model of the business process

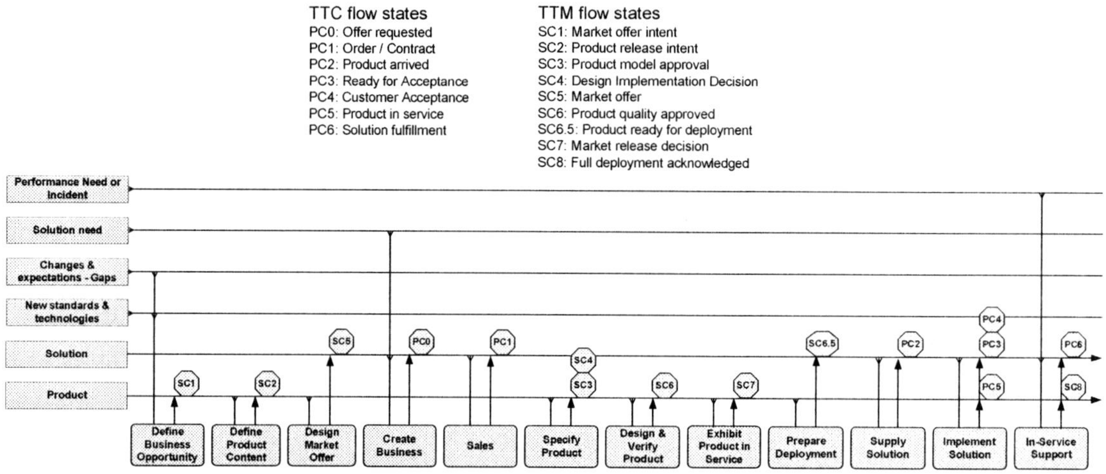

mation architecture as expressed by the information model in ADT. Of course, more details have to be provided (relations, state sets, attributes, etc.) to make the model complete.

The next step is to clarify the interdependencies between the spatial and temporal modalities. The natural candidate for modeling these interdependencies is the Information Interaction Model (IIM). An example of how such a model, derived from the dependency map in Figure 9, may look is given in Figure 10:

As can be seen, the spatialization modality is manifested as information elements lined up vertically to the left. The temporalization modality is manifested as activities lined up horizontally at the bottom. The interaction between these modalities is expressed in the "score" diagram showing the prerequisites and outcomes of each activity.

By inspecting the IIM in Figure 10, some concerns immediately become apparent. Some states (SC3 & SC4; PC3 & PC4) cannot be separated; they signify two state changes of the same element simultaneously. Thus, it can be concluded that the duplicated state changes are the result of two separate activities; however, these are for some reason not singled out in the original process model. Moreover, the state SC5 "Market offer" seems to belong to the TTC flow rather than the TTM flow. Likewise, the state PC5 "Product in Service" would more naturally appear in the TTM flow.

The appealing property of IIMs is that it provides a comprehensive specification of information management capabilities that the PLM system must provide. The information elements shown in the IIM diagram must be possible to perform basic management operations on: defining, storing, searching for, changing the state on, etc. Additional management capabilities may be needed such as access to rules for naming elements, support from workflows, making various reports, and so on.

If the main activity domains have been identified in the organization, it is (at least in principle) a straightforward task to analyze these in terms of dependency maps and IIMs, and consolidate all information management capabilities needed across all domains. The whole approach is sketched in Figure 11:

Figure 11. The ADT approach towards implementing PLM

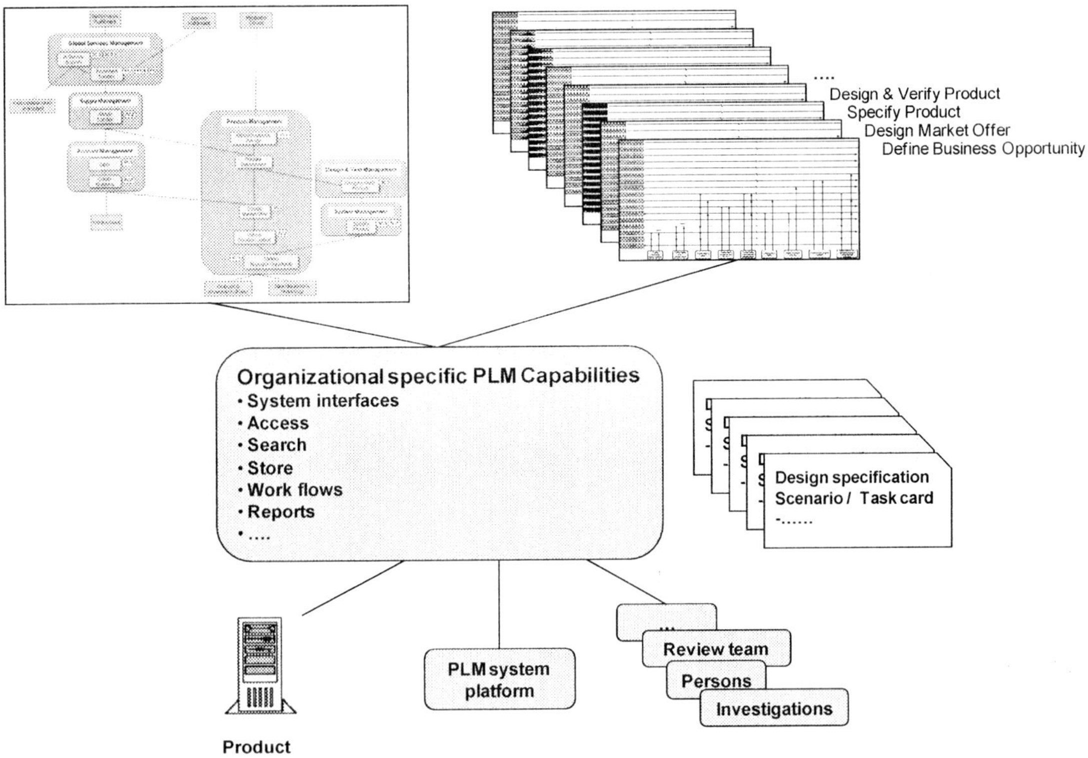

The business needs are expressed by the dependency maps and the IIMs. These are fulfilled by developing a PLM application providing organization specific PLM capabilities. These capabilities, which might be specified by scenarios, task cards, or other means, need the vendor's PLM system platform, the structure of the product, and various other capabilities provided by investigations, personnel, review teams, and more.

In summary, in this section I have indicated how an approach towards implementing PLM based on the ADT may be conceived. This approach, which of course must be put to test in realistic PLM initiatives, is based on the conviction that the complexity of such initiatives can best be managed if mediational means, such as images, are aligned with the activity modalities.

IMPLICATIONS

In this section I will discuss some implications of the ADT approach to PLM.

The Activity Domain as the Foundation

The use of the ADT as a guiding framework for PLM initiatives brings to the surface a number of issues which would otherwise have remained in oblivion. Analytical and constructive efforts can be focused on the interplay between information architectures, processes, and PM systems rather than on these elements in isolation. The all too common trap of focusing on the technology of the PLM system can be avoided. Furthermore, the notion of activity domains takes into consideration individual and social aspects of PLM, especially the issues of signification and communal meaning. These issues are of outmost importance and, in my opinion, more difficult to tackle than technological issues.

Conceiving the organization as a set of interacting activity domains provides a comparatively stable construct, which changes less frequently than formal organizational structures. Even if the label of the organizational unit changes, the work object and enacted capabilities of humans and means remain. The elements of an activity domain are the same regardless of whether the activity domains are confined to one and the same organization or in different organizations. Thus, the activity domain provides a unified way of conceiving the coordination of both inter-organizational and intra-organizational activities.

In summary, the activity domain provides a solid foundation for PLM initiatives, something which is particularly important in turbulent periods of organizational out-sourcing, mergers and acquisitions.

Conceptual Unveiling

When the PDM Core project was started, it was soon realized that the team members had to achieve some common understanding of key concepts used in the project. Another issue was to improve the communication to the organization outside the project by having a consistent structure in the description of the concepts. To this end, a set of *fundamental categories* for information management was defined in an internal report. These categories were illustrated as in Figure 12.

As can be seen, most of the elements in ADT are discernible, albeit differently structured. After some initial confusion, it was decided that the results of the PDM Core project would be based on the fundamental categories. Quite soon, an initial unveiling of abundant concepts and terminologies was achieved. Concepts that were perceived as separate could be seen as expressions of the same, underly-

Figure 12. The fundamental categories in PDM Core

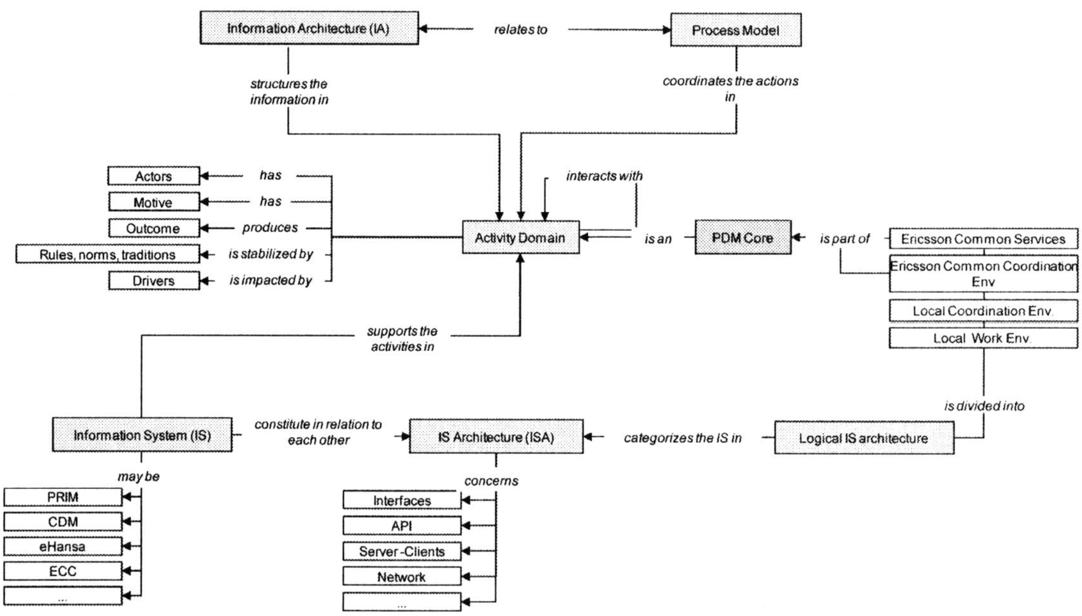

ing fundamental category. Gradually, these categories were internalized in the team and among people in the immediate surroundings of the project. The categories were constantly refined and continued to evolve during the project.

The fundamental categories and its derivatives in the ADT can be seen as a means to address the issue of conceptual disarray. Instead of seeing a multitude of seemingly unrelated things, it is possible to see something as "some-thing", i.e., a manifestation of a category in ADT. This means that locally defined concepts, which otherwise are confined to various domains, can be grounded in categories which are valid for all domains. This in turn makes it easier to allow locally defined dialects without losing control of how they are interrelated.

Balancing Central Control and Local Autonomy

The diversification of organizational functions has made it impossible for a central organizational unit to exercise complete control over the definition of information architectures, business processes, and information systems. This is especially valid for inter-organizational co-operations in the extended enterprise. However, it is not feasible for individual units to structure these resources entirely on their own, since that would prohibit coordination between units. Thus, a balance between central control and local autonomy has to be struck. This balance must comprise all constituents of the activity domain. For example, in the PDM Core project, it was proposed to specify a core set of object types, relations, etc., representing mandatory information entities that all units must adhere to. Other parts of the information architecture were left to each activity domain to decide on their own. It was anticipated that the core set would remain relatively stable over time while more frequent changes were anticipated for the domain specific parts.

An example of an information model for the Research & Development activity domain is shown in Figure 13. The grayish entities belong to the core set, while the other entities are specific for the R&D domain. The object in focus in this domain is the "DESIGN_PRODUCT" around which the other types are organized. Similar models may be constructed for the other activity domains.

The identification of core entities in the information architecture is an example of how centralized control can be balanced with local autonomy. The responsibility for the information, which is vital to the company, can be assigned to a central authority, thus creating a stable information core. Coordination is made possible without constraining the activity domains in ineffectual ways. A similar approach may be applied to the business processes and the ISs.

Domain Specific Product Structures

The traditional product structure at Ericsson exposes the functional and organizational partitioning of the product into various subsystems. A realization structure is defined for each subsystem, which defines the assemblies or parts making up the product. For each part there is an associated set of documents describing various views of the product such as marketing, development, manufacturing, installation, etc. The structure contains configuration and revision rules which are used, for example, to build configurations for manufacturing and installation purposes.

Basically, the traditional product structure reflects the needs of the R&D activity domain, where the needs from the other domains are, so to say, squeezed into this structure. However, each domain apprehends the product according to its own needs. For example, in the Marketing domain the products

Figure 13. An information model with core and supplementary entities (Taxén & Svensson, 2005. ©Inderscience Enterprises Limited. Used with permission)

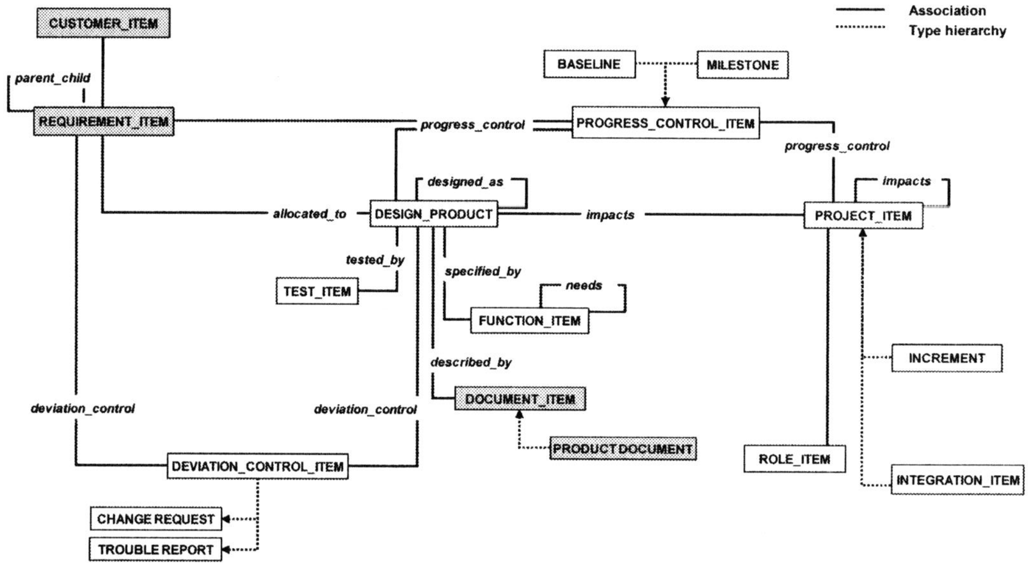

Figure 14. An alternative product structure (Taxén & Svensson, 2005. ©Inderscience Enterprises Limited. Used with permission)

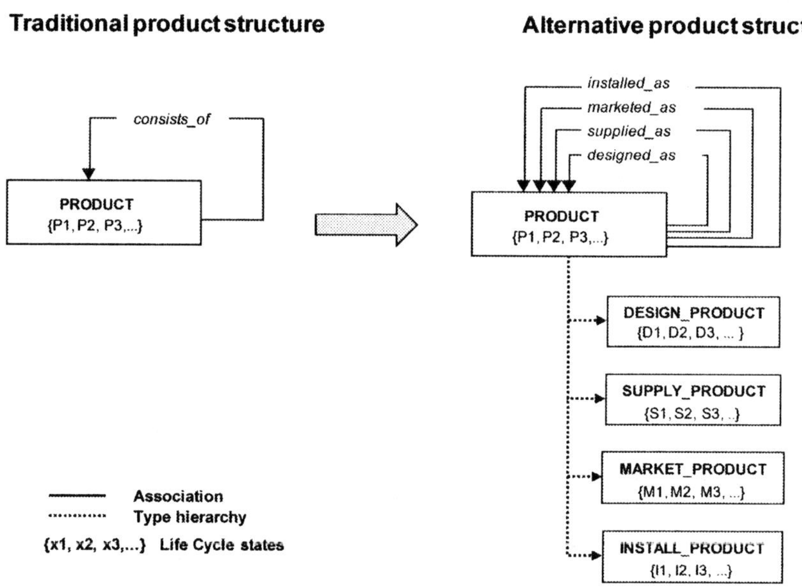

are conceived and managed from a marketing point of view. The most relevant aspect in this context is how the products are presented to the customer. As a result, the entities in the Marketing context need to carry marketing-specific attributes and be structured in line with marketing-specific structures. The same is valid for the other supplementary domains (see Figure 6).

Figure 14 shows one proposal of moving towards an alternative product structure.

The root entity is a product type that carries the properties which are common to all product entities, followed by activity domain specific product entity types. The entities have properties and lifecycles which are unique for each activity domain. These entities can be further specialized according to the needs of each activity domain.

The relationships between product entities are also domain-specific. The design-oriented structure "designed_as", is used to structure the product in the development domain, and likewise for the other domains. The transformation between lifecycle states in different domains (an example of the transition modality) is determined by rules which indicate how a certain state depends on other states. For example, the product state P1 may be triggered when the states D1 and M1 have been set, indicating that some design and marketing activities have been finished.

By allowing domain specific structures, each domain is given a certain freedom to construct structures aligned to their needs. This can be done without jeopardizing coordination between domains if the transition rules between the domains are upheld. As a consequence, the characterization of the product will be different in different domains, reflecting the contextualization modality. It is the ensemble of domain specific product related elements (attributes, documents, etc.) that together define the product. In this way a unified management of product structures is possible without hindering diversification.

The alternative product structure is a proposal which must be further detailed and verified in practical applications. Most certainly the current proposal will be modified. However, the general principle of several product structures aligned to the needs of the activity domains appears reasonable. This provides an 'unlocking' effect on the hitherto fruitless attempts to come up with one product structure for the entire organization. This in turn makes the effort to achieve common understanding about this structure substantially less, since the commonality can be limited to the bordering entities in the domain-specific structures. The price that has to be paid is that the different product structures have to be translated into each other when necessary. A similar proposal has been suggested by Brown (2001).

The Core Catalogue

A key component in achieving a balance between central control and local autonomy is the core catalogue, see Figure 15. The core catalogue is a central IS which interacts with activity domain-specific ISs. The core catalogue contains core data collected from all main activity domains. Only those objects, attributes and relationships classified as core ones are stored in the core catalogue.

The catalogue can be seen as a manifestation of the stabilization modality for the overall Ericsson activity domain, since it prescribes the form and content of Ericsson specific data. The catalogue enables a controlled communication of data between the activity domains. In doing so it may be necessary to map the information elements from these domains to the catalogue, which is an expression of the transition modality. An example of such a mapping is that the concept of "material" in the ERP system (SAP R/3) used in the Sales & Supply domain is mapped to the "product" concept in the core catalogue.

A prototype implementation of the core catalogue concept was done in the project. However, the classification of the information architecture into core and supplementary sets, which is crucial for the catalogue concept, was not consolidated. Moreover, the project did not address the question of what particular ISs are suitable as the core catalogue IS or the activity domain specific ISs. A general observation is that the activity domain specific ISs are more exposed to changes than the core catalogue IS and thus must be easy to modify.

Figure 15. The core catalogue concept (Taxén & Svensson, 2005. ©Inderscience Enterprises Limited. Used with permission)

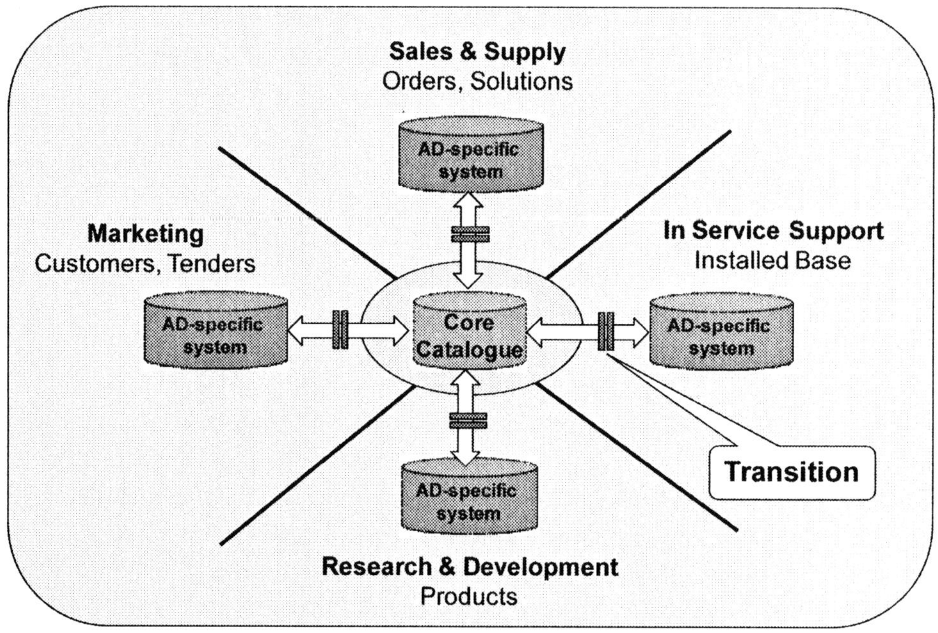

A central idea in the core catalogue concept is that the different activity domains shall be in control of the activity domain specific ISs. The enactment of these systems into domain-specific resources is best done by the actors in the domains. Of course, company-wide rules for these systems can be specified, for example, that the ISs should use the same IT-platform.

The core catalogue also provides a means to restructure the IS architecture in the organization. Some ISs used in the activity domains have acquired the status of "immutable mobile" or "black boxes" in the terminology of the Actor Network Theory (Latour, 1986; 1992). This means that the enactment of these ISs is firmly established among the actors. Although it would technically be possible to replace the ISs, it cannot be done without creating resistance and conflicts as long as the ISs are embedded in an existing activity domain. The core catalogue makes it possible to manage these conflicts by providing a structured exchange of core data between different ISs while they are in active use. If it is decided to phase out a certain IS, this can be done in a systematic way.

It can be noted that a similar approach has been embedded in the Product Life Cycle Support (PLCS) initiative developing the ISO standard ISO 10303-239 (PLCS, 2008). PLCS provides a mechanism called Data Exchange Specification (DEX), which is used to map various information sets to each other. The DEX entities are examples of manifestations of the transition modality.

PRODUCT LIFECYCLE MANAGEMENT: SUMMARY

In this chapter, I have made an inquiry into the potential for ADT to provide guidelines for implementing PLM systems. These systems represent a fundamental capability in product developing industries: maintaining product definition information during the whole lifecycle of the product. As such, PLM systems are at the core of organizational assets; if information about the product is lost or is of bad quality, the existence of the organization is endangered.

PLM from the ADT point of view is based on the activity domain as the foundation for PLM. From the definition of PLM as a business solution, I claim that such a foundation is capable of informing PLM initiatives, especially with respect to issues that are currently not up-front in such endeavors. Rather than the prevalent focus on the technology of PLM systems, the ADT perspective directs the attention to issues like achieving common understanding, balancing central control and local autonomy, and managing critical dependencies between PLM capabilities. These issues are particularly important to pay heed to in the turbulent and complex environments that product developing organizations face today.

Much remains to be done before it can be established that ADT indeed provides operational guidance for PLM-initiatives. The results from the PDM Core project are, promising as they might be, only a first start. The validity and transferability of the ADT based foundation on a large scale is not confirmed so far. However, it is my conviction that the ADT is a viable alternative approach to a more unified and complete understanding of various phenomena associated with management that, according to Martinsons & Davidson (2003), are much needed today.

REFERENCES

ARIS. (2008). Retrieved August 30, 2008, from http://www.ids-scheer.com/en/aris

Brown, D. H. (2001). *Fundamentals of shared product structure.* Port Chester, NY: D. H. Brown Associates, Inc.

Cao, G., Clarke, S., & Lehaney, B. (2001). A critique of BPR from a holistic perspective. *Business Process Management Journal, 7*(4), 332–339. doi:10.1108/EUM0000000005732

CIMdata. (2008). *Product Lifecycle Management (PLM) Definition.* Retrieved August 28, 2008, from http://www.cimdata.com/plm/definition.html

Eriksson, H. E., & Penker, M. (2000). *Business Modeling with UML - Business Patterns at Work.* New York: John Wiley & Sons, Inc.

Estublier, J., Favre, J.-M., & Morat, P. (1998). Toward SCM/PDM Integration? *System Configuration Management. SCM-8,* (LNCS 1439), 75–94.

Hallin, K., Zimmerman, T., & Malmqvist, J. (2004). Towards a Framework for Integrated Information Management in Mechatronic Product Development. In *NordDesign 2004 – Product Design in Changing Environment,* 18–20 August 2004, Tampere, Finland.

Latour, B. (1986). The powers of association. In J. Law (Ed.), *Power, action and belief. A new sociology of knowledge?* (pp. 264–280). London: Routledge.

Latour, B. (1992). Technology is society made durable. In J. Law (Ed.), *A Sociology of Monsters. Essays on Power, Technology and Domination* (pp. 103–130). London: Routledge.

Lindenthal, A. (2008). PDM and PLM. *2PLM John Stark Associates.* Retrieved Aug 28, 2008, from http://www.johnstark.com/2plm269.html

Martinsons, M. G., & Davidson, R. (2003). Cultural Issues and IT Management: Looking Ahead. Guest Editorial. *IEEE Transactions on Engineering Management*, *50*(1), 3–7. doi:10.1109/TEM.2003.808299

Persson Dahlqvist, A., Asklund, U., Crnkovic, I., Svensson, D., Ranby, J., Hedin, A., & Larsson, M. (2001). *Product Data Management and Software Configuration Management: Similarities and Difference.* Technical Report. The Association of Swedish Engineering Industries.

PLCS. (2008). *Product Life Cycle Support.* Retrieved Sept 4, 2008, from http://www.plcs-resources.org/

Svensson, D., & Crnkovic, I. (2002). Information Management for Multi-Technology Products. In *Design 2002, Dubrovnik, Croatia 14–17 May 2002.*

Svensson, D., Malmström, J., Pikosz, P., & Malmqvist, J. (1999). A Framework for Modeling and Analysis of Engineering Information Management Systems. In *Proceedings of the 1999 ASME Design Engineering Technical Conferences,* September 12–15, Las Vegas, NV.

Taxén, L. (2004). IS Design as Domain Construction. In O. Bertelsen, M. Korpela, A. Mursu, (Eds.), *ATIT 2004, Proceedings of the First International Workshop on Activity Theory Based Practical Methods for IT-Design* (pp. 111–122). September 2004, Copenhagen, Denmark. Report PB-574. Daimi, Denmark: University of Aarhus, Denmark. Retrieved Oct 6, 2008, from http://www.daimi.au.dk/research/papers-reports/reports

Taxén, L., & Svensson, D. (2005). Towards an Alternative Foundation for Managing Product Life-Cycles in Turbulent Environments. [IJPD]. *International Journal of Product Development*, *2*(1-2), 24–46. doi:10.1504/IJPD.2005.006667

Trauth, E. M. (1989). The Evolution of Information Resource Management. *Information & Management*, *16*(5), 257–268. doi:10.1016/0378-7206(89)90003-7

ENDNOTE

1 It is conspicuous that the project was baptized to "PDM Core" and not "PLM Core". The term "PLM" had not gained momentum at Ericsson around year 2000. PDM, which stands for "Product Data Management", was later to be included in PLM concept as the data management part of PLM.

Chapter 13
Alignment:
The Activity Domain in the Centre

Alignment, or fit, concerns how various components of an organization work in concert to meet the needs of its environment. The motivation for addressing alignment is that the performance of the organization is a consequence of the fit between these components. Maintaining the fit becomes a necessity for survival in a changing economy. Common components mentioned in connection with alignment are externally oriented ones such as strategies, goals and needs of the business, and internally oriented ones such as information technology (IT), processes and knowledge.

Although IT has changed dramatically since the 1980s, IT and business alignment is still the number one concern for IT executives (Luftman & McLean, 2004). In the era of globalization, ever escalating turbulence of the market, and increasing complexity of products, alignment pose immense challenges (e.g. Chan, 2002; Earl, 1996; Hackney, Burn, Cowan, & Dhillon, 2000; Opdahl, 1997; Regev & Wegmann, 2003). Some of the difficulties are:

- There is an ambiguity of how to define alignment, and what components are relevant for alignment.
- Central concepts in alignment such as "business goal", "business structure", "informal organization structure", etc., are inherently vague (Chan, 2002).
- Alignment spans several organizational boundaries. With increased organizational dynamics such as outsourcing, alliances formation, etc., inter-organizational aspects need to be considered.

DOI: 10.4018/978-1-60566-192-6.ch013

Outsourcing, for example, implies that the control of alignment concerning the outsourced functions will be lost.

- Alignment spans not only technical issues but also social ones such as how to align different informal structures and organizational cultures (ibid).
- There is an apparent lack of theoretical foundations from which alignment issues can be addressed. There is a lack of theories that can provide an integrative, socio-technical view on alignment and at the same time are possible to operationalize (e.g. Martinsons & Davidson, 2003).

A particular kind of alignment concerns the alignment between business strategies and what Zack has called "knowledge strategies":

Many executives are struggling to articulate the relationship between their organization's competitive strategy and its intellectual resources and capabilities. They do not have well-developed strategic models that help them to link knowledge-oriented processes, technologies, and organizational forms to business strategy, and they are unsure of how to translate the goal of making their organizations more intelligent into a strategic course of action. They need a pragmatic, yet theoretically sound model of what I call knowledge strategy. (Zack, 1999, p. 126, italics in original)

The importance of aligning business (B) and knowledge (K) strategies is well recognized (Abou-Zeid, 2008)[1]. In order to operationalize alignment, these strategies should be grounded in a common foundation from which general definitions or theories can be transformed into elements that can be manipulated, measured, or observed in practical situations. In particular, such a foundation must consider the socio-technical nature of alignment (Tuomi, 2002), i.e., the social and technological context in which alignment takes place, must be considered.

In Section 3, I discussed the inherent problems of using the concept of knowledge as a point of departure for operationalizing theories that aim at taking social and cognitive aspects into account. I suggested that the concept of "capability" is more productive for this purpose since capability is something that can be assigned to both human actors and mediational means – the objectivated and objectified aspects of capability respectively. The use of "capabilities" rather than "knowledge" indicates a deliberate intention to include mediational means as inseparable parts of actions. Wertsch (1991) maintains that action and mediational means are so deeply intertwined that it is more appropriate to speak of "individual(s)-acting-with-mediational-means" rather than individual(s) alone when referring to the agent of action. However, since the prevalent literature conceptualizes the B/K strategy alignment in knowledge terms, I will keep this terminology throughout this chapter. The reading of "K" should, according to my perspective, be read "capability" rather than "knowledge".

The purpose of this chapter is to permeate the alignment area from the ADT perspective. In doing so I will focus on the alignment of B and K strategies, since this can be apprehended as a wider scope than the more tangible alignment of, say B strategies and IT strategies. The reason is that knowledge according to ADT is directly related to the work object of the activity domain. It is in the activity domain that capabilities of both humans and mediational means are enacted, and mediational means are, among other things, ISs and IT. Thus, business strategies and IS/IT cannot be directly aligned since that would "short-cut" the enactment process in the activity domain.

DEFINITIONS

In order not to further confuse the conceptual disarray in alignment from the start, Goldkuhl (2002) provides appropriate advice. He emphasizes the importance of linguistic determinations of terms. In daily "language games", we do not have to reflect on the meaning of terms as long as their usage results in intended consequences. However, in scientific contexts we need to be more precise. Scientific conceptualization is use of language, and we need to be aware of how we define our concepts:

It is important to see that we use appropriate language forms when we label our concepts. We must be aware that an attribute, even if we use a substantive form, is a property and this means that it is a quality of something and not a separate entity in itself. (Goldkuhl, 2002, p. 9)

In this spirit I, will define some central concepts in alignment as follows.

Business Strategy

According to Porter, activities are the basic units of competitive advantage (Porter, 1996). Strategic positioning means "performing *different* activities from rivals' or performing similar activities in *different ways*" (ibid, p. 62, italics in original). From this follows that "strategy is the creation of a unique and valuable position, involving a different set of activities" (ibid, p. 68).

Business strategy has been defined as "the determination of the basic long-term goals and objectives of an enterprise, and the adoption of courses of action and the allocation of resources necessary for carrying out these goals" (Chandler, 1966, p. 16). A business strategy is unique to an organization, sometimes unique in time, and always shaped by the cultural values of the stakeholders, constituencies, the communities the organization serves, and by marketplace considerations (Bishoff & Allen, 2004).

Knowledge Strategy

A knowledge strategy is a plan that describes how an organization will manage its knowledge better for the benefit of the organization and its stakeholders. A good knowledge strategy is closely aligned with the organization's overall strategy and objectives. According to Zack (1999), a "knowledge strategy [...] describes the overall approach an organization intends to take to align its knowledge resources and capabilities to the intellectual requirements of its [business] strategy" (ibid, p. 135). This strategy "can be thought of as balancing knowledge-based resources and capabilities to the knowledge required for providing products or services in ways superior to those of competitors" (ibid, p. 131). In order to become operational, the strategy must be translated into an organizational and technological architecture to support knowledge creation, management, and utilization processes for closing those gaps (ibid, p. 142). In doing so, firms need some models, which "strategically guide their knowledge management efforts, bolstering their knowledge advantages and reducing their knowledge weaknesses" (ibid, p. 131).

Alignment

In the literature about alignment the concepts "alignment" and "fit" are used interchangeably. Other concepts used are linkage, congruence, match, adaptation, correspondence, and coherence (Regev &

Wegmann, 2004). Alignment is a noun defined by Merriam-Webster (2008) as follows:

[Alignment is] the act of aligning or state of being aligned; especially: the proper positioning or state of adjustment of parts (as of a mechanical or electronic device) in relation to each other.

"Fit" can stand for both a noun and a verb: "the fact, condition, or manner of fitting or being fitted." (noun); "to be in agreement or accord with." (verb); (Merriam-Webster, 2008). In this chapter, I will use "alignment" for the action of aligning and "fit" for the state of being aligned.

In a broad sense, alignment can be understood as the efforts of an organization to balance different stakeholder needs in order to survive in a changing environment (Regev & Wegmann, 2004). Such stakeholders may be external, for example, customers, competitors, share holders, suppliers, legislative authorities, etc. Examples of internal stakeholders are various business units providing some capability to the benefit of the organization, and the employees working at these units.

In order to maintain the balancing act, there is a need for some model of the organization that stakeholders can use as a means to understand what constitutes an optimal fit (ibid). The model represents "what the stakeholders consider as acceptable fit, to what extent this fit exists, what changes are likely to create this fit in the future, and what would be the consequences of a new fit on the stakeholders" (ibid, pp. 135-136).

RECONSTRUCTION OF STRATEGY ALIGNMENT

In order to position the ADT approach I will reconstruct my understanding of the B/K strategy alignment discourse. A suitable point of departure is provided by Tuomi (2002), who states that the sources of knowledge management (KM) can be separated into four intertwined clusters. The first one, *organizational information processing*, has its roots in the Artificial Intelligence community and is concerned with building corporate-wide information systems and expert systems. The core idea is that knowledge can be stored and shared with the help of computer systems. In this cluster, technology is in focus.

In the next cluster, *business intelligence*, the focus is on categorizing, searching and distributing information that is considered vital for the business. Knowledge sharing is a prime task for corporate librarians and intelligence professionals. This task is facilitated by the access to large databases and the Internet. However, information overload is an issue. This in turn brings the relevance of the information to the foreground. Ultimately, the problem of knowledge representation is reduced to the idea that "all knowledge can be represented as documents and associations between them" (Tuomi, 2002, p. 5). Making sense of the information is left to the reader.

In the third cluster, *organizational cognition*, organizational sense-making and the active process of knowledge construction are emphasized. The focus is on the effective use of human experts and the establishment of social and communicative networks. A more interpretative approach towards knowledge is taken, where tacit and situated aspects of knowledge are high-lighted.

The fourth cluster, *organizational development*, brings knowledge and social action to the foreground. The concept of "learning organization" is coined, and the knowledge creation process becomes subject to management. In this cluster, KM is linked to the B strategy, that is, the KM strategy is turned into a K strategy. Resource-based strategies, including analysis of competitive strengths and weaknesses, evolve to competence-based strategies. Knowledge is considered an asset in the balance sheet, and intellectual

Figure 1. The linking of K and B strategies

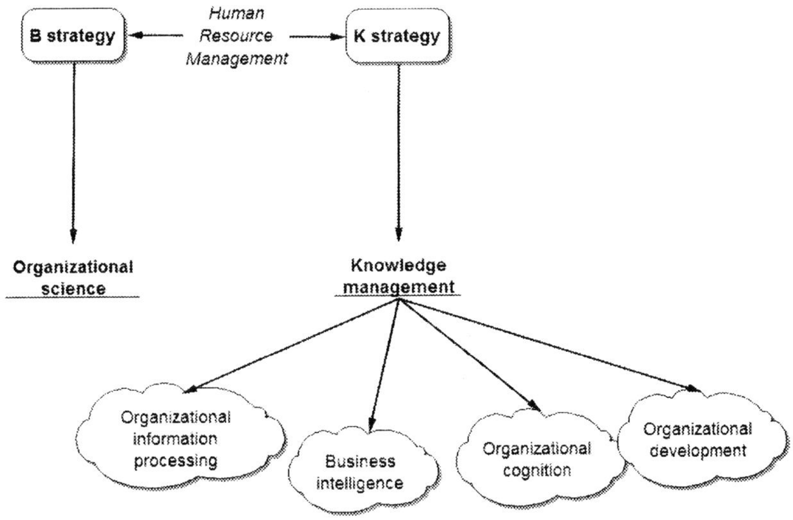

property is protected. The strategic needs of the organization are linked to the aggregation of individual skills by human resource (HR) management initiatives. The basic idea is to identify and fill in gaps in knowledge in order to execute the B strategy, which in turn is grounded in organizational sciences. Thus, the HR department becomes the link between K and B strategies as illustrated in Figure 1.

The clusters described emerged more or less sequentially between 1993-1996. Tuomi (2002) calls this period the first generation of KM. It is characterized by its focus on information sharing, repositories and intellectual capital management. In the second period, which started around 1997, companies included KM as part of their everyday organizational discourse. Specific KM positions and departments were established. Issues of tacit knowledge, social learning, situated and embedded knowledge, and communities of practice were in focus (ibid, p. 10).

Tuomi maintains that the first and second generations of KM will remain vital. A third generation of KM will in addition emphasize the role of information systems as support for knowledge construction and human sense-making. Knowledge will be viewed from a constructivist and pragmatic perspective. The action character of knowledge will be in focus as well as social aspects of knowledge. This will require a better understanding of the cultural basis of knowledge.

ALIGNMENT FROM THE ACTIVITY DOMAIN THEORY PERSPECTIVE

From the ADT perspective the outcome of an organization is achieved by the coordination of a constellation of activity domains. From this conceptualization alignment can be defined as follows:

Alignment is the act of constructing the constellation of activity domains in an organization in such a way that it fulfills the motive and social needs that motivate the existence of the organization.

Figure 2. A constellation of activity domains

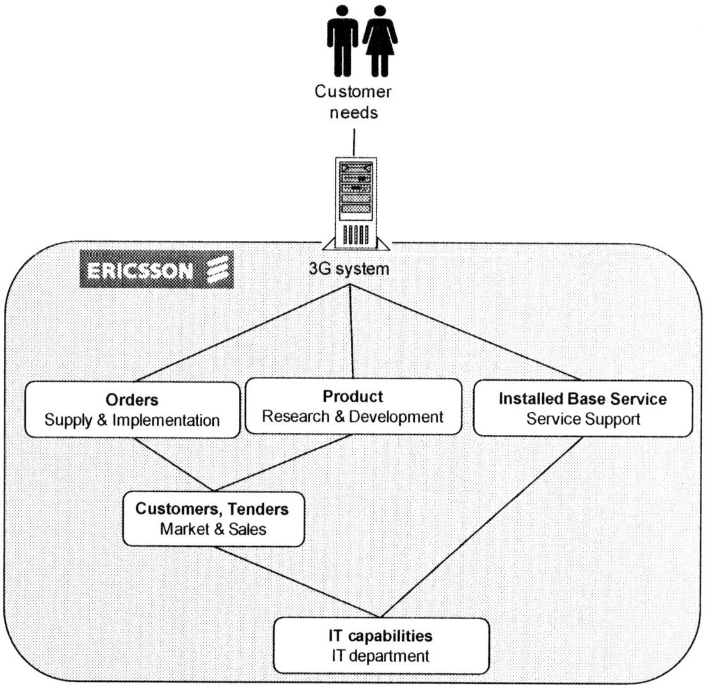

To illustrate this, consider the Ericsson example in Figure 29, p. 95 (repeated in Figure 2 for convenience).

The social need that motivates the existence of Ericsson is to provide telecom systems to customers. Each domain in the Ericsson context provides a certain capability, and these capabilities are dependent on each other.

With this general picture in mind, alignment can be seen from two perspectives: one from the inside and one from the outside of each domain. Both the internal and external aspects have to be considered in alignment: I will call these *internal* and *external* alignments respectively. The notions of internal and external are always relative to a certain activity domain: what is external for one domain may be internal to another domain. In line with the "client" – "server" concepts in IT, the external view of a domain can be seen as the server domain of the internal view of a client domain that needs the server domain capability. In the example above, Ericsson would be considered the client domain of the server domains Supply & Implementation, R&D, Service Support, Market & Sales, and the IT department.

By placing the activity domain in focus for alignment several things are achieved. First, a common pattern consisting of activity domains and their dependencies of each other can be used throughout the organization. This same pattern can be used in any relevant context, whether that is a team, group, business unit, organization or the entire enterprise. Second, a clear target for alignment can be identified in terms of capabilities. Since each domain produces an outcome that is needed elsewhere, the work object of each domain will be in focus. The constitution of the work object will determine what domain specific, internal capabilities are needed to achieve the outcome. Finally, the focal change between internal and external alignment is achieved through the transition modality, since this modality signifies how the internal and external world-views of a domain are related to each other. In external alignment, the external context of a domain is emphasized while internal alignment emphasizes its internal context.

The internal and external aspects of the activity domain enable us to conceive targets for the B and K strategies as follows.

Business Strategy Target

The main target for the B strategy is to manage the coordination of activity domains in such a way that the outcome of the organization is achieved. In this way, top management can focus on the macro structure of the organization. Some issues that need to be considered are:

- *The constellation of activity domains:* Which domains are needed to fulfill the strategic intents of the organization? How do the domains depend on each other?
- *Business level coordination*: How are the activity domains coordinated from the top-level domain, that is, the organization itself? In general, this coordination is signified by a business process model like the one in *The Ericsson Business Process around 2000*, p. 227).
- *In-house or out-sourcing*: Which domains should remain within the control of the business, that is, in-house? Which ones should be out-sourced? Are we lacking some basic capability that should be acquired through acquisitions or mergers?
- *Transitions between activity domains*: Are there business critical transitions between activity domains? How should these be managed? For example, an organization may choose to out-source the operation of its IT-platform to another organization. An issue that may appear in such a case is the translation between different organizational languages. Other issues concern the pragmatic functions of language. Which assignments, contracts, agreements, responsibilities, etc., are needed to regulate the co-operations between organizations?
- *Central versus local control*: Each domain enacts a unique world-view. The concrete manifestations of the activity modalities are all different between domains. However, the coordination of activity domains calls for some communal meaning across domains. This raises the question of maintaining an optimal balance between what is centrally controlled and what can be left to each domain to control locally. This balance affects all the activity modalities.
- *IT architecture*: How should the IT architecture of the organization be designed in order to support the coordination of activity domains?
- *Business rules:* What business rules concerning identification, core concepts, correct ways of working, etc., are valid throughout the organization?

Knowledge Strategy Target

So far, most KM initiatives have taken as a point of departure the knowledge needed for an actor to carry out a certain task, for example, writing software code in C++. This is certainly valid also in the ADT approach. However, with the introduction of the activity domain, individual actions are immersed in a social fabric where these actions make sense only in relation to the motive of the domain. Thus, the communal and situated aspects of individual knowledge are brought to the fore. This means that knowledge about how to coordinate individual actions becomes crucial. Based on these considerations, two targets for K strategies can be discerned: the transformative capability needed to transform the work object of the domain, and the coordinative capability needed to coordinate transformative actions in the domain. Thus, K strategies should be centered on securing transformative and coordinative capabilities in each activity domain.

Transformational and coordinative capabilities are needed in every activity domain, including the organization itself. The actors in this domain, for example, the CEO and his or her steering group, need to acquire the particular competence needed, and coordinate their actions. Thus, the same targets for K strategies apply also to this domain.

OPERATIONALIZATION

A tentative sketch of an alignment process based on the ADT approach would look as follows. The B strategy would most likely benefit from using some kind of dependency diagram like the one in Figure 2, showing dependencies between activity domains. Such a diagram would be the common means by which management can discuss various scenarios and foresee consequences of different actions.

Concerning the K strategy, the management of transformational capabilities needs to be related to the motive and object of the activity domain. Actors in a domain producing printed circuit boards need very different kinds of capabilities as compared to actors in a domain producing software in C++. The strategy needs to consider the enactment view of capabilities: it is only when working with means in the activity domain that meaningful resources are brought about. This does, of course, not exclude general training out of context. The ultimate capabilities, however, can only emerge in the activity domain.

Strategies for constructing coordination capabilities should be based on the operational guidelines presented in Section 3. A key element is the domain construction strategy (DCS), which is focused on continuous iteration between reflection and action. All activity modalities should be attended, and the interdependencies between these modalities need to be managed as well as the transitions between activity domains. The operationalized elements – the information model, the IS, and so on – are ceaselessly modified in the construction process.

The result will be, as the name on the strategy suggests, an ongoing construction of the entire coordinative mode of the activity domain. In this way, common understanding is gradually established among the actors. The enactment process unfolds as long as the domain exists. The elements in the domain, such as the information systems, will never be "finalized".

EMPIRICAL GROUNDING

The 3G project related previously in Section 1 was coordinated at several sites. Two of the main ones were the A-site in Aachen, Germany, and the S-site in Stockholm, Sweden. Both of these sites, which in the ADT terminology should be considered as activity domains, coordinated a number of other development sites around the globe (see Figure 3).

The goal of the project was to deliver 3G systems to the customers, which in this case were operators running the 3G network. All the main activity domains at Ericsson were involved. The Market & Sales domain negotiated requirements from the operators. These were forwarded to the Research & Development domain, in which the A-domain was one domain. The A-domain developed customer oriented software, and had the overall responsibility to coordinate and integrate deliveries from all other sites.

Figure 3. Two activity domains in the 3G development project: the A- and S-domains

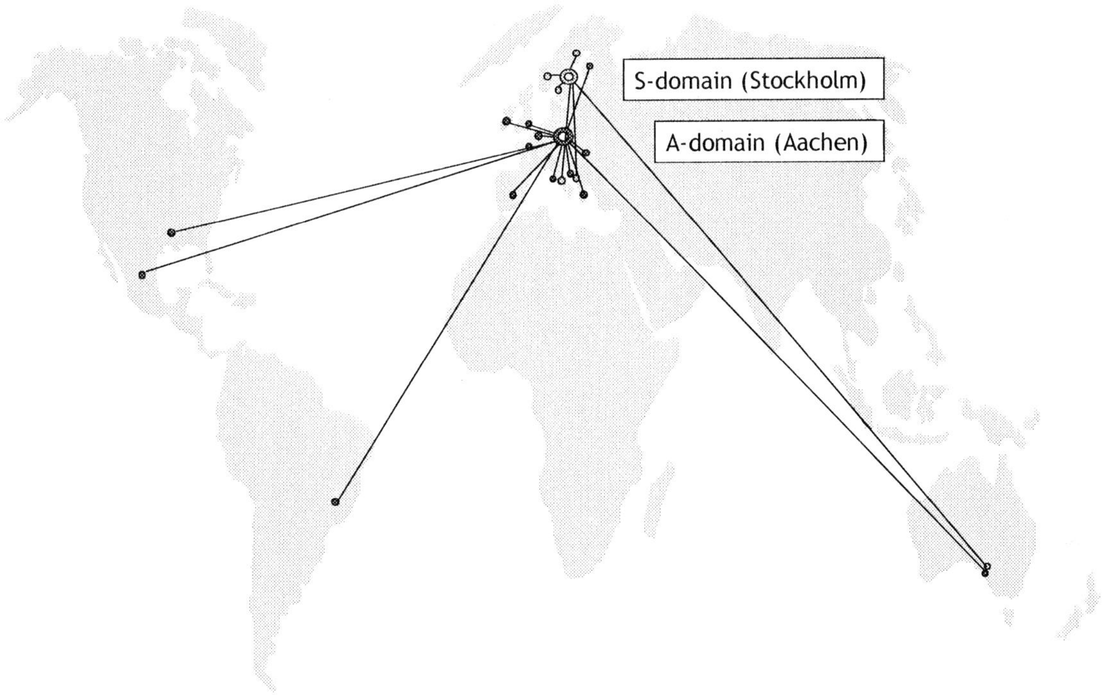

The S-domain was responsible for hardware heavy platform parts of the system such as switching equipments and processors. As such, the applications being developed in software were dependent on the S-domain. The dependencies between the main domains are illustrated in Figure 4.

The 3G project is a concrete illustration of the alignment problem. The B strategy was more or less settled when the project started. The transition from the 2nd generation GSM-based systems to 3rd generation of UMTS-based systems was a necessity for telecom system suppliers to survive. New applications, such as Multimedia Messaging Service (MMS), access to the Internet and mobile data communication demanded higher bit-rates in the communication network; bit-rates that only 3G systems could provide.

The K strategy was, to the extent it existed as an outspoken strategy at that time, quite naturally focused on the transformative capabilities needed to develop the 3G system parts. These capabilities varied depending on the specific technology used. Some capabilities could be reused from previous projects, but in general, quite new ones had to be enacted. For example, one ASIC (Application Specific Integrated Circuit) used in the switching core was at that time in the absolute fore-front of what was technically possible. Moreover, the mediational means used to design this chip were also state of the art and, to aggravate things further, modified during the project. The situation for the transactional K strategy could be summarized as experimental, explorative and learning on the spot.

The situation for the coordinative K strategy was similar in the sense that the established way of coordinating projects could not be pursued further. As related earlier, the anatomy-centric approach had a rambling history at Ericsson since the early 1990s, but the application of this method for managing project of the complexity faced in the 3G system had never been tried before. In addition, the IS used

Figure 4. Dependencies between activity domains

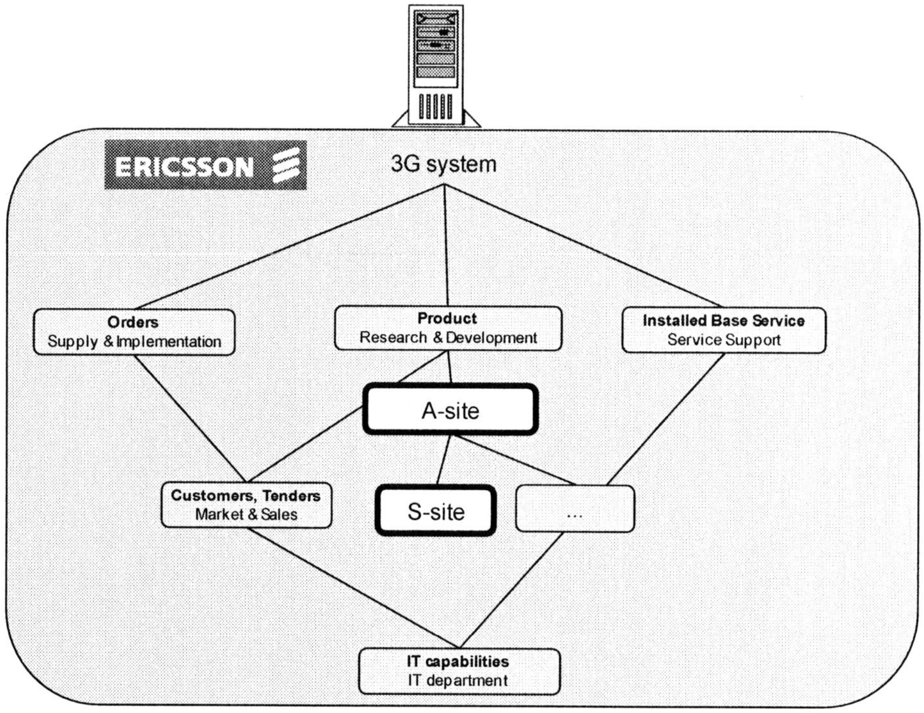

– Matrix – was new to Ericsson, and there was no accumulated experience of how the coordination should be carried out.

From an alignment perspective, the situation can be summarized as follows. The internal alignment of the A- and S-domains had to be done more or less from scratch. A new paradigm, so to say, had to be established. A major challenge in both the A- and S-domains was to arrive at a common understanding about the meaning of coordination, which comprised a multitude of items such as work-packages, products, product related documents, requirements, engineering change orders, baselines, milestones, etc. These items had to be characterized in terms of what attributes they had, what states they could pass through during their life cycle, how they were related to each other, etc. Thus, an overwhelming number of phenomena had to be defined by the actors, hence the difficulties in achieving common understanding.

The alignment of the domains was carried out by a small group of actors in each domain. These actors represented various stakeholders in the project such as project managers, requirement managers, configuration managers, etc. Other actors were application developers from the vendor of the Matrix IS. Thus, the same information system platform was used to implement coordinative capabilities in both domains.

As it turned out, the alignment of the A- and S-domains followed quite different trajectories. In Figure 5, the information model of the A-domain around 2001 is depicted.

The model signifies various items that the actors in the A-domain found to be relevant. For example, “WP” signifies “Work Package”, which is the focus of the model (encircled). A work package is similar

Figure 5. The information model in the A-domain (2001) (Taxén, 2005. ©Wiley Inter Science. Used with permission)

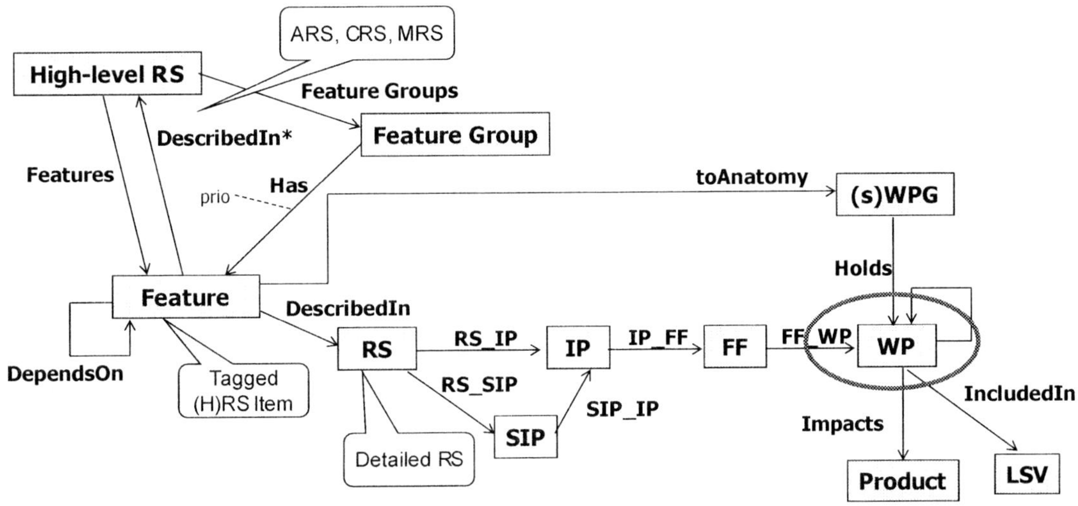

to "increment" and signifies a logically coherent capability that can be delivered for integration with other work packages according to the integration plan (see Figure 1, page 13).

The alignment of the A-domain was carried out at about the same time as the alignment of the S-domain. In Figure 6 the corresponding information model of the S-domain is illustrated.

As can be seen, the A- and S-domains were internally aligned very differently. The only element in common is the Work Package. However, when zooming in on the details characterizing the Work Package, it turned out that even this central element was constructed differently in terms of attributes, state sets, icons, etc.

In spite of, or rather due to, these differences the internal alignment at each site turned out to be highly successful in providing coordination capabilities for the projects. Each site was aligned in terms of common understanding about coordination and the coordination support provided by the applications built on the Matrix platform. In addition, the internal alignment in both domains was constantly adjusted during the project.

In contrast to the successful internal alignment of both the A- and S-domains, the external alignment between these sites was a failure. In order to coordinate these two sites, at least the Work Package should have been defined the same way (see Figure 7).

Some initiatives were taken to achieve this alignment but no real progress was made. The necessary interaction between the domains had to be improvised on a personal basis by e-mail, personal interaction, etc.

In summary, the application of the principles in the ADT enabled the internal alignment of the A- and S-domains. However, the external alignment between them was not achieved. This could have been achieved by mandating the definition of the Work Package item. However, the power structure to achieve this was not in place in the organization during the progress of the 3G project. In other words, the external R&D client domain was unable to coordinate the A and S server domains.

Figure 6. The information model in the S–domain (2001) (Taxén, 2005. ©Wiley Inter Science. Used with permission)

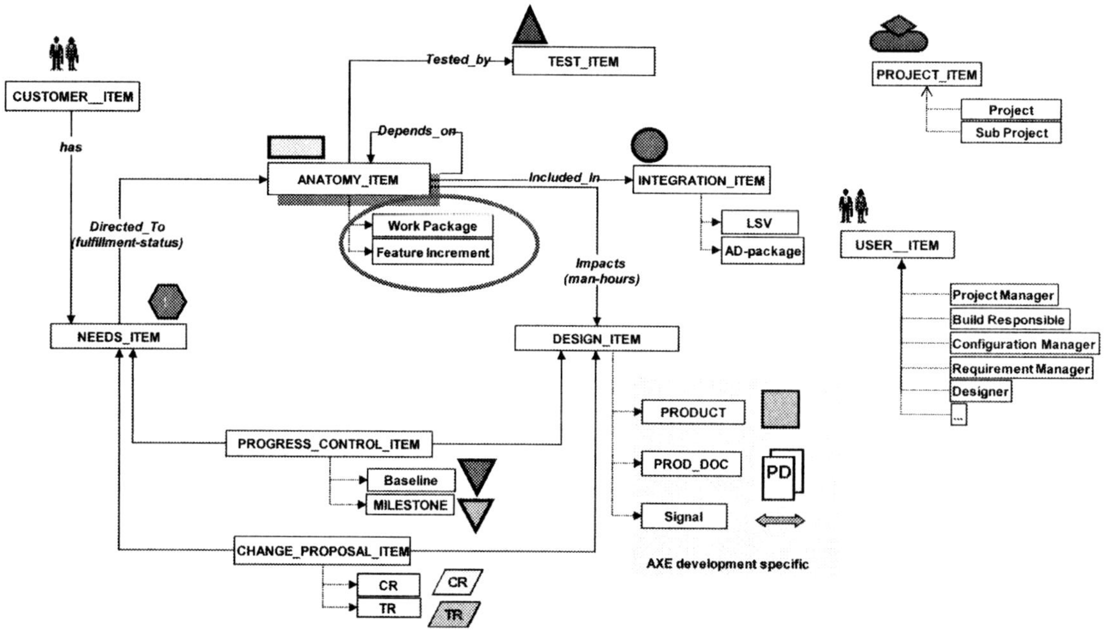

Figure 7. The external alignment of the S- and A-domains

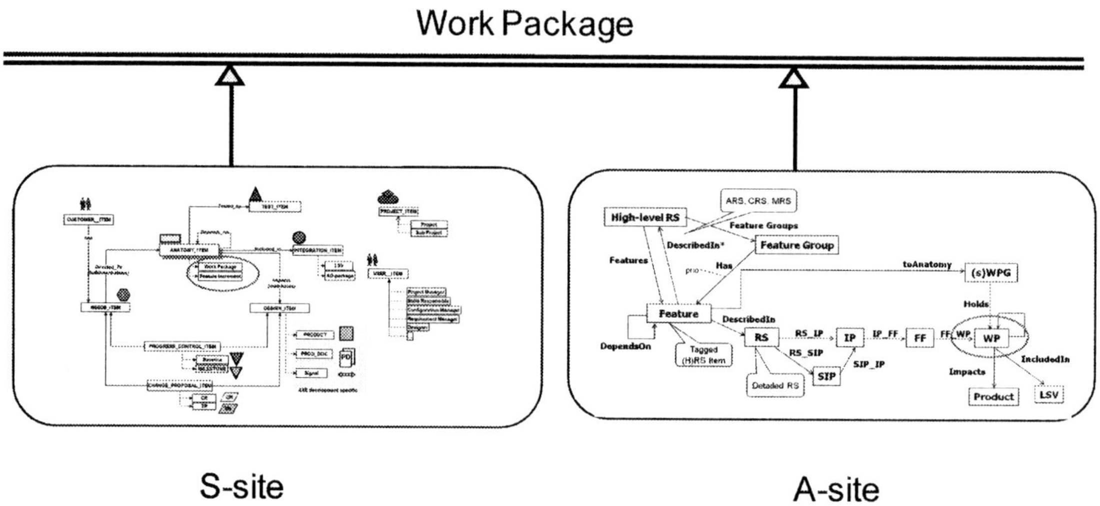

IMPLICATIONS

The gist of the ADT approach towards alignment is the introduction of the activity domain as a common foundation for B and K strategies, suggesting that the proper point of departure for knowledge and organizational discourses is the activity domain (see Figure 8).

Figure 8. The activity domain as the foundation for B and K strategies

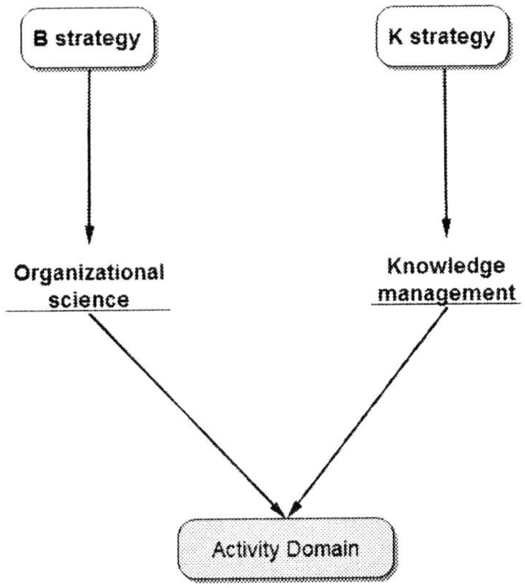

In essence, the HR department is replaced with the activity domain as the link between B and K strategies. The HR focus on individual knowledge is re-focused to the social unit of the activity domain. The individual is of course still important, but individual capabilities are meaningful only in the context of the domain where capabilities become resources. Although tentative, some implications of the ADT approach towards alignment of B and K strategies can be envisaged.

General Implications

The different outcomes in the 3G project with respect to external and internal alignment indicate that internal alignment is easier to address since this concerns the actors own "backyard" and deals with imminent, concrete issues. External alignment is, as the term indicates, considered as someone else's problem. However, what is external from the perspective of one activity domain is internal from another domain's point of view. Unless the whole picture is seen, alignment efforts will most likely turn into a patchwork.

A further implication of the results is that alignment of large, culturally and geographically diverse organizations is hard to achieve if the unit of analysis is the organization. Some intermediate construct, like the activity domain, is needed in order to capture significant differences between organizational units. A slogan like "one company – one process" is, in spite of its attractive simplicity, an oversimplification that most likely causes more problems than it solves. The ADT approach enables a balanced strategy where common, imperative measures may live side by side with local autonomy that acknowledges the uniqueness of each activity domain.

Furthermore, the results suggest that all modalities need to be considered in alignment. Aligning process and IT systems are necessary but not enough. Information structures, rules, transitions to other domains, and common understanding have to be included. Moreover, the interdependencies between these modalities must be attended.

Concerning the development of alignment, this is often staged in a creation and maintenance phase. However, such a distinction is problematic (see, for example, Regev & Wegmann, 2004; Dietz, 2004).

This observation is also supported by the results. Alignment is continuously enacted, and it is hard to discern a borderline between creation and maintenance. It seems more appropriate to conceive alignment in the mathetic, consolidation and pragmatic phases suggested by the DCS.

The ADT approach towards alignment must overcome some substantial obstacles in order to make a persistent impact within an organization. First, the activity domain construct must gain acceptance as a viable way of conceiving organizations. The formal organization structure is important but inadequate as a basis for alignment. Second, a structure of cooperating activity domains must be identified. It is not always clear what should be considered as an activity domain in a particular organization. In line with the pragmatist epistemology in the ADT, this is something that must ultimately be worked out in praxis. Third, it must be realized that the fully centralized and decentralized strategies towards alignment are not the only options towards alignment. It is possible to balance the imperative and the optional.

An Integrated View on Knowledge

Using ADT as a foundation makes it possible to chisel out a distinct position of knowledge. As been suggested in Section 3, the concept of "capability" can be seen as a dialectical unity in which enactment results in two types of manifestations: objectivated capabilities in the human mind and body, and objectified capabilities of mediational means in the domain. This view of knowledge can be related to prevailing concepts in KM as follows:

- *Commodification of knowledge, embedded knowledge*: the tendency to focus on objectified manifestations of enactment, e.g. tools, drawings, models, and other mediational artifacts.
- *Mentalistic view of knowledge*: The tendency to focus on objectivated elements in the mind such as communal meaning, concepts, language, tacit knowledge, proficiencies, and the like.
- *Organizational memory*: The objectified manifestations of enactment. Since an organization can employ different types of activity domains, depending on the type of object, various types of organizational memory will exist in an organization.
- *Organizational learning*: The enactment of capabilities of actors and mediational means that takes place in various activity domains in accordance with the overall motive of the organization. It can be noted that the concept of learning has an anthropocentric "flavor" that I seek to avoid with the dialectical unity of actors and means embedded in the enactment concept.

Moreover, the ADT approach makes it possible to address problems found in existing KM practices. For example, Fahey & Prusak (1998) listed "The Eleven Deadliest Sins of Knowledge Management":

1. Not developing a working definition of knowledge.
2. Emphasizing knowledge stock to the detriment of knowledge flow.
3. Viewing knowledge as existing predominantly outside the heads of individuals.
4. Not understanding that a fundamental intermediate purpose of managing knowledge is to create shared context.
5. Paying little heed to the role and importance of tacit knowledge.
6. Disentangling knowledge from its uses.
7. Downplaying thinking and reasoning.
8. Focusing on the past and the present and not the future.

9. Failing to recognize the importance of experimentation.
10. Substituting technological contact for human interface.
11. Seeking to develop direct measures of knowledge.

Point 1 – a working definition of knowledge is given in Section 3 and is recapitulated above. Points 2, 3, 7, and 9 – objectified ("knowledge stock") and objectivated ("in the head") manifestations are enacted in the activity domain. This also implies that it is futile to develop measures of knowledge (point 11). Point 4 – "shared context" is enacted in the alignment of the domain towards fulfilling its motive. Point 5 – tacit knowledge is apprehended as enacted objectivated elements in the mind and body of actors. Point 6 – enactment of capabilities is inherently related to usage. In ADT, a disentangled view of knowledge is inconceivable.

The Practice Turn in KM

From a knowledge management point of view, the ADT approach can be seen as one line of inquiry belonging to a broader KM strand that takes the *practice* as its point of departure (e.g. Brown & Duguid, 1991; Gherardi, 2000; Gorelick & Tantawy-Monsou, 2005; Hildreth & Kimble, 2002; Peltonen & Lämsä, 2004; Tuomi, 2002). The practice construct has gained an increased interest in contemporary social science (e.g. Schatzki, Knorr Cetina, & von Savigny, 2001). Practices are considered to be "embodied, materially mediated arrays of human activity centrally organized around shared practical understanding" (Schatzki, 2001, p. 2)

One reason for the practice turn in KM is a growing discontent with the disentangled views of knowledge as either a commodity or residing in the brain. One line of inquiry centers on the Community of Practice (CoP) approach (Wenger, 1998). A CoP is defined as:

A group of people who share a concern, a set of problems, or a passion about a topic, and who deepen their knowledge and expertise in this area by interacting on an ongoing basis. (Wenger, McDermott, & Snyder, 2002, p. 4)

An example of a CoP is engineers who design a certain kind of electronic circuits, for example, so called phase-lock loops. They "find it useful to compare designs regularly and to discuss the intricacies of their esoteric specialty" (Wenger, McDermott, & Snyder, 2002, p. 4). Thus, a CoP is primarily a context outside the activity domain, where actors can share experiences and knowledge. As such, a CoP may become important in the K strategy as a means for knowledge transfer.

However, the CoP approach misses the central element in ADT: the primacy of the activity domain and its associated work object. Some contributions in the KM literature have exploited this line of inquiry (Blackler, 1995; Virkkunen & Kuutti, 2000). For example, Blackler states that "Central to activity theory is the idea that collective action is driven by the conceptions people have of the object of their activities" (Blackler, 1995, p. 1041). In line with this I argue that the work object, as determinant for the formation of the activity domain, is crucial for the alignment of B and K strategies.

KM Systems

The storing, distribution and retrieving of information in a KM system should be structured from an activity domain perspective. Some conceivable capabilities of such systems are:

- Listing activity domains and their characteristics, for example, their motives, what needs they fulfill, and what kinds of work objects are manipulated.
- Keeping track of the dependencies between activity domains.
- Matching activity domains with similar characteristics where actors with similar capabilities may be found.
- Listing individuals with expert knowledge related to a particular domain.

It has been noted that the predominant use of KM systems is to capture, store and transmit "commodified knowledge" in the form of patents, documents, experiences, etc. (e.g. Hildreth & Kimble, 2002). From the ADT point of view, this kind of management concerns only the objectified manifestations of activity in various activity domains. In order to manage the objectivation aspect, KM systems need to support the entire spectrum of meaning and knowledge implicit in the activity domain construct. This is in line with Tuomi when he suggests that:

[Information] systems for knowledge management and organizational memory should be seen as media that is used as an interpersonal cognitive artifact. A critical factor in designing such artifacts is to consider those knowledge stocks that are needed to make sense of the information stored in the system. (Tuomi, 1999, p. 9)

If two domains have similar motives and objects, the objectivated manifestations of activity modalities, that is, what is embodied in the minds of the actors will be similar. Communities of Practices (CoPs) (Wenger, McDermott, & Snyder, 2002), consisting of actors from different domains, can be cultivated based on these considerations. In CoPs, actors can exchange experiences and look for solutions to similar problems.

It can also be noted that systems, which are used in manipulating the work object, for example, configuration management systems, contribute to the knowledge construction in the domain. They are an intrinsic part of the construction of the domain, and should be designed to facilitate the construction of communal meaning. This implies, for example, that the semiotic aspects of such systems should be given a high priority when designed. The action character of cues, symbols, and help texts should be made as evident as possible.

Comparison with Other Alignment Approaches

In this section, I compare the ADT approach with some other alignment approaches reported in the literature. This comparison can only be superficial, given the width and depth of the subject area. I will use the following categories as reference:

- *Integrative perspective*: Is the approach based on an integrative socio-technical perspective?
- *Practice based:* Is there a practice construct akin to the activity domain in the approach?

- *Communal meaning*: Is the issue of communal meaning salient?
- *Dialectical view of knowledge*: Is there a dialectical view on the construction of objectivated elements in the mind and objectified elements in the work context?
- *Emphasis on the work object*: Does the approach emphasize the work object as a focus for B/K strategy alignment?
- *Coordination*: Is coordination an essential theme in the approach?

Earl (2001) has made a thorough investigation of various approaches to KM. He reports on seven schools of KM and suggests how these can be used as points of departures for alignment initiatives. These schools are related to the ADT approach as follows[2].

Systems School

The focus of the systems school is to capture specialist knowledge. Domain specific knowledge is codified and stored in knowledge databases. Technical know-how is provided to those qualified to use it. The KM systems have virtually the same role as ISs. Knowledge is generated from objective data and experience through practice.

In this school, there is a practice touch since the domain specificity of the knowledge is emphasized. The objectification aspect of knowledge is in focus. However, there is less focus on the actual construction of knowledge.

Cartographic School

This school tries to survey the knowledge of the organization: who in the organization knows what? This information is stored in knowledge directories, similar to the "yellow pages" in a phone directory. The individual and tacit aspects of knowledge are communicated to other individuals. IT supports the connections of people.

The domain aspect is present in this school since individual knowledge acquired in one practice is supposed to be transferred to other, similar practices. The objectivation aspect of knowledge is emphasized.

Process School

In the process school contextual and best practice knowledge related to tasks are emphasized. Learning from experience is shared, based on similarity of tasks in key knowledge areas. Improvements made in particular practices are collected and distributed within the organization. The role of IT is the unrestricted provision of knowledge by shared data bases.

This school is related to the ADT approach in several ways. The constructive aspect of learning is emphasized. Contextual and best practice aspects as well as "knowledge areas" are related to the activity domain construct. Moreover, different KM strategies are advocated, which is an indication of differentiation based on type of process or practice. It appears that the process school tries to "balance" several of the activity modalities in ADT. However, the emphasis on process indicates that the temporalization modality is in focus.

Organizational School

The organizational school nurses knowledge communities in which participants can exchange and share knowledge interactively. This takes place outside their daily practice. The communities are organized as networks of domain specific knowledge across business units, sites, and countries. Groupware IT support is heavily employed.

This school is similar to the ADT approach in the sense that the community is in focus. However, in the knowledge community the work object in the activity domain is only indirectly present. The knowledge that is commonalized in a knowledge community has been constructed elsewhere, in the activity domain.

Strategic School

In the strategic school, knowledge is the key resource. KM is the essence of the B strategy. Intellectual capital and a learning organization are both heavily stressed. Knowledge achieved through systems, processes, and people is converted into knowledge-based products or services. The domain specificity of knowledge is recognized as captured in the slogan "multi-local, multi-national".

As in ADT, the strategic school takes an integrative view of knowledge. Knowledge is needed in every practice and is situated in nature. However, in these practices, the work object is subdued in the favor of knowledge itself as the essence of the organization.

In Table 1 below a qualitative mapping of the ADT approach to the schools above is shown. More stars indicate a stronger relation.

Although the mapping is indeed crude, some observations can be made. First, none of the schools can be directly mapped onto the categories of ADT. The process school is the one that has most in common with the ADT approach. It seems that the various schools highlight one or several of the ADT categories. This may be a consequence of the fact that the schools are grounded in different views or "philosophies" of knowledge (Earl, 2001, p. 217). Second, only the strategic school appears to take an integrative view of knowledge. However, the relation of knowledge to the organization's competitive products or services is not salient. Third, the categories of meaning, work object, and coordination are by and large absent in the different schools.

Zack (1999) suggests that the link between K and B strategies has been widely ignored. There is a need for pragmatic and theoretically sound models that enable executives to relate the firm's competi-

Table 1. Mapping the ADT approach to KM schools according to Earl (2001)

	Systems	Cartographic	Process	Organizational	Strategic
Integrative			*		***
Practice	*	**	***	***	**
Meaning			*		
Objectification	***		**		
Objectivation		***	*		
Work object			**	*	
Coordination			*		

tive strategy to capabilities and intellectual resources. According to Zack, a knowledge strategy should identify "which knowledge-based resources and capabilities are valuable, unique, and inimitable as well as how those resources and capabilities support the firm's product and market" (ibid, p. 131).

This position goes well with the ADT approach. The coordination of activity domains provides the products or services of the firm. By analyzing the knowledge needed to coordinate the outcomes and to produce the outcome of each individual activity domain, a strategy for knowledge management can be devised that matches the B strategy. In doing so the classification of knowledge into core, advanced and innovative knowledge suggested by Zack (1999) can be applied to each activity domain as well as to coordination knowledge. Moreover, Zack advocates that B and K strategies should be simultaneously aligned (ibid, p. 135). In ADT this is provided by targeting both strategies to the activity domain construct.

Managerial Implications

As indicated earlier in the Section *Complex systems theory and ADT*, p. 135, the ADT view on organizations retains many properties of complex adaptive systems. As a consequence, traditional targets for top level and middle level managers need to be reconsidered. Managers "should facilitate, guide, and set the boundary conditions within which successful self-organization can take place" (Lewin, 1999, p. 215). Anderson (1999) underlines that "the task of those responsible for the strategic direction of an organization is not to foresee the future or to implement enterprise-wide adaptation programs [...]. Rather, such managers establish and modify the direction and the boundaries within which effective, improvised, self-organized solutions can evolve" (ibid, p. 228).

With the activity domain as the core organizational construct, these recommendations can be interpreted as follows. The role of top managers is to align the organizational architecture within which local self-organization takes place. The architecture can be interpreted as a constellation of dependent activity domains. This implies that espoused views on organizations as a homogenous entity need to be revised. In this process, top management has a key role in "spreading the word".

A natural role for middle managers is to take on the responsibility for the local self-organization in activity domains. One side of this responsibility is directed towards the interior of the domain. The manager must secure that capabilities of actors and mediational means become resources to produce the outcome. This amounts to implementing the K strategy in the domain. The other side is externally oriented towards the coordination with other domains. As such, the manager is involved in implementing the B strategy of the organization as a whole.

Further Inquiries

Since the ADT approach introduces a partly new conceptualization on B/K alignment, it is of course at present not possible to estimate the full potential of this approach. However, some guidelines for further inquires that might strengthen the theory can be identified.

Mediating Construct Between the Individual and the Organization

In order to model the relationship between an organization's competitive B strategy and its K strategies I suggest that there is a need for constructs that mediate the actions of individuals with the multitude of different kinds of knowledge needed in large and possibly globally distributed organizations. In ADT, this mediating construct is the activity domain. Without such a construct the task of coordinating the specialist knowledge of individuals becomes overwhelming. Thus, the unit of analysis should be neither the individual, nor the organization, but some intermediate, practice oriented construct like the activity domain.

Dynamics of Alignment

The dynamics of alignment needs a practice-oriented construct that mediates between the K and B strategies. Alignment implies that the constellation and coordination of activity domains must be rearranged to reflect changes in the organization's strategic positioning on the market. Some domains may become obsolete and others may have to be constructed. The domain construction strategy provides guidelines of how to construct the new domains in such a way that both objectified elements (such as artifacts, tools, and institutions), and objectivated elements (communal meaning among actors about coordination) are manifested according to the motive of the domain.

Enablers of B and K Strategies Alignment

The indications from ADT in researching this issue are as follows. A first enabler is to take the activity domain, or any similar practice based construct, as the main Unit of Analysis. Without such a unit the complexity of alignment cannot be mastered. Next, an integrative perspective of knowledge is needed where objectivated ("in the head") and objectified ("commodified") forms of knowledge are seen as a dialectical unity. These forms of knowledge are constructed by actors working on a common work object. Thus, a second enabler is to emphasize the work object, that is, the target for activities in the domain. A third enabler is to bring coordination into the KM discourse as one way to relate B and K strategies.

Inhibitors of B and K Strategies Alignment

The ADT approach indicates that the main inhibitor of alignment is the disjoint views on knowledge represented by the commodification and the mentalistic views. If this persists, the fundamental dialectics between objectified and objectivated forms of knowledge will remain unattended. Consequently, the dynamics of knowledge construction is not considered, which in turn implies that the full scope of alignment cannot be mastered.

The Impacts of Culture (Organizational and National) on Alignment Process

Here, the indications from the ADT approach is that culture cannot be seen as a detached element that can be managed separately. On the contrary, culture is constituted by meaningful activity:

Socially meaningful doings constitute cultures (social semiotic systems): [...] cultures are systems of interlinking, socially meaningful practices by which we make sense to and of others, not merely in explicit communication, but through all forms of socially meaningful action. (Lemke, 1993)

Consequently, culture expressed as various communal meanings needs to somehow be reconciled if different domains shall be able to coordinate their work. An indication of how to research this issue is given by the transition modality in ADT. Manifestations of this modality signify ways that actors enact in order to coordinate domains harboring different communal meanings.

ALIGNMENT: SUMMARY

In this chapter, I have proposed the ADT as a theoretical framework for informing the alignment of business and knowledge strategies. The approach addresses a major problem in alignment – how to operationalize an integrated B/K alignment strategy. As such, the approach is well suited to address the so called third generation of knowledge management, where knowledge construction, tacit and situated knowledge, and a social understanding of technology are emphasized.

The main tenet of the approach is that the activity domain is seen as a mediator between B and K strategies. The elaboration of business and knowledge strategies as well as their alignment cannot be considered as separate tasks. On the contrary, these tasks are highly interdependent.

The positions taken with respect to knowledge and the nature of the organization can certainly be contested. However, given that they are accepted, the line of argumentation needs to be further articulated and grounded in both literature and empirical research. This said and done, I claim that the ADT approach indicates new directions for researching the B/K alignment problem. An integrative, socio-technical perspective of grounding knowledge is necessary. I suggest that the activity domain provides such grounding.

REFERENCES

Abou-Zeid, E. (2008). Preface. In E. Abou-Zeid (Ed.), *Knowledge Management and Business Strategies: Theoretical Frameworks and Empirical Research* (pp. xii–xvii). Hershey, PA: Information Science Reference.

Anderson, P. (1999). Complexity Theory and Organization Science. *Organization Science*, *10*(3), 216–232. doi:10.1287/orsc.10.3.216

Bishoff, L., & Allen, N. (2004). *Business Planning for Cultural Heritage Institutions*. Washington, DC: Council on Library and Information Resources.

Blackler, F. (1995). Knowledge, Knowledge Work and Organization: An Overview and Interpretation. *Organization Studies*, *16*(6), 1021–1046. doi:10.1177/017084069501600605

Brown, J. S., & Duguid, P. (1991). Organizational Learning and Communities of Practice: Towards a Unified View of Working, Learning, and Innovation. *Organization Science*, *2*(1), 40–57. doi:10.1287/orsc.2.1.40

Chan, Y. (2002). Why haven't we mastered alignment? The importance of the Informal Organization Structure. *MIS Quarterly Executive*, *1*(21), 76–112.

Chandler, A. (1966). *Strategy and Structure*. Garden City, NY: Doubleday & Company.

Dietz, J. L. G. (2004). Basic Notions Regarding Business Processes and Supporting Information Systems. In *the 5th BPMDS Workshop on Creating and Maintaining the Fit between Business Processes and Support System* (pp. 160–168). Riga, Latvia, 2004.

Earl, M. J. (1996). An Organizational Approach to IS Strategy Making. In M. J. Earl (Ed.), *Information Management - The Organizational Dimension* (pp. 136–170). Oxford, UK: Oxford University Press.

Earl, M. J. (2001). Knowledge Management Strategies: Toward a Taxonomy. *Journal of Management Information Systems, 18*(1), 215–233.

Fahey, L., & Prusak, L. (1998). The eleven deadliest sins of knowledge management. *California Management Review, 40*(3), 265–276.

Gherardi, S. (2000). Practice-based Theorizing on Learning and Knowing in Organizations . *Organization, 7*(2), 211–223. doi:10.1177/135050840072001

Goldkuhl, G. (2002). Anchoring scientific abstractions - ontological and linguistic determination following socio-instrumental pragmatism. In *European Conference on Research Methods in Business and Management (EcRM2002)* (pp. 147–156). Reading, 29–30 April 2002. Draft retrieved Feb 4, 2006, from http://www.ida.liu.se/~gorgo/engpub.html

Gorelick, C., & Tantawy-Monsou, B. (2006). For performance through learning, knowledge management is the critical practice. *The Learning Organization, 12*(2), 125–139. doi:10.1108/09696470510583511

Hackney, R., Burn, J., Cowan, E., & Dhillon, G. (2000). Challenging Assumptions for Strategic Information Systems Planning: Theoretical Perspectives. [CAIS]. *Communications of the Association for Information Systems, 3*(9), 2–24.

Hildreth, P. M., & Kimble, C. (2002). The duality of knowledge. *Information Research, 8*(1). Retrieved May 2, 2006, from http://informationr.net/ir/8-1/paper142.html

Lemke, J. L. (1993). Discourse, Dynamics, and Social Change. *Cultural Dynamics, 6*(1), 243–275. In M. A. K. Halliday (Ed.) *Language as Cultural Dynamic.* Leiden, the Netherlands: Brill. Retrieved January 6, 2006, from http://academic.brooklyn.cuny.edu/education/jlemke/cult-dyn.htm

Lewin, A. Y. (1999). Application of Complexity Theory to Organization Science. *Organization Science, 10*(3), 215. doi:10.1287/orsc.10.3.215

Luftman, J., & McLean, E. R. (2004). Key Issues for IT Executives. *MIS Quarterly Executive, 3*(2), 89–104.

Martinsons, M. G., & Davidson, R. (2003). Cultural Issues and IT Management: Looking Ahead. Guest Editorial. *IEEE Transactions on Engineering Management, 50*(1), 3–7. doi:10.1109/TEM.2003.808299

Merriam-Webster On Line. (2008). Retrieved October 5, 2008, from http://www.merriam-webster.com/

Opdahl, A. L. (1997). A model for comparing approaches to IS-architecture alignment. In B. Irgens, K. Ellingsen, E. Mathisen, N. M. Nielsen, & Z. Uddin Quazi (Eds.), *Proceedings of "Norsk konferanse om organisasjoners bruk av IT"* (pp. 67–79). Bodö, Norway, June.

Peltonen, T., & Lämsä, T. (2004). 'Communities of Practice' and the Social Process of Knowledge Creation: Towards a New Vocabulary for Making Sense of Organizational Learning. *Problems and Perspectives in Management, 2004*(4), 249–262.

Porter, M. (1996). What Is Strategy? *Harvard Business Review*, (November-December): 61–78.

Regev, G., & Wegmann, A. (2003). Why Do We Need Business Process Support? Balancing Specialization and Generalization with BPS Systems. Introductory note to *the 4th BPMDS Workshop on Requirements Engineering for Business Process Support*. Velden, Austria 2003.

Regev, G., & Wegmann, A. (2004). Remaining Fit: On the Creation and Maintenance of Fit. Introductory note to *the 5th BPMDS Workshop on Creating and Maintaining the Fit between Business Processes and Support System* (pp. 131–137). Riga, Latvia, 2004.

Schatzki, T. R. (2001). Introduction: Practice theory. In T. R. Schatzki, K. Knorr Cetina, E. von Savigny (Eds.), *The practice turn in contemporary theory* (pp. 1–14). London: Routledge.

Schatzki, T. R., Knorr Cetina, K., & von Savigny, E. (Eds., 2001). *The practice turn in contemporary theory*. London: Routledge.

Taxén, L. (2005). A Socio-technical Approach Towards Alignment. *Software Process Improvement and Practice*, *10*(4), 427–439. doi:10.1002/spip.245

Tuomi, I. (1999). Data Is More Than Knowledge: Implications of the Reversed Knowledge Hierarchy for Knowledge Management and Organizational Memory. In *Proceedings of the 32nd Hawaii International Conference on System Sciences* (pp. 1–12). Volume Track 1, 5–8 Jan. 1999.

Tuomi, I. (2002). The Future of Knowledge Management. [LLinE]. *Lifelong Learning in Europe*, *VII*(2), 69–79.

Virkkunen, J., & Kuutti, K. (2000). Understanding organizational learning by focusing on "activity systems." *Accounting . Management and Information Technologies*, *10*(4), 291–319. doi:10.1016/S0959-8022(00)00005-9

Wenger, E. (1998). *Communities of Practice. Learning, Meaning and Identity*. Cambridge, UK: Cambridge University Press.

Wenger, E., McDermott, R., & Snyder, W. M. (2002). *Cultivating Communities of Practice: A Guide to Managing Knowledge*, Boston, MA: Harvard Business School Press. Excerpt retrieved Oct 8, 2008, from http://hbswk.hbs.edu/archive/2855.html

Wertsch, J. V. (1991). *Voices of the mind: a sociocultural approach to mediated action*. Cambridge, MA: Harvard University Press.

Zack, M. (1999). Developing a Knowledge Strategy. *California Management Review*, *41*(3), 125–145.

ENDNOTES

[1] In the literature "knowledge strategy" is more or less used synonymously with "knowledge management strategy". In this chapter, I refer to knowledge strategy as a strategy that is strongly linked to the business strategy in order to emphasize knowledge as a strategic resource (Zack, 1999). By knowledge management strategy, I indicate strategies for managing knowledge without a direct coupling to the business strategy.

[2] The commercial and spatial schools are excluded since these are less relevant for the comparison.

Chapter 14
In Search of an Integrating Construct

Coordination, communication, business processes, business rules, information architectures, Information Systems (IS), and Information Technology (IT) are fundamental organizational realms subject to intense organizational inquiry. A typical passage from the vast body of organizational literature is the following:

A business process is a sequence of activities aimed at producing something of value to the business [...] business processes display a similar set of features [...] have at least one start point and at least one end point [...] can be broken down into a set of tasks or subactivities [...] a process has an objective [...] requires resources such as people and technology [...] spans several organizational units, each of which delivers some value to the process [...] there's a natural ordering of activities in a process because of internal dependencies [...] a process handles one or more things which could be physical objects or pieces of information [...]. (Morgan, 2002, pp. 35-36)

DOI: 10.4018/978-1-60566-192-6.ch014

This excerpt demonstrates several problematic positions:

- The definitions of basic realms in organizational inquiry are imprecise. A business process is "sequence of activities"; yet it seems to be inextricably associated with objective, people, technology, organizational units, and "things".
- There is a "natural ordering" of activities because of "internal dependencies". The problematic concepts here are of course "natural" and "internal". What do these concepts stand for? What does "natural" mean and what is "internal" the interior of?
- The concept of "business process" has an integrating character; although defined as a "sequence of activities" it is distorted into an entity that has precedence over objects, resources, people, technology, information, and so on.

From an ADT point of view, these positions demonstrate several things. First, there is a need for some integrating construct, but it is unclear what this construct is. Second, in the absence of such a construct the most near at hand, the business process, is assigned the role of the integrating construct. This has the effect that the original meaning of the concept "process" as a sequence of activities becomes overloaded with connotations that obscure its essence: signifying a temporal order. Third, the conceptual confusion creates insecurity about how basic organizational realms are interrelated. In general, different organizational units are responsible for business processes, information architectures, business rules and IS/IT. Moreover, it is not unusual today to outsource the IS/IT responsibility since this is not considered "core business". The result is a focus on separate organizational realms and less attention on how they mutually affect each other. Neglecting these interdependencies may cause severe misalignment in organizational development programs.

The purpose of this chapter is to analyze prevalent conceptions of basic organizational realms such as coordination, communication, business processes, etc., using ADT as a guiding framework. Based on this analysis, I will suggest the activity domain as the fundamental integrating construct in organizational inquiry, and discuss the implications of this proposal. I argue that the persistent confusion about the definitions of organizational realms is due to the lack of an integrating perspective. Taking any of these realms as a basic perspective for inquiry implies that a multi-dimensional problem is "compressed" into a one-dimensional one where other dimensions are concealed or unfocused. A consequence of this superficial way of approaching a multi-dimensional problem is that interdependencies are veiled. Without an integrative perspective that lays bare these interdependencies, and thus makes them accessible to operationalization, it is likely that the inherent difficulties in managing organizational change programs will be even more aggravated. This calls for a "decompression" approach where relevant dimensions are recognized per se, however dialectically related to each other:

When analyzing processes it is important to recognize that such analysis must be based [...] on a, necessarily, "multi-perspective" representation [...]. As regards the problem of intervention, [...] it is clear that the problem of redesigning [business processes] is [...] no different from the more general problem of organisational change, which latter consists of the complex search for mutual adaptations between tasks, structure (communications, authority, and workflow systems), people and technology. Thus it is also clear that (re)designing a process poses a socio-technical problem. (Biazzo, 1998, pp. 1012-1013)

In line with Goldkuhl, Röstlinger & Braf (2001), I claim that meaning is the key enabler of operationalizable, integrative perspectives:

If we look at BPR [Business Process Reengineering] and KM [Knowledge Management] together we can see that these approaches contribute with a claim for an integrated development of information technology, business processes and humans, and their knowledge and competences. How this integration should be done in practical development and change has seldom been thoroughly described. We think this is due to an insufficient theoretical integration of these different aspects. [...] We think that one key to the lack of such a theoretical integration in BPR and in KM literature is the lack of semiotic and linguistic awareness. (Goldkuhl, Röstlinger, & Braf, 2001, p. 2)

BUSINESS PROCESS RECONSTRUCTION

The definition of business process usually adopted is formulated by Bititci & Muir as "a collection of various tasks which produce an output" (Bititci & Muir, 1997, p. 366). This definition implies a transformational view on a process where an input is transformed into a product or service. Moreover, the definition also indicates a specific temporal dimension associated with processes.

The view on a process as a structured chain of activities has a direct coupling to coordination as defined by Malone & Crowston (1994). According to their definition, coordination is simply the management of the dependencies between these activities. This implies that coordination is an activity in itself carried out by some actors. The work object of the coordination activity is coordination manifested as various tangible and intangible elements in the organisation. For example, a business process model would be seen as a result of a coordination activity.

With the advent of the Business Process Reengineering (BPR) initiative, it soon became clear that it was necessary to include other organizational dimensions in BPR besides the temporal ordering of activities. For instance, Deakins & Makgill (1997) identify eighteen different themes related to BPR in the literature. However, disagreement persists about the definition of business processes and related concepts. For example, Biazzo reports on several definitions of BPR found in the literature (Biazzo, 1998). These definitions see BPR alternatively as modeling of old and new business processes, as alignment between IT and business processes or as the advancement of organizational performance by improving business process performance.

A recurrent theme in the literature is that the temporal dimension is considered insufficient to define a process. For example, Davenport & Short suggest that processes may be based on three dimensions (Davenport & Short, 1998, p. 53-56):

- *Organisational units*[1]*:* Processes take place between interorganizational, interfunctional and interpersonal business units. Interorganizational processes take place between two or more business organizations. Interfunctional processes exist within the organization, but cross several functional or divisional boundaries. Interpersonal processes involve tasks within and across small work groups, typically within a function or department.
- *Objects*: Processes manipulate objects that can be physical or informational. In physical object processes, real, tangible things are either created or manipulated as in, for example, manufacturing. Informational object processes create or manipulate information.

- *Activities*: Processes involve two types of activities: managerial (e.g. a budget process) and operational (e.g. fulfilling a customer order). Operational processes involve the day-to-day carrying out of the organization's basic business purpose. Managerial processes help to control, plan, or provide resources for operational processes.

Thus, there is a clear tendency to include other dimensions beside temporality in the definition of business processes. This is revealed by the linguistic determination used. For example, Chan states that "IT can be an initiator, a facilitator, and an enabler in [emphasis added] a business process" (Chan, 2000, p. 235). Other expressions, such as "coordination in the process", "information in the process", both abundant on the Internet, indicate a similar shift in meaning.

Another aspect of business processes is discussed by Marjanovic (2005). In dynamic business environments, the process may evolve during the process execution as a result of new needs. Customer demands may change as well as competitors and partners. This places new focus on KM. The dynamic, evolving and knowledge intensive processes have been termed "emergent business processes". In such processes, it is natural to "include general, specific and tacit knowledge distributed across experts and non-experts" (ibid, p. 477). In this conceptualization, knowledge is considered to be part of the process.

The transformational view on business processes has been challenged for neglecting the communicative aspects of coordination: "Coordination means an orientation towards the communicative interaction between different parties such as the customer and the producer" (Goldkuhl, Röstlinger, & Braf, 2001, p. 7). It is important to realize that the execution of a business process is dependent on commitments and agreements exchanged between the actors executing the process. Thus, from this perspective a process is considered as something that has two facets: a transformational one and a communicative one.

Analysis

It is customary to illustrate business processes with workflow diagrams as the one in Figure 1.

To the left and right, the input and output objects to the process are shown. These correspond to the object process dimension according to Davenport & Short (1998). The arrow-shaped figures in the center represent activities, that is, the activity process dimension. At the bottom, the organizational units – the organizational unit dimension – are shown. In addition, octagonal icons show the states of objects at various positions in the process. The arrows between activities indicate the flow of physical or informational objects between activities. Finally, arrows from organizational units to activities signify which organizational unit is responsible for which activity.

From the ADT perspective organizational units would most likely be interpreted as activity domains. The crucial question is whether there is a clearly identifiable motive and work object for these units, or if they are grouped together for some other reason. The "activities" in the model would probably also be conceived as activity domains depending on the granularity in the model. For example, the Account Management activity domain would be conceived as utilizing two other domains: "Define Business Opportunity" and "Sales". A natural criterion for when the recursive chain of activity domains shall be stopped is when all activities are entirely internal for an activity domain, in which case the activities would be renamed "actions".

Two activity modalities are visible in the process model: temporalization signified by the flow of activities from left to right, and spatialization signified by the input and output objects. However, these represent a limited view of spatialization, since there is no indication of how these objects are related to

Figure 1. A workflow model of a business process from Ericsson (Taxén & Svensson, 2005. ©Inderscience Enterprises Limited. Used with permission)

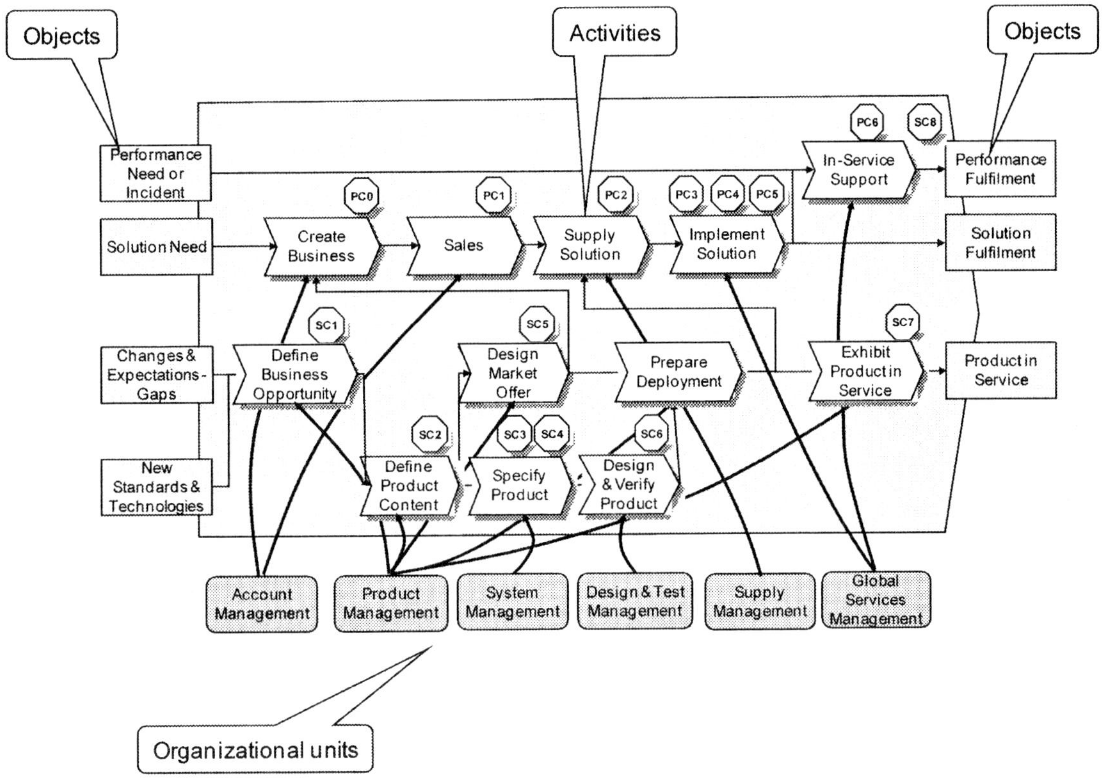

each other. This can only be shown in a model of the information architecture in the context considered. In Figure 2 the activity domains and activity modalities in the process model are shown.

The most severe problem with the process model in Figure 1 is that the interdependencies between the modalities are virtually impossible to cognize. First, there is no easily comprehended flow between the activities. The arrows go from left to right, bottom to top and the other way around without any visible logic. Moreover, the progression of the objects from input to output is not conceivable other than at the states signified by the octagons. From the ADT point of view, models like the one in Figure 1 "compress" too many things into one image, and without any appreciation of aligning the tokens along the modalities. The lateral direction displays mainly one modality – spatialization, while the longitudinal direction displays two modalities – temporalization and contextualization (the activity domains). In short, the construction of an actable communal meaning about the process model is difficult, to say the least.

Recently, a modeling notation called "swimlanes" has been used widely in business process modeling (see Figure 3). A swim lane (or swimlane) is a visual element used in process-flow diagrams that depicts what or who is working on a particular subset of a process. Swim lanes are arranged vertically and are used for grouping the sub-processes according to the responsibilities of those swim lanes (Wikipedia, 2008d).

In a swimlane diagram, the lateral divisions show responsible organizational units such as departments or functions. Along the longitudinal direction, "swimlanes" show which sub-processes or activities are executed by the units in lateral division. The lateral direction represents the sequence of events in the

Figure 2. Activity domains and activity modalities in the business process from Ericsson (Taxén & Svensson, 2005. ©Inderscience Enterprises Limited. Used with permission)

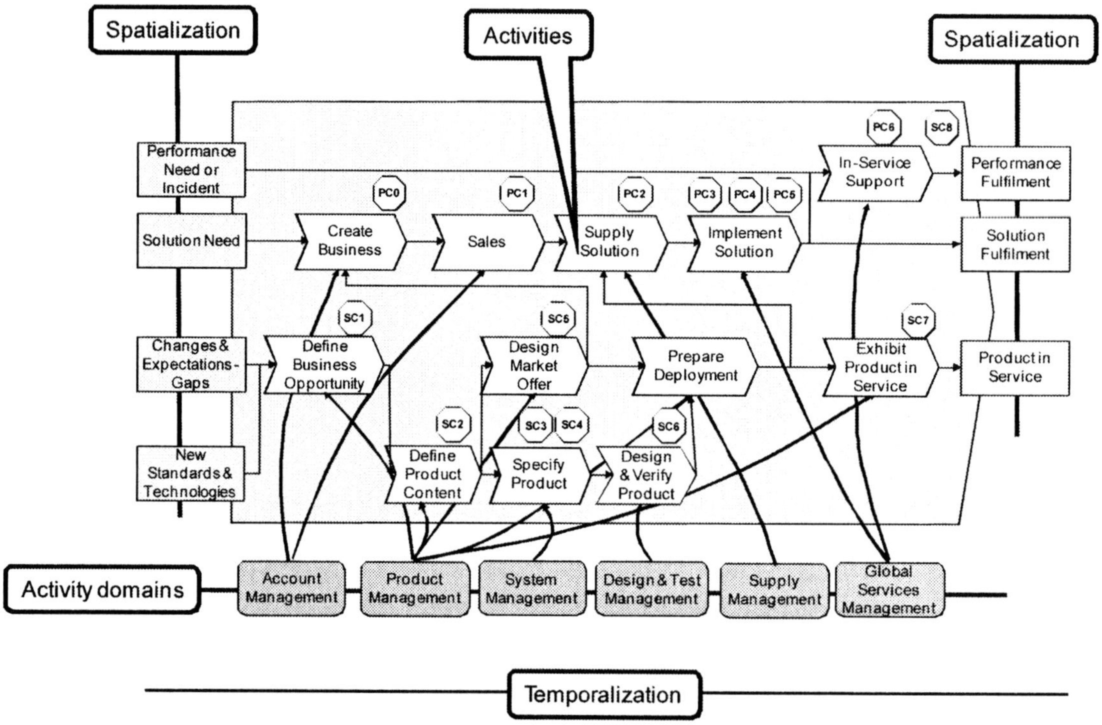

Figure 3. The swimlane

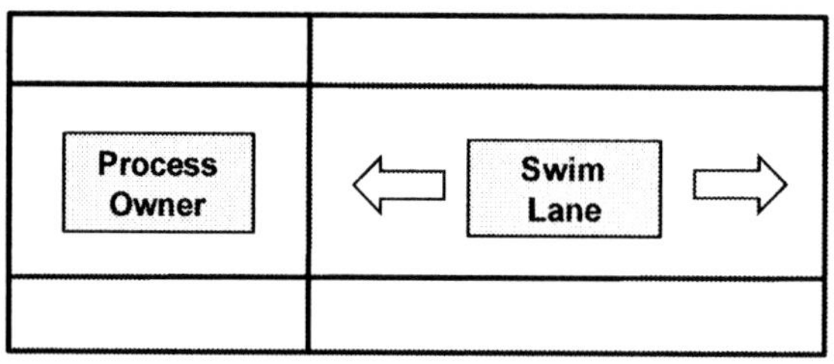

overall process, and depicts a time orientation from left to right. Arrows between the lanes represent how information or material is passed between the units and activities. The swimlane diagram can clarify who is responsible for each step and what information is passed between organizational units. Swimlanes are used in Business Process Modeling Notation (BPMN) and Unified Modeling Language (UML) Activity diagram modeling methodologies (Wikipedia, 2008d).

A swimlane corresponding to the business process in Figure 1 is shown in Figure 4.

From the ADT perspective, the swimlane is aligned with two major modalities and their interdependencies: the longitudinal – contextualization and the lateral – temporalization. The longitudinal division shows activity domains. A particular swimlane shows the order between activities for which the activity domain in the lateral division is responsible. Thus, the lateral direction is clearly a manifestation of temporalization. Moreover, each arrow crossing the border between two swimlanes indicates

Figure 4. The Ericsson business process as swimlanes

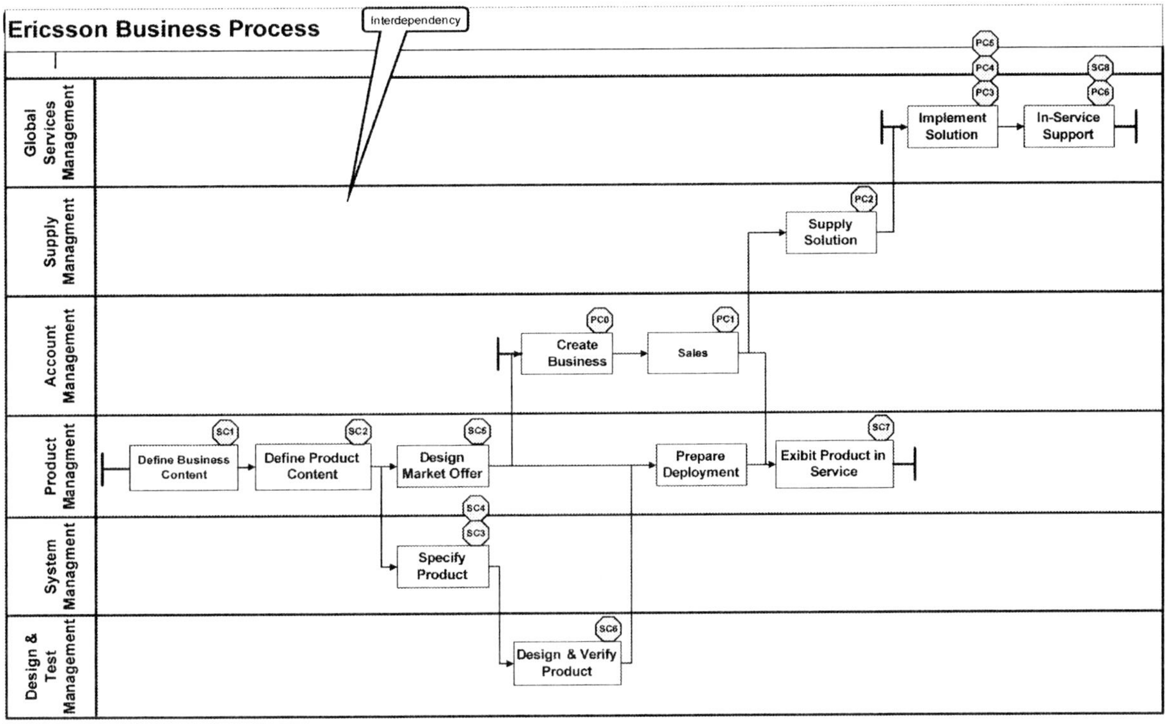

a manifestation of transition. At these crossings, potential transitions between activity domains might occur. It is evident that the swimlane layout is much easier to cognize that the model in Figure 1, since the lateral and longitudinal directions display only one modality each. Thus, the swimlane diagram is, from an activity modality point of view, better aligned with the modalities.

Still, there are some problems associated with swimlanes. The first is the ambiguity of the elements in the lateral and longitudinal directions. From the definition of swimlanes above, the longitudinal division shows "responsibility", while the lateral division contains "a particular subset of a process". However, from the ADT point of view, both may be activity domains that produce something and are responsible for their outcomes. From the social perspective inherent in the activity domain construct, it is not particularly relevant to single out the aspect of "responsibility" in a diagram. The responsibility is, so to say, built into the very concept of the activity domain. The second issue is that the spatialization modality is not visible in the swimlane diagram other than indirectly as text in the activities, for example, "Supply Solution". "Solution" is obviously an entity that has its place in the information architecture.

The absence of visualization of the spatialization modality makes the swimlane unsuitable for modeling information management capabilities in an organization. In contrast, such a representation is provided by the previously treated Information Interaction Model (IIM). In Figure 5, the same process as in Figure 1 and Figure 4 is depicted.

The scope of an IIM is an entire activity domain. The spatialization modality is aligned along the longitudinal direction as information entities and the temporalization modality along the lateral direction as activities. The interdependencies between the modalities are signified by the "score" element in

Figure 5. The Ericsson business process as Information Interaction Model (IIM)

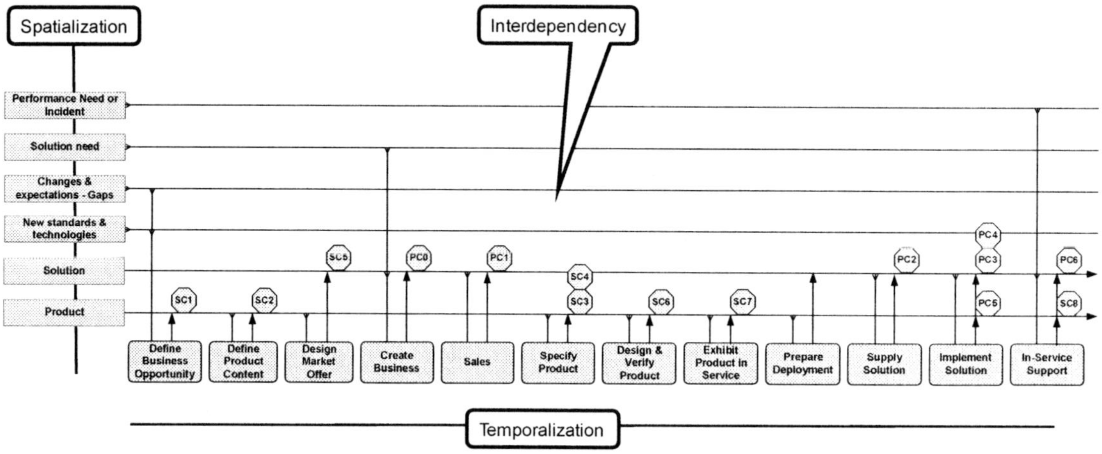

the diagram. Due to the repetitive nature of the activity domain each activity may in itself be an activity domain that, when opened up, displays the same structure.

The IIM has several appealing qualities. First, the work object of the domain is emphasized by the information carrying entities in the longitudinal direction. The transformative activities modifying the work object are laid out along the lateral direction. Also, the state progression of each information entity is clearly visible. This means that the IIM is well suited to specify requirements on information management projects such as implementing Enterprise Resource Planning (ERP) or Product Lifecycle Management (PLM) systems in an organization.

In summary, in the absence of an integrative construct, prevalent conceptualizations of business processes take on this role. In spite of acknowledging the inherent temporal dimension of the process concept, additional modalities are incorporated under the banner of "process" until the process concept becomes diluted and overloaded. Moreover, visualizations of processes are in general not aligned with the activity modalities, which make them hard to comprehend and consequently unsuitable for achieving common understanding. This in turn derails their potential as means of concerted actions in organizational change endeavors.

The modeling nomenclature in the swimlane concept is a step in the right direction, since the lateral and longitudinal directions are aligned with two modalities. Moreover, the interdependencies between the modalities are clearly shown in the visual element of the swimlane. However, as pointed out, from an information management initiative, there are other drawbacks with the swimlane model. In such initiatives, the IIM diagram layout appears to be more appropriate.

The main implications from the analysis in this section are as follows. The activity modalities represent fundamental dimensions by which coordination is cognized. In order to make sense of complex coordination situations, these modalities must be represented by artifacts such as diagrams in such a way that it is easy to cognize each modality, and, in addition, how these are interdependent. Prevailing process representations such as the one in Figure 1 do not acknowledge this.

FROM INFORMATION SYSTEM DEVELOPMENT TO DOMAIN CONSTRUCTION

In the previous section several types of business processes models were analyzed from the ADT perspective. I argued that the models aligned with the activity modalities and their interdependencies are powerful means for achieving common understanding of the process. In this section, I will pursue the same line of inquiry for IS/IT implementation projects, or IS development (ISD) projects for short[2]. The purpose of this inquiry is to substantiate my claim for the activity domain as an integrating construct; now with the focus on the IS as a mediational means for the coordination of transformative intra- or interorganizational actions.

IT is a key enabler for BPR (Davenport, 1993). In their extensive examination of the BPR literature, Deakins & Makgill found that IT and its implementation in organisations dominated the research agenda (Deakins & Makgill, 1997) rather than human aspects. In spite of documented importance, the operationalization of such aspects seems to be under-researched:

Of particular concern [...] is the lack of attention given to so-called "people issues". [One] might have expected that HR [Human Resource] and change management issues would have featured in the literature as strongly as implementation and IT issues. However, this paper has clearly illustrated that the clarion call for researchers to concentrate on people issues in BPR has so far gone unheeded. (Deakins & Makgill, 1997, p. 104)

With the advent of ERP systems such as SAP, Baan, Oracle, and People-Soft in the early 1990s, the focus on information technology was even more accentuated. These systems promise, according to Davenport, "the seamless integration of all the information flowing through a company – financial and accounting information, human resource information, supply chain information, customer information" (Davenport, 1998, p. 121).

Issues

According to The Gartner Group, 70 percent of all ERP projects fail to be fully implemented, even after three years (Gargeya & Brady, 2005, p. 501). Davenport claims that the main reason is the failure of companies to reconcile the technological imperatives of the ERP system with the business needs of the enterprise itself. The system drives the company towards full integration and generic processes even when customized processes and some level of business unit segregation may be a competitive advantage (Davenport, 1998, p. 122).

The core of the dilemma is how to operationalize the coordination of business units in order to fulfill enterprise goals. There is a need to arrive at a communal vision of joint processes, information sharing, ways of working, etc. This might require efforts that are more costly than the up-front purchase of the ERP package (Gosain, Lee, & Kim, 2005). Moreover, while the ERP system may contribute to the internal coherence of the company, it may also create problems in the supply chain due to sub-optimization (Cumbie, Jourdan, Peachey, Dugo, & Craighead, 2005). This problem is aggravated due to company mergers, acquisitions, etc. Thus, the effect of ERP on supply chains is a major topic for future research (ibid).

Communal Meaning

The need to establish communal meaning is a persistent theme throughout this book. Achieving communal meaning becomes increasingly difficult when generic and decontextualized models are to be implemented in a particular organisation:

A more contentious issue is whether a set of generic business processes can be defined with universal applicability. [At] an abstract level, some consensus may be achieved over a generic set of business processes. However, it is also becoming evident that as the level of detail increases, disagreements begin to surface. Since most enterprises are concerned with detailed operational models rather than abstract models, the value of a top-down, i.e. generic, approach to business process definition is becoming increasingly questionable. (Bititci & Muir, 1997, p. 366)

Moreover, what actors hold for true about coordination is socially determined. This means that every organization has to work out how to operationalize coordination and how it is to be supported by IS/IT:

In order to analyze a situation in terms of coordination, it is sometimes important to explicitly identify the components of coordination in that situation. [...] It is important to realize that there is no single 'right' way to identify these components of coordination in a situation. (Malone & Crowston, 1994, p. 101)

Thus, in order to implement IS support for coordination it is necessary to include the construction of communal meaning in the project, regardless of whether that is done in a systematic way or just ad hoc, which seems to be the prevalent strategy in many projects.

Communication

An essential aspect of coordination is communication: "The capacity of an organisation to manage complex patterns of interdependent activities is closely related to its capacity to manage the communication required for coordination" (March & Simon, 1958, p. 183). Of course, communication is a vast topic in itself, which is far beyond the scope of this book. Here, it suffices to realize that the actors need to have a common language to enable coordination:

A practice is coordinated through communication. Different linguistic actions are necessary in order to coordinate actions so that the intended result can be produced. This is necessary within a practice in which several producers cooperate. (Goldkuhl & Röstlinger, 2005, p. 6, my translation)

March & Simon also point out that the language needed for communication is well developed when it refers to concrete things or things that have acquired a state of "facts" in the organisation. When this is not the case, the communication is much harder:

[It] is extremely difficult to communicate about intangible objects. Hence, the heaviest burdens are placed on the communications system by the less structured aspects of the organization's tasks, particularly by activity directed towards the explanation of problems that are not yet well defined. (March & Simon, 1958, p. 186)

The importance of communication for establishing communal meaning in coordination situations is stressed by Melin:

Communication about other actions (of coordinating and coordinated character as well as executed or planned) is important to the establishment of a common understanding in the actual coordination situation. (Melin, 2002, p. 402, my translation)

Thus, we can talk about at least two principal functions of language: communication in order to establish communal meaning and communication for coordination of actions when communal meaning is established, preferably through various speech acts. As has been pointed out earlier, these two functions are called mathetic and pragmatic by Halliday (1975).

The distinction between mathetic and pragmatic functions of language has also been discussed by Habermas. In his Social Action Theory, Habermas (1984) coined the notion of communicative action, which is usually defined as "actions towards mutual understanding (Verständigung) whose goal is the coordination of the actions of the participants" (Weigand, van der Poll, & de Moor, 2003).

According to Weigand & Dignum (1997), Habermas's definition of communicative action consists of two elements: conversational action and consensual action. In conversational action, the emphasis is on achieving common understanding. The actors exchange messages without a particular coordination goal in mind. Conversational action contributes to a mutual definition of a situation at hand. In this type of communicative action, the mathetic function of language is emphasized. Consensual action has a clear coordination objective and is based on an established, common definition of the situation. In this type of communication, the pragmatic function of language is emphasized.

IS Development from the Activity Domain Theory Perspective

From the ADT perspective an information system is a mediational means supporting the coordination of transformative actions in a particular activity domain. According to the experience from Ericsson, at least the following issues need to be resolved:

- What items are subject to coordination? In a product development organisation such items are typically products and documents, drawings, CAD-models, requirements, engineering change orders, error reports, software builds, test cases, etc.
- What are the relationships between these items?
- What properties characterize an item?
- What is the status of an item at any particular instant of time?
- How should the responsibilities for managing items (creating, deleting, relating, etc.) be allocated?
- What guidelines, rules, norms, etc., are valid in the domain?
- What are the dependencies between activities operating on the items?

In order to resolve these issues, the IS needs to be enacted, which means that capabilities of both the actors using the IS and capabilities of the IS itself must co-evolve into a mutual resource. Thus, it is not enough to focus on the technical capabilities of the IS. The objectivation component, i.e., the manifestation in the minds and bodies of engaging with the IS is necessary to consider in ISD. An inherent aspect of objectivation is, as pointed out earlier, communal meaning about how to implement the IS.

An important corollary of this stance is that there is no such thing as COTS (Commercial, off-the-shelf) systems, i.e., ready-made and available IS applications for immediate use without modification. First, the objectivation component, the establishing of a human capability, must always be enacted. Second, it is very unlikely that the IS can be used without configuration or adaptation in some way, since that would imply a one-sided, objectivation adaptation, the burden of which falls entirely on the actors and the organization. In other words, a COTS system taken literally would imply a complete submission of the organization to the COTS system, something that would be extremely awkward. Even though a commercial system is sold as "out-of-the-box" or as a "vanilla installation"[3], the objectivation component implies that every system is always domain specific.

A second corollary concerns the so called "best practice" methodologies that come with commercial ERP systems like SAP R/3. A SAP package comprises more than 5000 different parameters that may have to be configured to the needs of an organization. In order to alleviate this process, "reference models" are included in the package. Reference models "benefit the customer by utilizing business process knowledge and best practices" (Scheer & Habermann, 2000, p. 59). "Best practice" is derived from analyzing implementations in a variety of sectors such as automotive, chemicals, industrial machinery, pharmaceuticals, etc., and in so called cross-industry packages such as business planning and consolidation, business process management, supply chain management, and strategy management. Reference models are documented as Event-driven Process Chains (EPC).

What do the reference models actually represent? From the ADT perspective, these models originate from a large sample of EPCs enacted in various organizations. When the EPCs are taken out of context and consolidated into a reference model, this model has in fact never been enacted in any organization. It represents the alienated expression of a multitude of enacted models that, when in use, include both the objectivated human component, and the objectified external component – the EPC. In essence, a reference model is just a diagram as alien to actors in an organization as the hieroglyphs were before the Rosetta stone and the deciphering done by Jean-François Champollion. A reference model has nothing to do with knowledge. It may or may not be a proper point of departure for enacting a resource in an organization. That, however, requires that the actors need to understand the nomenclature in EPC diagrams, modify the reference model to their needs, and configure or adapt the SAP system. Only then will the modified reference model become an objectified component in the enactment process. Thus, the value of reference models must be critically assessed:

While at the abstract level the idea of "universal" best practices may be seductive, at the detailed process level these mismatches create considerable implementation and adaptation problems. (Kumar & Van Hillegersberg, 2000, p. 25)

Constructing the Activity Domain

In ISD it is possible to distinguish several "tendencies". The 1st tendency concentrated on the technical aspect of the IS (Iivari & Lyytinen, 1998). A clear separation was made between users and designers. In the 2nd tendency, which was largely influenced by Scandinavian researchers, the use context of the IS became more pronounced, for example, in design approaches such as the socio-technical, the trade-unionist, the professional work practice approach and others (ibid). However, in all these approaches, the IS was still the target of design, and the context, the activity domain, was more or less taken for granted.

The ADT approach implies a focal change from the IS to the activity domain as the development target. It is not only the IS that is developed/constructed, but *the entire activity domain,* including the construction of communal meaning about how coordination should be conceived. This approach might be regarded as a 3rd tendency in ISD. Thus, ISD is conceptualized as an enactment process where both objectified manifestations in the form of IS capabilities, and objectivated manifestations in the form of user capabilities to use the IS, are constructed. This means that all actors performing coordinative acts are contributing to the IS design, some more, and some less. The users are but one of several groups of actors participating in the co-construction of the domain and accordingly, the IS.

In ADT, enactment continues along the activity modalities, which implies that the target of the construction is the objectification of all modality dimensions, i.e., spatialization, temporalization, stabilization, and transition. Contextualization is inherent in the prerequisites: that the IS should be enacted into a resource in the context framed by the activity domain. In ordinary business language, this means that the IS should be capable of managing the elements in the information architecture (spatialization), the activities in the business process (temporalization), the business rules (stabilization), and the boundary objects signifying the interactions between domains (transition). Since the activity modalities are assumed to be dialectically related, the IS must also implement the interdependencies between the modalities.

In practice, the means used in the construction are usually a subset of the models representing the activity modalities: the spatialization model, the temporalization model, the stabilizing core, and the transition model. A general property of these models is semiotic expressiveness. Sign-tokens like icons, accompanying text, etc., should be easy to understand and unambiguous. Care should be taken to use the same graphical layout of icons in the models and the IS to alleviate meaning construction.

As described in the Section "*The domain construction strategy*", p. 131, the DCS can be roughly divided into three distinct phases: the mathetic, the consolidation, and the pragmatic ones. In the mathetic phase, an initial, tentative communal meaning is constructed. Information models, rules, etc., are suggested and tried out frequently. Thus, in this phase, the primary property needed from the IS is flexibility in terms of ease of changing the implementation of models. In the consolidation phase, a "scaffolding" domain is constructed that eventually will become operational. A key issue is to make actors confident with the IS and the data in it. Important IS properties in this phase are acceptable performance worldwide, and interfaces to ISs in other domains. In the pragmatic phase the prime property is the ability to manage instances of all items and to change the implementation according to new needs encountered.

Coordination in the conceptualization of the ADT implies that an activity domain is regarded as a resource that can used in other domains. Even if the same IS platform is used in these domains, the enactment of IS capabilities will differ depending on the motive of the domain, its history, the pre-understanding of its actors, etc. Thus, if an item is relevant in several domains, it will be characterized differently. For example, a product in marketing will emphasize different characteristics of the product than the same product in production. This means that the in-built information model of the IS platform should recognize *contexts* in order to discriminate what properties are relevant. Such an information model has been proposed by Parsons (1996). Established modeling languages, such as UML, do not appear to recognize context as a key modeling construct.

Enabling Mechanisms

As pointed out earlier the practical results achieved by applying the DCS in extraordinarily complex development tasks were unparalleled in the history of Ericsson in terms of functionality and effort spent. Several IS applications were developed that supported coordination on a global basis. Based on these experiences the following enabling mechanisms can be identified:

- *Focusing on construction of communal meaning*: develop the system in many iterations ("daily build"); fast design and test cycles; engage main users as partners in the development; iconic compliance between the IS and the spatialization model; provide "active" communication to users globally through event triggered information such as e-mail notification of major events; provide "passive" communication by enabling user access to the same information everywhere at all times; implement "discussion group" mechanisms regarding engineering change orders; etc.
- *Taking contextual aspects seriously*: allow the construction of several coordination domains on the same IS platform; provide users with means to define context specific views and reports; take measures to implement cross-domain coordination.
- *Pragmatic approach*: usefulness rather than "correctness" for evaluating IS functionality; prepare for discovering new ways of working as the project proceeds; allow failures, create an experimental atmosphere.
- *Technical qualities of the IS*: ease of adapting the IS platform (changing types, attributes, relations ...); fast global access through Java Servlet based web-techniques to improve performance worldwide; changes "on the fly", no need to close the IS database at changes; interpretative programming language to speed up prototyping.

CONCLUSIONS: THE ACTIVITY DOMAIN AS AN INTEGRATING CONSTRUCT

In this chapter, I have investigated the activity domain as an integrating construct. Integration, "the act of combining into an integral whole", is a necessity in complex development tasks. If the multitude of seemingly unrelated things and events in such tasks cannot be conceptualized from an integrating perspective, the management of such tasks becomes overwhelming.

In the first section, I analyzed various process models from the ADT perspective and found that the ways such models are illustrated indeed make a difference. If such models try to capture too many activity modalities in one diagram, they become virtually incomprehensible, which diminish their capability to mediate consorted actions. Conversely, if the process model diagrams are structures in such a way that modalities and their interdependencies are clearly recognized, the potential of such diagrams to mediate actions increases.

In the second section, the searchlight is directed to IS development of applications supporting coordination. The IS is conceived as a mediational means that provides capabilities to manage all activity modalities and their interdependencies. The enactment of these capabilities is done according to the domain construction strategy, which is geared towards the mutual development of ISs and human capabilities for coordinating transformative actions in activity domains. In addition to these two areas, the anatomy-centric approach provides further evidence for the activity domain as an integrating construct. The anatomy, the increment plan, and the integration plan are well aligned with different activity modalities and their interdependencies.

In conclusion, there are significant arguments for the viability of the activity domain as an integrating construct.

REFERENCES

Biazzo, S. (1998). A critical examination of the business process re-engineering phenomenon. *International Journal of Operations & Production Management, 18*(9/10), 1000–1016. doi:10.1108/01443579810225586

Bititci, U., & Muir, D. (1997). Business process definition: a bottom-up approach. *International Journal of Operations & Production Management, 17*(4), 365–374. doi:10.1108/01443579710159950

Chan, S. (2000). Information technology in business processes. *Business Process Management Journal, 6*(3), 224–237. doi:10.1108/14637150010325444

Cumbie, B., Jourdan, Z., Peachey, T., Dugo, T. M., & Craighead, C. W. (2005). Enterprise Resource Planning Research: Where Are We Now and Where Should We Go from Here? [JITTA]. *Journal of Information Technology Theory and Application, 7*(2), 21–36.

Davenport, T. (1993). *Process Innovation: Reengineering Work through Information Technology.* Boston: Harvard Business School Press.

Davenport, T. (1998). Putting the Enterprise into the Enterprise System. *Harvard Business Review, 76*(4), 121–129.

Davenport, T., Short, J. (1998). The New Industrial Engineering: Information Technology and Business Process Redesign. *IEEE Engineering Management Review, Fall*, 46–60.

Deakins, E., & Makgill, H. (1997). What killed BPR? Some evidence from the literature. *Business Process Management Journal, 3*(1), 81–107. doi:10.1108/14637159710161594

Gargeya, B., & Brady, C. (2005). Success and failure factors of adopting SAP in ERP system implementation. *Business Process Management Journal, 11*(5), 501–516. doi:10.1108/14637150510619858

Goldkuhl, G., & Röstlinger, A. (2005). *Praktikbegreppet - en praktikgenerisk modell som grund för teoriutveckling och verksamhetsutveckling.* [The practice notion – a practice generic model as foundation for theory development and development of organizations]. Department of Computer and Information Science, Linköping University (in Swedish). Retrieved Feb 4, 2006, from http://www.ida.liu.se/~gorgo/engpub.html

Goldkuhl, G., Röstlinger, A., & Braf, E. (2001). Organisations As Practice Systems - Integrating Knowledge, Signs, Artefacts And Action. In *IFIP 8.1 Working Conference Organisational Semiotics: evolving a science of information systems*. July 23–25, 2001, Montreal, Canada. Draft retrieved Feb 6, 2006, from http://www.ida.liu.se/~gorgo/engpub.html

Gosain, S., Lee, Z., & Kim, Y. (2005). The management of cross-functional inter-dependencies in ERP implementations: emergent coordination patterns. *European Journal of Information Systems, 14*, 371–387. doi:10.1057/palgrave.ejis.3000549

Habermas, J. (1984). The Theory of Communicative Action, (Volume 1, *Reason and the Rationalization of Society)*. Boston, MA: Beacon Press.

Halliday, M. (1975). *Learning How To Mean*. London: Edward Arnold.

Iivari, J., & Lyytinen, K. (1998). Research on Information Systems Development in Scandinavia - Unity in Plurality. *Scandinavian Journal of Information Systems, 10*(1&2), 135–186.

Kumar, K., & van Hillegersberg, J. (2000). ERP experiences and evolution. *Communications of the ACM, 43*(4), 23–26. doi:10.1145/332051.332063

Malone, T., & Crowston, K. (1994). The Interdisciplinary Study of Coordination. *ACM Computing Surveys, 26*(1), 87–119. doi:10.1145/174666.174668

March, J. G., & Simon, H. A. (1958). *Organizations*. Cambridge, MA: Blackwell Publishers.

Marjanovic, O. (2005). Towards IS supported coordination in emergent business processes. *Business Process Management Journal, 11*(5), 476–487. doi:10.1108/14637150510619830

Melin, U. (2002). *Koordination och informationssystem i företag och nätverk.* [Coordination and Information System in Companies and Networks]. Linköping Studies in Information Science, Dissertation No. 6, Institutionen för datavetenskap, Linköpings universitet (in Swedish).

Morgan, T. (2002). *Business Rules and Information Systems: Aligning IT with Business Goals*. Boston: Addison-Wesley.

Parsons, J. (1996). An Information Model Bases on Classification Theory. *Management Science, 42*(10), 1437–1453. doi:10.1287/mnsc.42.10.1437

Scheer, A.-W., & Habermann, F. (2000). Making ERP a Success. Using business process models to achieve positive results . *Communications of the ACM, 43*(4), 57–61. doi:10.1145/332051.332073

Taxén, L., & Svensson, D. (2005). Towards an Alternative Foundation for Managing Product Life-Cycles in Turbulent Environments. [IJPD]. *International Journal of Product Development, 2*(1-2), 24–46. doi:10.1504/IJPD.2005.006667

Weigand, H., & Dignum, F. (1997). Formalization and rationalization of communication. In *Proceedings of the Second International Workshop on Communication Modelling, the Language/ Action Perspective (LAP'97)* (pp. 71–86). Veldhoven, June 9–10, The Netherlands.

Weigand, H., van der Poll, F., & de Moor, A. (2003). Coordination through Communication. In *Proceedings of the 8th International Working Conference on the Language-Action Perspective on Communication Modelling (LAP 2003)* (pp. 115–134). Tilburg, The Netherlands, July 1–2, 2003. Retrieved Feb 6, 2006, from http://infolab.uvt.nl/research/lap2003/

Wikipedia (2008d). Retrieved Aug 15, 2008, from http://en.wikipedia.org/wiki/Swim_lane

ENDNOTES

1. Davenport & Short (1998) use the term "entities" instead of "units". However, I wish to reserve the term "entities" for other purposes in the book, and have therefore changed "entities" to "units" when referring to Davenport & Short.
2. I will use the term "IT" to denote infrastructural capabilities, including the IS platform as it is provided by the vendor. "IS" I will take as a generic term for computer based applications on the platform, aiming at supporting coordination.
3. Vanilla installation of software: installing the initial version without adaptations or configurations, usually by accepting all the default options.

Chapter 15
In Conclusion

It is now time to look back and recapitulate what has been achieved. The book can be seen as a parable from practice to practice over troubled water (see Figure 1).

The story begins with a confused group of people, faced with the problem of figuring out how to coordinate very complex development projects in the development of the 3rd generation of mobile systems at Ericsson. Due to a number of historical and contemporary coincidences – the early projects in the 1990s using the integrated centric development approach; the encounter I had with the Matrix information system in 1996; the conviction of one project manager to use this system for supporting coordination; and my personal role in developing the coordination support during the early phases in the 3G projects – a quite new way of managing coordination evolved during the years 1996 – 2003. This story is described in Section 1 of the book.

During that time there was not much time for thinking about what happened; things just had to be done, and the projects did not wait. However, I slowly began to realize that the course of events might have wider and more general consequences for how to conceive coordination of complex development tasks. Again, due to quite an unexpected event – the threat of ousting the Matrix system from Ericsson – I began my postgraduate studies at Linköping University simultaneously with my work at Ericsson. From the sometimes chaotic events that took place on the worried surface, I began to look beneath, trying to find the essence behind seemingly unrelated phenomena.

DOI: 10.4018/978-1-60566-192-6.ch015

Figure 1. The journey from practice to theory and back to practice

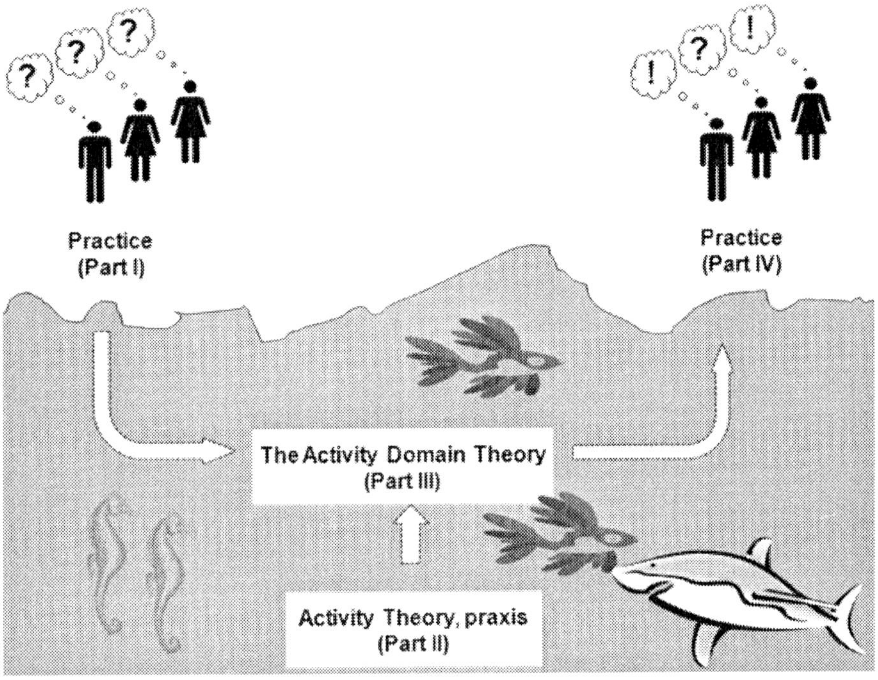

At the university, I gradually became acquainted with the theoretical instruments that was necessary to understand what was happening at Ericsson. A guiding searchlight, hinting at where to search for such instruments, was my own lifelong interest in the thinking of young Marx and the praxis philosophers that followed. I slowly became aware of a long historical tradition of philosophers and scholars working in the action oriented paradigm, and the depths and breaths of their thinking. Above all the Russian Theory of Activity appeared to be in concordance with my own thinking. The story of how the theoretical foundation of ADT was shaped is recapitulated in Section 2 of the book.

A prosperous interaction between the Ericsson practice and theoretical insights took place from 1998 to 2003; the year I put forward my Ph.D. thesis, and, more or less simultaneously, left Ericsson. The result of this period became the main contribution to the book: the Activity Domain Theory, described in Section 3.

Armed with the ADT, it was time to return to the surface and take a look at practice again; this time with the ADT as new glasses. The findings from this work are reported in Section 4 of the book. If these findings are convincing, which I hope they are, ADT might illuminate some issues and point to new ways of managing complex projects and systems. This, however, remains to be seen.

So, what has been achieved? Which are the main findings from the endeavor related in this book? I believe that these can be summarized as follows.

THE ACTIVITY DOMAIN AS AN INTEGRATING CONSTRUCT

There seems to be an endless flow of information, tools, consulting services, and research efforts aimed at improving organizational capabilities to meet increasingly tougher demands. These capabilities center around areas we have encountered over and over again in this book: business processes, information systems, enterprise architectures, PLM systems and so on. Yet, there are serious doubts if these efforts really make a difference in practice:

Numerous frameworks have emerged to support large scale organizations and government entities but to date there has been no empirical support to determine if they meet the needs of their users. [It] is not clear whether these frameworks [...] meet the needs of their users to identify areas for change, help to specify computer systems to implement those changes, and whether the changes actually result in organizational performance improvement. (Cane & McCarthy, 2007, p. 437)

The message in this book is that these shortcomings may be overcome by taking the activity domain as the core organizational entity. The activity domain frames the social context in which actors provide outcomes fulfilling social needs. It integrates transformative and coordinative actions, communal meaning, and the enactment of human and mediational capabilities. Moreover, the activity domain is a recurrent structure that can be scaled from teams all the way to the extended enterprise. From these characteristics, the activity domain might be considered as the "DNA of the organization"; the building block from which the coordinative properties of all organizational units can be derived.

COMMUNAL MEANING TO THE FOREFRONT

The experience from Ericsson as well as observations reported in the literature clearly indicates that common understanding must be treated more systematically than today. The construction of communal meaning needs to be considered as an engineering task which must be informed by strategies, methods, and support systems. However, the intangible nature of the target implies that this construction task is inherently more difficult than ordinary construction tasks. In ADT, the domain construction strategy is suggested as one way to achieve communal meaning.

THE ACTIVITY MODALITIES: BRIDGING INDIVIDUAL COGNITION AND SOCIAL REALITY

A recurrent theme in the organizational discourse is how to conceive the relation between the individual and the organization. A common conceptualization is to see the individual as the "micro" level and the organization as the "macro" level (e.g. Wiley, 1988). However, as Wiley (ibid) points out, this conception is problematic since the notion of "level" is not unambiguously defined. Other conceptualizations regard the organization in an anthropocentric light; organizations are "actors", they have "memories", they "learn" things, and, in the extreme, are considered as living entities (Örtenblad, 2005).

Weick views organizations as "collections of people trying to make sense of what is happening around them" (Weick, 2001, p. 5). The view presented in this book share this view, but from a different perspective. The activity modalities are seen as phylogenetically inherited treats that humans make use of in coordinating actions, whether in isolation or together with others. In order to act, we frame a situation by focusing on some object; we filter out relevant phenomena in the situation, and how these are related to each other; we figure out a sequence of actions; we make use of tried-out ways of acting; and we switch between situations when needed. These are basic dimensions of coordinating actions in every situation; dimensions that I call activity modalities: contextualization, spatialization, temporalization, stabilization, and transition.

The concrete manifestation of these modalities takes place in activity domains. By engaging with available means in the domain, congruent objectified impression in the domain and objectivated impressions in the minds and bodies of the actors are developed, regardless of the scope of the domain. Two people working on some shared object will leave the same modality specific type of traces behind as a number of organizations cooperating in the extended enterprise.

Thus, the issues of "levels" and "humanized" organizations as actors disappear. The same construct, the activity domain, frames situations where the activity modalities bridge individual cognition and the social reality constructed in the domain.

PAY HEED TO THE UNITY OF OPPOSITES

When we conceive two objects as related in some way, it is customary to analyze this situation in either of two ways. We may look at one object at a time, paying less attention to their relationship – a reductionist approach. Alternatively we may see the two objects as a unity and downplay the individual identity of each object – a holistic approach. In ADT, a third way is suggested – the unity of opposites. In this approach, each object maintains its identity, but one cannot exist without the other and they mutually constitute each other.

"Unity of opposites" is the central category of dialectics, suggesting that the existence or identity of a thing (or situation) depends on the co-existence of at least two conditions which are opposite to each other, yet dependent on each other and presupposing each other, within a field of tension (Wikipedia, 2009d):

[Every] actual thing involves a coexistence of opposed elements. Consequently to know, or, in other words, to comprehend an object is equivalent to being conscious of it as a concrete unity of opposed determinations. (Hegel, 1830)

The most conspicuous occurrences of the unity of opposites in ADT are the activity modalities. These modalities are distinct, yet dialectically interrelated. They all constitute each other within a concrete totality, the activity domain. I have discusses the harmful consequences of modality "compression", which is the tendency to include all modalities under the temporalization modality in the shape of the business process.

The practical consequence of this way of conceiving the modalities is that manifestations of them, such as information architectures (spatialization), business processes (temporalization), business rules (stabilization), and business support systems (mediational means) cannot be developed in isolation from each other. The interdependencies between these organizational artifacts must be given more attention than what is customary today, where separate and detached organizational responsibilities are allocated to each modality (cf. the role of "process owner"). The obvious way to allocate responsibility is to the activity domain. Such an arrangement would indicate that the interdependencies between modalities need to be managed in addition to each modality on its own.

Another example of the unity of opposites is the enactment of capabilities employed as resources in the activity domain. As described, enactment is a dialectical process where human capabilities – what I have called objectivation – and capabilities of means – called objectification – are manifested in the domain. This means that these two aspects, objectivation and objectification, cannot exist in isolation. In practice, this is quite obvious. Take, for example, the case when a new information system is introduced into an organization. The inherent capabilities in the system are ineffectual until actors have learnt how to use it for the benefit of the organization. Conversely, an employee who is proficient in using the system is virtually a lame duck if the system is not in place in the organization.

A final example is the interdependencies between practice and reflection. In the ADT perspective these aspects of action cannot exist in isolation. Unfortunately, we have institutionalized such a separation on the institutional level in business/industry on the one hand, and academia/university on the other. My own experiences are quite clear on this point. After working for many years in industry, I acquired a deep understanding of the complexity of the practical problems in coordinating large, globally distributed telecom projects. This kind of understanding I did not find at the university, where the focus was either on analyzing such projects or to apply theories on very simplified, practical problems. The real issue at stake, to operationalize theories in order to inform complex development projects, was not addressed.

On the other hand, the interest for theory development, which could illuminate the often chaotic situation in the industrial practice, is not highly honored in industry. So, we are left with the highly unsatisfactory situation with a practical focus in industry, underestimating reflection, and a reflective focus in academia, underestimating practice. One way to overcome this dilemma might be to encourage careers with alternating periods of working in industry and academia, in line with my own experiences.

CONCLUDING WORDS

Men make their own history, but they do not make it as they please; they do not make it under circumstances chosen by themselves, but under circumstances directly encountered, given and transmitted from the past. (Marx, 1852, p. 398)

Although seemingly embedded in sophistic and esoteric language, the message of ADT is in fact very simple: apply the same experiences we have in everyday life to the management of complex systems! We make use of contextualization, spatialization, temporalization, stabilization and transition every moment of our life without paying attention to it. The ADT is in essence only an application of this perspective onto the organizational arena. But, as in many other situations, seeing the well-known and familiar from a different perspective is not easy. It demands abandoning of the previous taken for granted, letting loose established truths, and reorientation to the new paradigm. Once done, however, new things and relations appear on the arena; a new paradigm emerges that over time becomes as self-evident as the old one once was.

In a way the meeting between a controversial 19th century philosopher and the industrial reality of the 20th century, which resulted in the ADT, is a true serendipity. It will never happen again exactly the same way. However, as Marx once said: "Hegel remarks somewhere that all great, world-historical facts and personages occur, as it were, twice. He has forgotten to add: the first time as tragedy, the second as farce" (Marx, 1852, p. 398). It is my hope that the ADT can be revived somewhere else than at Ericsson, this time hopefully not as a farce, but as one building block in managing complex systems in a better way. Or, to paraphrase Marx: "Scholars have only *interpreted* the organization, in various ways; the point, however, is to *change* it!"

REFERENCES

Cane, S., & McCarthy, R. (2007). Measuring the impact of enterprise architecture. *Issues in Information Systems*, *VIII*(2), 437–442.

Hegel, G. F. W. (1830). *Encyclopaedia of the Philosophical Sciences (1830), Part One*. Retrieved Sept 25, 2008, from http://www.marxists.org/reference/archive/hegel/works/sl/sl_iv.htm#SL48n

Marx, K. (1852). The Eighteenth Brumaire of Louis Bonaparte. In *Karl Marx and Frederick Engels, Selected Works, Vol. 1* (pp. 394–487). Moscow: Progress Publishers, 1966.

Örtenblad, A. (2005). Of course organizations can learn! *The Learning Organization*, *12*(2), 213–218. doi:10.1108/09696470510583566

Weick, K. (2001). *Making Sense of the Organization*. Oxford: Blackwell Business.

Wikipedia (2009d). Retrieved Jan 30, 2009, from http://en.wikipedia.org/wiki/Unity_of_opposites

Wiley, N. (1988). The Micro-Macro Problem in Social Theory. *Sociological Theory*, *6*(2), 254–261. doi:10.2307/202119

About the Author

Lars Taxén, associated professor, received his M.Sc. from the Royal Institute of Technology in Stockholm in 1968. Between 1968 and 2003 he was employed at the Ericsson Telecommunication Company, where he held several positions related to processes and information systems for hardware and software design. From 1995 on he was engaged in the development and implementation of incremental development methods for large, globally distributed software development projects. The experiences from this work were reported in his PhD thesis *A Framework for the Coordination of Complex Systems Development* (2003). In 2007 he became an associated professor at Linköping University in Sweden. He has published in various conference proceedings and journals and is now active as a researcher and consultant.

Index

B

C

D

E

F

G

H

I

T

U

V

W

Z